The ARRL

SATELLITE HANDBOOK

Steve Ford, WB8IMY

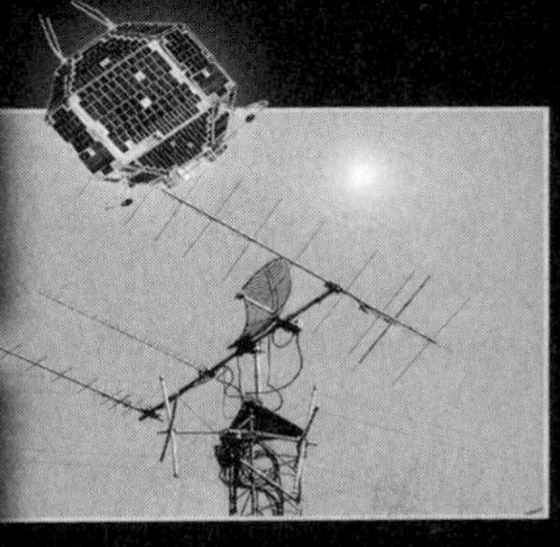

Production: Jodi Morin, KA1JPA
Michelle Bloom, WB1ENT

Technical Illustrations: David Pingree, N1NAS

Cover Design: Sue Fagan, KB1OKW

About the Cover

From upper left, clockwise: The OSCAR 40 satellite undergoes final inspection prior to launch; the Japanese Fuji-OSCAR 20 satellite; the satellite antennas at ARRL Headquarters station W1AW; SuitSat, a Russian spacesuit that was outfitted with a telemetry transmitter and launched from the International Space Station.

Published by:

ARRL *The national association for AMATEUR RADIO*

225 Main Street • Newington, CT 06111-1494

Printed in USA

ISBN: 978-0-87259-985-7

First Edition
Third Printing

Table of Contents

Acknowledgements

This book was inspired by the work of Dr Martin Davidoff, K2UBC. For many years Dr Davidoff was the author of the ARRL's *Satellite Experimenter's Handbook* and *Radio Amateur's Satellite Handbook*, the "gold standard" references in the amateur satellite community. You'll find that some of Dr Davidoff's work from these editions shows up in this book as well (some things simply cannot be improved upon!).

In Chapter One you'll find excerpts from an interesting story titled "Space Satellites from the World's Garage — The Story of AMSAT." This vital part of Amateur Radio history originally appeared in a paper presented to the National Aerospace and Electronics Conference in Dayton, Ohio in May 1994. It comes to us courtesy of Keith Baker, KB1SF, and Dick Jansson, KD1K.

Speaking of Keith Baker, this former AMSAT-NA president also served as my editor and fact checker. His experience in the satellite world far exceeds my own and I deeply appreciate his kind assistance.

Steve Ford, WB8IMY
QST Editor and ARRL Publications Manager
Newington, CT
October 2008

Foreword

Amateur Radio became a partner in space exploration a mere four years after the launch of *Sputnik*, the first artificial satellite. Orbiting Satellites Carrying Amateur Radio, or *OSCARs* as they are better known, have been riding rockets into space ever since. When this book was written, the OSCAR fleet was comprised of more than a dozen active spacecraft.

Amateurs have made huge strides in space technology. Hams pioneered Microsat technology, which clearly demonstrated that small, affordable satellites could carry out a variety of useful missions. The commercial satellite world watched with interest and soon emulated our successes. At the other end of the scale, hams have built and launched large, highly complex spacecraft such as AMSAT-OSCAR 40. The ARRL has often supported the Radio Amateur Satellite Corporation (AMSAT) and the rest of the amateur satellite community in these efforts and we're confident that more amazing technology awaits on the horizon — no pun intended!

The purpose of the *ARRL Satellite Handbook* is to make you familiar with all that amateur satellite technology has to offer, and to give you the tools to become active in this fascinating aspect of Amateur Radio. Satellite operating is unlike anything you've ever done before. Not only are the hardware requirements different, the operational dimension is unique as well. The reward is the thrill of realizing that you're communicating through a spacecraft traveling thousands of miles per hour through the emptiness of space. And you are accomplishing this feat not through a multimillion dollar communication network, but with your own bare hands. The only thing between you and the satellite is the distance spanned by your signals.

David Sumner, K1ZZ
ARRL Chief Executive Officer
Newington, CT
October 2008

A Brief History of Amateur Radio Satellites

This chapter includes excerpts from Space Satellites from the World's Garage — The Story of AMSAT *by Keith Baker, KB1SF, and Dick Jansson, KD1K.*
The author also gratefully credits historical resources provided by AMSAT-NA.

In the Beginning: The '60s

Many people are astonished to discover that Amateur Radio satellites are not new phenomena. In fact, the story of Amateur Radio satellites is as old as the Space Age itself.

The Space Age is said to have begun on October 4, 1957. That's the day when the Soviet Union shocked the world by launching **Sputnik 1**, the first artificial satellite. Hams throughout the world monitored Sputnik's telemetry beacons at 20.005 and 40.002 MHz as it orbited the Earth. During Sputnik's 22-day voyage, Amateur Radio was in the media spotlight since hams were among the only civilian sources of news about the revolutionary spacecraft.

Almost four months later, the United States responded with the launch of the Explorer 1 satellite on January 31, 1958. At about that same time, a group of Amateur Radio operators on the West Coast began considering the possibility of a ham satellite. This group later organized itself as Project OSCAR (OSCAR is an acronym meaning Orbiting Satellite Carrying Amateur Radio) with the expressed aim of building and launching amateur satellites. (See the sidebar "When Does a Satellite Become an OSCAR?")

The Soviet Union began the Space Age with the launch of Sputnik 1 in 1957.

After a series of high-level exchanges with the American Radio Relay League and the United States Air Force, they secured a launch opportunity. The first Amateur Radio satellite, known as **OSCAR 1**, would "piggyback" with the Discoverer 36 spacecraft being launched from Vandenberg Air Force Base in California. Both "birds" (as satellites are called among their builders and users) successfully reached low Earth orbit on the morning of December 12, 1961.

OSCAR 1 weighed in at only 10 pounds. It was built, quite literally, in the basements and garages of the Project OSCAR team. It carried a small beacon transmitter that allowed ground stations to measure radio propagation through the ionosphere. The beacon also transmitted telemetry indicating the internal temperature of the satellite.

OSCAR 1 was an overwhelming success. More than 570 amateurs in 28 countries forwarded observations to the Project OSCAR data

OSCAR 1 was the first Amateur Radio satellite.

collection center. OSCAR 1 lasted only 22 days in orbit before burning up as it reentered the atmosphere, but Amateur Radio's "low tech" entry into the high tech world of space travel had been firmly secured. When scientific groups asked the Air Force for advice on secondary payloads, the Air Force suggested they study the OSCAR design. What's more, OSCAR 1's bargain-basement procurement approach and management philosophy would become the hallmark of all the OSCAR satellite projects that followed, even to this day.

OSCAR 2 was built by the same team, and although it was similar to OSCAR 1 there were a number of improvements. One such upgrade modified the internal temperature sensing mechanism for improved accuracy. Another improvement modified the external coating of the satellite to achieve a cooler internal environment. Yet another modification lowered the beacon transmitter output to extend the battery life of the satellite. Thus, the "continuous improvement" strategy that has also become a central part of the amateur satellite approach was set into place very early in the program. On June 2, 1962 it blasted to orbit from Vandenberg Air Force Base, California aboard a Thor Agena B rocket.

OSCAR 2 was followed by **OSCAR 3** on March 9, 1965. OSCAR 3 would become the first Amateur Radio satellite to carry a *linear transponder* to allow the satellite to act as a communications relay. The transponder was designed to receive a 50 kHz-wide band of uplink signals near 146 MHz and then retransmit them near 144 MHz. This would allow amateurs with relatively modest Earth stations to communicate over much longer distances at these frequencies.

OSCAR 3's transponder operated for 18 days, during which time about 1000 amateurs in 22 countries were heard operating through it. The satellite was the first to clearly demonstrate that multiple stations could successfully use a satellite simultaneously, a technology that is largely taken for granted in satellite telecommunications today.

The fourth Amateur Radio satellite, **OSCAR 4**, was targeted for a geostationary circular orbit 22,000 miles above the Earth. OSCAR 4 would ride into space aboard a Titan III-C rocket on December 21, 1965. Unfortunately, despite a valiant effort on the part of the hams and others involved, (most of whom were members of the TRW Radio Club of Redondo Beach, California), the top stage of the launch vehicle failed and OSCAR 4 never reached its intended orbit. Despite this apparently fatal

When Does a Satellite Become an OSCAR?

While worldwide AMSAT organizations are largely responsible for the design and construction of the modern day Amateur Radio satellites, the original "OSCAR" designation is still being applied to many satellites carrying Amateur Radio. However, most Amateur Radio satellites are not usually assigned their sequential OSCAR numbers until *after* they successfully achieve orbit and become operational. Even then, an OSCAR number is only assigned after its sponsor formally requests one.

For example, let's make up a satellite and call it ROVER. The ROVER spacecraft won't receive an OSCAR designation until (1) it reaches orbit and (2) its sponsor submits a request. Now let's presume that ROVER makes it into orbit and the OSCAR request is made and granted. ROVER is now tagged as OSCAR 99 and its full name becomes ROVER-OSCAR 99. You'll find, however, that many hams will abbreviate the nomenclature. Some will simply call the satellite ROVER, or OSCAR 99. They may even abbreviate its full name to just RO-99.

If a satellite subsequently fails in orbit, or it re-enters the Earth's atmosphere, its OSCAR number is usually retired, never to be issued again.

blow, OSCAR 4 operated long enough for amateurs to successfully develop innovative workaround procedures to salvage as much use out of the satellite as possible.

The Birth of AMSAT

In 1969, the **Radio Amateur Satellite Corporation** (AMSAT) was formed in Washington, DC. AMSAT has participated in the vast majority of amateur satellite projects, both in the United States and internationally. Now, many countries have their own AMSAT organizations, such as AMSAT-UK in England, AMSAT-DL in Germany, BRAMSAT in Brazil and AMSAT-LU in Argentina. All of these organizations operate independently but may cooperate on large satellite projects and other items of interest to the worldwide Amateur Radio satellite community. Because of the many AMSAT organizations now in existence, the North American AMSAT organization is frequently designated AMSAT-NA.

Since the very first OSCAR satellites were launched in the early 1960s, AMSAT's international volunteers have pioneered a wide variety of new communications technologies. These breakthroughs have included some of the very first satellite voice transponders as well as highly advanced digital "store-and-forward" messaging transponder techniques. All of these accomplishments have been achieved through close cooperation with international space agencies that often have provided launch opportunities at significantly reduced costs in return for AMSAT's technical assistance in developing new ways to launch paying customers. Such spacecraft design, development and construction efforts have also occurred in a fiscal environment of individual AMSAT member donations, thousands of hours of volunteer effort and the creative use of leftover materials donated from aerospace industries worldwide.

AMSAT's major source of operating revenue is obtained by offering yearly or lifetime memberships in the various international AMSAT organizations. Membership is open to radio amateurs and to others interested in the amateur exploration of space. Modest donations are also sought for tracking software and other satellite related publications at Amateur Radio gatherings. In addition, specific spacecraft development funds are established from time to time to receive both individual and corporate donations to help fund major AMSAT spacecraft projects.

From a personnel standpoint, AMSAT-North America is a true volunteer operation. The only person in the entire organization drawing a regular paycheck is the office manager at their headquarters near Washington, DC. She conducts the day-to-day business of membership administration and other key organizational tasks. The rest of AMSAT-NA, from the President of the Corporation, on down to the workers designing and building space hardware, all donate their time and talents to the organization.

The '70s

The 1970s was a prolific decade for amateur satellites. The decade began with the launch of **OSCAR 5**. Also known as Australis-OSCAR-5, it was designed and built by students in the Astronautical Society and Radio Club at the University of Melbourne, Australia. OSCAR 5 launched on January 23, 1970, from Vandenberg Air Force Base, California and achieved a 925-mile-high polar orbit aboard a Delta rocket ferrying an American weather satellite.

OSCAR 5 had telemetry beacons transmitting data at 29 and 144 MHz. It was the first amateur satellite to be controlled from the ground; it contained a command receiver that allowed ground stations to control its 29 MHz beacon transmitter. OSCAR 5 did not have a transponder, so it didn't function as a communications relay. However, it did have an

AMSAT-OSCAR Satellite "Phases"

The many spacecraft designed and constructed by radio amateurs since 1961 can be classified by their design and flight characteristics into three *Phases*. Phase I designs comprised the low Earth orbit (LEO), short lifetime, predominantly beacon-oriented satellites such as OSCARs 1 through 4 and the Russian Iskra-1 and -2 series spacecraft.

Phase II OSCARs are also LEO birds, but are launched into somewhat higher orbits and are designed for much longer lifetimes. These AMSAT satellites included OSCARs 6, 7 and 8, as well as UoSAT OSCARs 9 and 11, both of which were built by a team of AMSAT members and students at the University of Surrey in England. These satellites have since been followed by a series of analog and packet radio satellites that were launched by a variety of AMSAT groups from several countries into similar orbits. One large subset of Amateur Radio satellites are called *Microsats* since these birds consist of small cubes or rectangles. These satellites make up the bulk of the AMSAT satellites currently in orbit. There are also smaller *CubeSats* and even *picosats*. Some of these are Phase II birds as well.

Phase III satellites are designed to be launched into highly elliptical Molniya-type orbits first pioneered by the Soviet Union. These satellites, which included OSCARs 10, 13 and 40, offer their users much longer access time, higher power and more diverse communication transponders. What's more, these so called "high altitude" satellites offer their users far larger communications footprints than the LEO satellites. In some cases, Phase III satellites can "see" nearly an entire hemisphere of the Earth at one time, allowing users the luxury of simultaneous contacts on one or more continents.

The Holy Grail of amateur satellite phases is Phase IV —geosynchronous orbit. A Phase IV satellite would be "parked" 22,000 miles above the Earth, providing continuous communication coverage over a large portion of the globe. Hams have yet to deploy a Phase IV satellite, but as this book was being written, AMSAT-NA was involved in discussions with Intelsat that may result in an Amateur Radio transponder module "piggybacking" with a commercial Intelsat satellite in geosynchronous orbit.

innovative magnetic attitude-stabilizing system.

OSCAR 6, which reached orbit on October 15, 1972, was the first Phase II satellite. (See the sidebar, "AMSAT-OSCAR Satellite 'Phases'".) This bird carried a two-way communications transponder that received signals from the ground on 146 MHz and repeated them at 29 MHz with a transmitter power of 1 W.

OSCAR 6 had a sophisticated telemetry beacon that reported information about many parts of the spacecraft, including voltages, currents and temperatures. OSCAR 6 also had an elaborate ground-control system.

Another innovation in OSCAR 6 was *Codestore*, a digital store-and-forward message system. Ground controllers in Canada sent messages to the satellite that were stored and repeated later to ground control stations in Australia.

Static noise plagued OSCAR 6, mimicking signals that the onboard computer interpreted as a command to shut down. To overcome the problem, controllers sent a continuous stream of **ON** commands to the satellite to keep it turned on. The trick worked and OSCAR 6 continued operating for 4.5 years, receiving 80,000 **ON** commands per day!

OSCAR 7, better known as AO-7, was launched on November 15, 1974 by a Delta 2310 booster from Vandenberg Air Force Base, California. OSCAR 7 had two transponders; one received at 146 MHz and repeated what it heard at 29 MHz, while the other listened on 432 MHz and relayed those signals on 146 MHz. The latter transponder, built by radio amateurs in West Germany, had an 8-W downlink transmitter.

In the first satellite-to-satellite link-up in history, a ham transmitted a signal to OSCAR 7, which relayed the signal to OSCAR 6, which then repeated it to a different station on the ground.

Australians built a telemetry encoder for the satellite and Canadians built a 435 MHz beacon. Other beacons were at 146 and 2304 MHz. The 2304 MHz beacon, with a transmitter power of 100 mW, was built by the San Bernardino Microwave Society of

California. Unfortunately, the FCC denied the OSCAR 7 team permission to turn on their 2304 MHz beacon so it never was tested in space.

OSCAR 7's radio system worked for 6.5 years until being declared dead in mid 1981 due to battery failure. However, after more than two decades of silence, OSCAR 7 came back to life in 2002. Both operating modes still function, but the satellite's control system is not working. Even though it can't be controlled from the ground, OSCAR 7 supports conversations on most daylight passes.

OSCAR 8, the third Phase II amateur satellite, was launched March 5, 1978, on a Delta rocket from Vandenberg Air Force Base, California to a circular 570-mile-high polar orbit. It had two transponders, including one designed by Japanese radio amateurs. It listened at 146 MHz and repeated what it heard through a transmitter on 435 MHz. American, Canadian and West German amateurs built the rest of OSCAR 8's flight hardware. OSCAR 8 functioned for more than 5 years until its batteries died in 1983.

Russia entered the Amateur Radio satellite community when an F-2 rocket blasted off on October 26, 1978 from the Northern Cosmodrome at Plesetsk carrying a government satellite and the first two *Radiosputniks*: **RS-1** and **RS-2**. They were both deployed in an elliptical orbit 1000 miles above Earth.

Each satellite had a 145-to-29 MHz transponder. The satellites, sometimes referred to as Radio-1 and Radio-2, circled the globe every 120 minutes. They transmitted telemetry beacons in Morse code, relaying temperature and voltage data. These hamsats (short for "ham satellites") had solar cells as well as a Codestore message store-and-forward mailbox. Ground control stations were at Moscow, Novosibirsk and Arseneyev near Vladivostok.

RS-1 and RS-2 had very sensitive receivers and overload breakers designed to disable the receivers whenever someone used excessive transmitter power on the uplink. (Similar equipment would fly much later on OSCAR 40). The breaker could be reset from the ground when the satellites were over the USSR. However, Western hams, sometimes (needlessly) transmitting with hundreds of watts of power, kept tripping the systems and turning the Radiosputniks off. The Russian ground controllers kept resetting the breakers, but most operation ended up being over the Soviet Union since Western hams kept shutting off the transponders when the satellites were over North America and Western Europe.

RS-1 lasted only a few months, but RS-2 was heard until 1981.

The 1980s

Amateur satellite construction and deployment increased dramatically in the 1980s. In fact, 1981 was a record launch year. Eight hamsats blasted to space that year — a tie with 1990 for the most Amateur Radio satellites launched in a single year.

Figure 1.1—A Molinya orbit slingshots a satellite high above the Earth at apogee, and then allows it to plunge back toward the planet for a relatively low pass at perigee.

The '80s saw the debut of the Phase III birds — complex communications satellites designed to be deployed in highly elliptical *Molniya* orbits that take the craft more than 40,000 km into space at the "high" end (apogee) before plunging back toward Earth, skimming a few hundred km above the surface at the perigee (see **Figure 1.1**). At apogee the satellites appear to "hover," offering coverage over entire hemispheres for several hours.

The **Phase 3A** satellite was launched on the second flight of Europe's new Ariane rocket on May 23, 1980 from a site outside Kourou, French Guiana, on the northeast coast of South America. Unfortunately, the rocket failed soon after lift off, sending Phase 3A into the Atlantic Ocean.

Meanwhile, the Russians were busy with projects of their own. Although most Soviet Amateur Radio satellites were called Radiosputniks, there was a series of birds christened *Iskra*, meaning "spark" in Russian. Students and radio amateurs at Moscow's Ordzhjonikidze Aviation Institute built the 62-lb Iskras, each powered by solar cells. Both satellites carried a transponder, telemetry beacon, ground-command radio, Codestore message bulletin board and computer. The Iskra transponders received at 21 MHz and transmitted at 28 MHz. Their telemetry beacons were near 29 MHz. Controlled by ground stations at Moscow and Kaluga, Iskras were intended for communication among hams in Bulgaria, Cuba, Czechoslovakia, East Germany, Hungary, Laos, Mongolia, Poland, Romania, USSR and Vietnam.

Iskra-1 was launched July 10, 1981 on an A-1 rocket from the Northern Cosmodrome at Plesetsk to a 400-mile-high polar orbit. After 13 weeks, it burned up while re-entering the atmosphere on October 7, 1981.

The first British amateur radio satellite, **UoSAT-OSCAR 9**, also called UO-9, was designed and built by students at the University of Surrey. In fact, "UoSAT" is short for "University of Surrey Satellite." The 115-lb science and education satellite was blasted to a 340-mile-high polar orbit on October 6, 1981 on a US Delta rocket from Vandenberg Air Force Base, California.

Although OSCAR 9 did not have a communications transponder, it transmitted data and had a television camera that sent pictures back to Earth. The satellite had one of the earliest two-dimensional charge-coupled device (CCD) arrays, forming the first low-cost CCD television camera in orbit. The resulting images transmitted from space were spectacular, considering the early technology. UO-9 was not a stabilized Earth-pointing satellite so the areas covered by its photos were random.

UO-9 had a magnetometer and radiation detectors. In addition, two on-board particle counters measured the effect solar activity and auroras had on radio signals. UO-9 also carried a synthesized radio voicebox with a 150-word vocabulary to announce spacecraft condition reports.

The beacons transmitted at 145 and 435 MHz. For propagation studies, there were additional beacons at shortwave frequencies near 7, 14, 21 and 28 MHz and microwave frequencies near 2 and 10 GHz.

In 1982, a software error mistakenly activated both the 145 and 435 MHz beacons at the same time, preventing the satellite's receiver from hearing command signals from controllers. Surrey hams called on radio amateurs at Stanford University, California to override the jamming. Stanford hams used a 150-foot dish antenna to transmit power equal to 15 megawatts toward the satellite. The gambit worked and satellite control was later recovered.

After seven more years of reliable service, OSCAR 9 reentered the atmosphere and was destroyed on October 13, 1989.

In the early 1980s, the Amateur Radio club at the University of Moscow was busy building a covey of new Radiosputniks. Like RS-1 and RS-2, the six satellites each weighed 88 lbs and were housed in cylinders 17 inches in diameter and 15 inches long.

The new birds were launched to 1000-mile-high orbits on December 17, 1981 on one C-1 rocket from the Northern Cosmodrome at Plesetsk. At that time, it was the largest group of Amateur Radio satellites ever orbited at one time.

Designated **Radiosputnik-3 (RS-3) through Radiosputnik-8 (RS-8)**, they were in orbits similar to those used by RS-1 and RS-2. The six satellites had transponders receiving at 145 MHz and transmitting at 29 MHz. They also had store-and-forward mailboxes, solar cells and Morse code temperature and voltage data beacons.

Some of the new Radiosputniks carried the first *autotransponders*. Hams could call a satellite on CW and the robot would respond with a greeting and signal report.

An Ariane rocket carried AMSAT-OSCAR 10 to orbit on June 16, 1983.

Each RS satellite died as its batteries failed. RS-5 and RS-7 were able to stay on the air until 1988.

The USSR's *Salyut-7* space station was launched to Earth orbit April 19, 1982, with the second 62-lb Iskra satellite inside. Cosmonauts Anatoli Berezovoi and Valentin Lebedev blasted off from Baikonur Cosmodrome on May 13 in a Soyuz transport and docked at *Salyut-7* two days later. The cosmonauts unwrapped **Iskra-2** and pushed it out an airlock on May 17 at an altitude of 210 miles.

Iskra-2's telemetry beacon was at 29 MHz. Since it started life in such a low orbit, the satellite was able to remain in space only about seven weeks before burning up in the atmosphere on July 9, 1982.

Yet another Iskra, **Iskra-3**, was hand launched from the *Salyut-7* airlock at an altitude of 220 miles on November 18, 1982. Even though Iskra-3 was much like Iskra-2, it suffered from internal overheating and didn't work as well. Iskra-3's telemetry beacon also was at 29 MHz. The third Iskra remained in space only four weeks before descending into the atmosphere and burning on December 16.

After the destruction of Phase 3A in 1980, AMSAT immediately began work on Phase 3B. The 200-lb clone was built mostly by German hams and launched on an Ariane rocket on June 16, 1983. It was later named **AMSAT-OSCAR 10**.

Like its predecessor, however, OSCAR 10 had a run of bad luck. Seconds after OSCAR 10's deployment, it was struck by the final stage of the Ariane booster. The collision damaged an antenna and sent the satellite spinning wildly. AMSAT had to wait for the satellite to stabilize before firing an internal thruster to change the orbit on July 11.

The main kick-motor firing did not go well, either. It didn't shut off as ordered and the satellite shot into an exaggerated orbit taking it nearly twice as far away from Earth as planned. Another motor firing was attempted July 26, but, by that time, helium had leaked from the satellite (probably as a result of the Ariane booster collision) which, in turn, most likely caused a failure of the kick-motor's helium-activated fuel valves to operate properly. As a result, OSCAR 10 wound up in an odd orbit ranging from 2390 to 22,126 miles.

AMSAT had a crippled satellite on its hands. The damaged antenna wouldn't work right and the orbit exposed the satellite to excessive radiation. The incorrect attitude kept the solar panels from orienting toward the Sun so the batteries couldn't charge properly. AO-10's transponders worked, but the broken antenna and low orbital inclination made it less useful. Its signals were weak and access time was limited. Even so, hundreds of radio amateurs used the satellite.

What's more, because radiation-hardened computer chips were prohibitively expensive at the time of its construction and launch, AO-10s builders were forced to use non-radiation-hardened computer chips in OSCAR 10's onboard computer. Unfortunately, AO-10s odd orbit was causing it to endure a continuous, intense bombardment of subatomic particles trapped in the Earth's magnetic field. This, in turn, caused a slow destruction of OSCAR 10's non-radiation-hardened computer memory chips. By 1986, AO-10's command computer had deteriorated to the point that mysterious data bits were regularly turning up in the telemetry stream and the satellite was becoming ever harder to stabilize and control. AO-10's transponders would switch off from time to time as

The second science and education satellite built by students at England's University of Surrey was UoSAT-OSCAR 11, launched March 2, 1984 from California to a 430-mile-high polar orbit.

voltage dropped when sunlight was low. The satellite required solar illumination 90% of the time, but sometimes received only 50%. When this happened, AO-10 would turn itself off and a command station would be required to transmit a reset order. OSCAR-10 continued this erratic operation until it finally became silent in the 1990s.

The second science and education satellite built by students at England's University of Surrey was UoSAT-B, launched March 2, 1984 from California to a 430-mile-high polar orbit. The 132-lb satellite was eventually renamed **UoSAT-OSCAR 11** (UO-11). It also has been called UoSAT-2.

UO-11's beacons transmit at 145, 435 and 2401 MHz. It handled messages while photographing aurora over the Poles with a sensitive camera which stores the images in memory. Digital telemetry beacons relayed news bulletins from AMSAT and UoSAT, which is headquartered at the Spacecraft Engineering Research Unit at the University of Surrey. When this book was written, OSCAR 11 was operating only occasionally.

Japanese amateurs reached space in 1986 with the launch of the Japan Amateur Satellite (JAS-1a) on August 12 from Japan's Tanegashima Space Center. AMSAT labeled it OSCAR-12; Japanese hams called it Fuji. It came to be known as **Fuji-OSCAR-12** or simply FO-12.

FO-12 had a transponder that received at 145 and retransmitted at 435 MHz. Primarily a packet radio satellite, or *PACSAT*, Fuji's transponder could be used either as a message bulletin-board or as a voice repeater. Fuji's telemetry beacon sent data in 20-words-per minute Morse code.

The mailbox in the sky received typewritten messages from individual ham stations and stored them in a 1.5 megabyte RAM memory. This electronic message center permitted amateurs to place messages on the satellite's bulletin board to be read by others.

Users were disappointed when FO-12's solar generator was unable to produce sufficient electricity for Fuji's battery. Japanese controllers were forced to turn the satellite off on November 5, 1989.

Soviet hams planned to return to space with **Radiosputnik-9** in the mid-1980, but its launch was delayed repeatedly. Finally, the flight was canceled and the number RS-9 retired permanently.

That disappointment notwithstanding, Soviet hams delighted the amateur satellite world on June 23, 1987 with the launch of a combo package of hamsats, **RS-10 and RS-11**, aboard one large government spacecraft. Radiosputnik-10/11 went to a 621-mile-high circular orbit as part of the Russian navigation satellite Cosmos 1861, which circles the globe every 105 minutes. The two satellites were, in fact, communications modules riding piggyback on Cosmos 1861 and sharing its considerable electric power budget.

RS-10 and RS-11's telemetry beacons transmitted near 29 and 145 MHz. They had identical shortwave and VHF transponders, but the specific frequencies they used were different. Hams on the ground sent signals to RS-10 and RS-11 on frequencies near 21 and 145 MHz. Downlink signals from the satellites were at 29 and 145 MHz. RS-10 and 11 also featured robot autotransponders that responded to Morse code calls with greetings and contact numbers. QSL cards were even issued for contacts with the autotransponders.

RS-10/11 became two of the most popular amateur satellites in history. Anyone with an HF transceiver and the ability to send or receive SSB or CW on 2 meters could access the birds. Simple wire antennas were all that were required. On some weekends, the transponders would be crowded with signals as hams took advantage of the 15 minute contact windows.

When the Russians decommissioned Cosmos 1861 in the 1990s that spelled the end of RS-10/11 as well.

As the saying goes, the third time is the charm. That was certainly the case for AMSAT's third attempt at a Phase III satellite, **AMSAT-OSCAR 13**, which safely reached orbit on June 15, 1988. In its Molniya orbit, OSCAR 13 reached an apogee of 22,000 miles before sweeping back around the Earth at an altitude of 1500 miles.

OSCAR 13 was the most complex Phase III satellite of its time. It was a project lead by AMSAT-DL with AMSAT teams from a number of other countries also contributing to its success. AMSAT-DL. OSCAR 13 offered four transponders for packet, facsimile, slow-scan television, voice (SSB), radioteletype (RTTY) and Morse code (CW). Transponders received at 435 and 1269 MHz and retransmitted at 145, 435 and 2400 MHz. The satellite's computer followed a schedule to manage transponder activation and use.

With its near-hemispheric coverage at apogee, OSCAR 13 quickly became famous as a "DX satellite." Hams in different continents could communicate for hours at a time. There were even scheduled roundtable chats via OSCAR 13.

All good things must come to an end, and so did OSCAR 13. The satellite re-entered the Earth's atmosphere on December 6, 1996.

The Last Satellites of the 20th Century

With the dawn of the 1990s, space enthusiasts at England's University of Surrey had several new satellites ready to fly. At about the same time, another international group of AMSAT experimenters had devised a new, far more radical satellite design approach to take advantage of rapid advances in solar cell efficiency. Rather than using bulky spaceframes, these new birds were small boxes only 22.6 × 22.6 × 22.3 cm on each side and weighing only 13 kg. Known as *Microsats*, these satellites represented a pioneering approach to satellite design, one that commercial satellite builders would soon follow.

On January 22, 1990, two UOSATS and an entire fleet of Microsats were placed in orbit by an Ariane 4 rocket. The lineup included UoSAT-OSCAR 14, UoSAT-OSCAR 15, AMSAT-OSCAR 16, DOVE-OSCAR 17, WEBERSAT-OSCAR 18 and LUSAT-OSCAR 19.

On January 22, 1990, an entire fleet of Microsats (along with two new UOSATS) were placed in orbit by a single Ariane 4 rocket. The lineup included **UoSAT-OSCAR 14**, **UoSAT-OSCAR 15**, **AMSAT-OSCAR 16**, **DOVE-OSCAR 17**, **WEBERSAT-OSCAR 18** and **LUSAT-OSCAR 19**.

UoSAT-OSCAR14 spent its first 18 months in orbit operating as a packet radio store-and-forward satellite. It received electronic mail and stored it for later reading by ground stations in other parts of the globe. In early 1992, all amateur operations were moved from AO-14 to UoSAT-OSCAR 22. AO-14 operations were then dedicated for use by VITA (Volunteers In Technical Assistance) who used it for sending and receiving e-mail messages in Africa.

Several years later, the computer used for store-and-forward communications became non-operational. In March 2000 UO-14 was returned to amateur use and reconfigured as a single-channel FM repeater. This move dramatically increased its popularity in the ham community. For the first time, hams were able to enjoy satellite communication using nothing

The LUSAT-OSCAR 19 satellite is typical of Microsat design.

more than common dual-band FM transceivers.

OSCAR 14 remained in service until November 2003 when it was declared officially dead. The Mission Control Centre at the Surrey Satellite Technology Ltd reported that the venerable and popular bird "had reached the end of its mission after nearly 14 years in orbit."

OSCAR 15 made it to orbit with its companions, but its lifetime was short. Within hours after being deployed, it fell silent.

AMSAT-OSCAR-16 was designed to be a dedicated store-and-forward file server in space. Using 1200 bit per second radio links, AMSAT-OSCAR-16 interacted with ground station terminal software to appear as a packet radio bulletin board system to the user. Anyone wishing to download files and personal e-mail from anywhere in the world could request that information be "broadcast" to everyone under the footprint of the spacecraft, or directed specifically to a particular ground station. This broadcast protocol differs from terres-trial packet radio communications, but allows a greater number of ground stations access to the spacecraft's resources during the often limited time a satellite is in view. OSCAR 16 enjoyed a long life, but finally went off the air in 2008.

DOVE-OSCAR 17 was sponsored by AMSAT-Brazil. The project was lead by Dr Junior Torres DeCastro, PY2BJO. DOVE, an acronym for Digital Orbiting Voice Encoder, carried hardware capable of reproducing digitized speech, or controlling a Votrax speech synthesizer. However, due to hardware failures that occurred after launch, the primary mission of broadcasting voice messages of world peace was not fully realized.

DOVE operated sporadically on a downlink frequency of 145.825 MHz FM, transmitting AX.25 protocol packet radio telemetry. Using 1200 bps Bell 202 style AFSK emissions, DOVE-OSCAR-17 could be copied with packet radio equipment in wide use on VHF at the time. Today DOVE is no longer operational.

In March 2000 UO-14 was configured as a single-channel FM repeater. It remained in service until 2003.

Sporting a full-color CCD camera, **WEBERSAT-OSCAR 18** digitized Earth images and downlinked them as an AX.25 serial data stream. WEBERSAT-OSCAR 18 was a product of the efforts of the Center for Aerospace Technology (CAST) at Weber State University in Ogden, Utah. WO-18's CCD camera had a resolution of 700 pixels by 400 lines and could be viewed with *Weberware* software running on a personal computer having adequate graphics display capability. Digitized NTSC video from the camera was assembled into packets that were sent as unnumbered information UI frames. Ground stations had to receive this data over several passes to capture a complete image. Each image contained about 200 kilobytes of data.

The final member of the 1990 Microsat fleet was **LUSAT-OSCAR 19**. OSCAR 19 is coordinated by AMSAT Argentina and is a packet radio store-and-forward spacecraft much the same as AMSAT-OSCAR-16. The only difference between the two satellites is that AO-16 supports an S-band beacon in addition to the mailbox, while LO-19 has a 70-cm CW beacon. Today only the CW beacon remains operational.

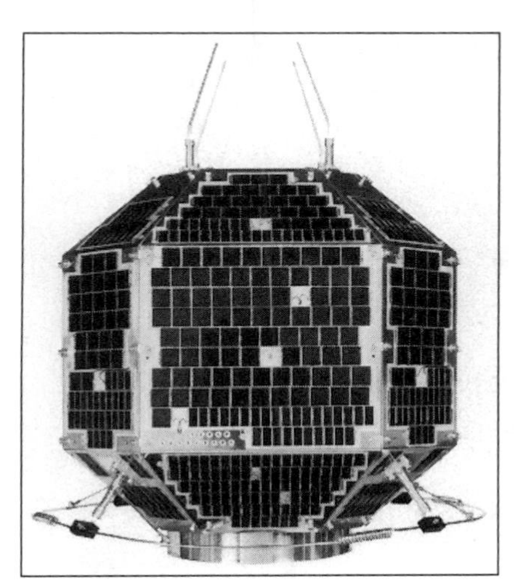
The Japanese launched Fuji-OSCAR 20 on February 7, 1990 from the Tanegashima Space Center on an H-1 two-stage rocket.

In the wake of the Microsat successes, the Japanese launched their next Amateur Radio satellite, **Fuji-OSCAR 20**, on February 7, 1990 from the Tanegashima Space Center on an H-1 two-stage rocket. Also known as FO-20, its orbit differed slightly from most OSCAR satellites, being slightly elliptical with a high inclination. This assured that the satellite would remain in sunlight for the majority of its orbit.

The physical structure of FO-20 was that of a 26 sided polyhedron with a weight of approximately 50 kg, so it was much larger than the Microsats. Although Fuji-OSCAR-20 used AX.25 packet radio communications links as the Microsats did, one big difference between FO-20 and the Microsats was that FO-20's packet radio could be accessed without the need for special Microsat terminal software. Any computer or terminal that could be used to access terrestrial packet radio bulletin board systems (BBSs) could be used to access the FO-20 mailbox.

The other difference between FO-20 and the Microsat satellites was that, in addition to the packet mailbox features of the satellite, FO-20 also supported an analog linear transponder for SSB and CW communications. FO-20 suffered from a declining power generation capacity over the years and eventually became silent.

The year 1990 would see one more amateur satellite launch. It would also be the first from Pakistan. Known as **Badr-1**, the satellite was launched July 16, 1990 by China on one of its Long March rockets to a 375-mile-high circular orbit. The 150-lb Badr-1 was constructed by engineers who were hams at the Space and Upper Atmosphere Research Commission (SUPARCO) at the University of the Punjab at Lahore. Several had completed Masters Degrees in engineering at England's University of Surrey. Back home, they used their new knowledge to build the satellite with support from the Pakistan Amateur Radio Society.

Badr-1's glory was brief. It reentered the atmosphere on December 9, 1990.

Radiosputnik-14/AMSAT-OSCAR 21 was the first satellite of 1991, taking to the skies on January 29, 1991 as a joint venture between AMSAT-U (Russia) and AMSAT-DL (Germany). It was essentially a module riding inside a larger Russian government satellite known as INFORMATOR-1. Amateur Radio lost a valuable asset when the Russian government ran out of funds for the project and turned the entire satellite off on September 16, 1994. RS-14/AO-21 had been a popular satellite because it was easy to use. The hamsat was an FM repeater that also transmitted recorded messages commemorating events like the 25th anniversary of the first landing on the Moon.

Another Radiosputnik, **RS-12/13**, was launched on February 5, 1991. Like the popular RS-10/11 combo, RS-12/13 module rode piggyback on a Russian COSMOS navigation satellite. Each Radiosputnik had a 40 kHz-wide linear transponder allowing for many simultaneous CW and SSB contacts. They also carried autotransponders that acknowledged CW calls with a greeting and contact number.

RS-12/13 was quite popular because of its HF and low VHF transponders. Anyone with simple station equipment could work them. The satellites enjoyed an 11-year lifespan until a solar flare disabled the parent COSMOS satellite in August 2002.

The British Microsat UoSAT-5, built by the University of Surrey, was launched July 17, 1991, to become **UoSAT-OSCAR 22**. OSCAR 22 was designed to serve several missions. One mission was to carry out experiments originally slated for UoSAT-OSCAR-15, whose on-board electronics failed shortly after the spacecraft reached orbit. The primary purpose

Manned Spacecraft and Amateur Radio

In addition to unmanned spacecraft, Amateur Radio has a rich history aboard manned spacecraft.

It all began in the fall of 1983 with space shuttle mission STS-9. According to a history compiled by the Johnson Space Center Amateur Radio Club, on November 28, 1983 STS-9 was launched carrying Mission Specialist Owen Garriott, who was licensed with Amateur Radio call sign W5LFL. For 10 days the space shuttle *Columbia* streaked through the skies and for the last 7 of those days, hams around the world heard Dr Garriott's voice break their 2-meter FM squelches calling earthbound stations.

In addition to the possibility that earthbound hams could make random contacts with the ham-astronauts aboard orbiting space shuttles, beginning with STS-35 in 1990 Amateur Radio took on a new role with the creation of the Space Shuttle Amateur Radio EXperiment (SAREX). SAREX brought the opportunity of scheduled radio contacts between the orbiting ham-astronauts and schools, putting astronauts and the space program in direct contact with school children around the world. The SAREX program finally ended in 1998.

The former Soviet Union was quick to incorporate Amateur Radio into its manned space program as well. In 1986 the first module of the *Mir* space station was placed in orbit. Over the next 15 years *Mir* housed Amateur Radio gear and hosted several Amateur Radio operators as crew members, who often used the RØMIR call sign. Countless earthbound hams, including many students as part of SAREX, got the chance to speak directly with *Mir's* crew or access *Mir's* packet messaging system. Pictures transmitted via an SSTV experiment installed aboard *Mir* also delighted amateurs.

On November 28, 1983 space shuttle STS-9 was launched carrying Mission Specialist Owen Garriott, W5LFL. It was the first use of Amateur Radio aboard a manned spacecraft.

The Soviet *Mir* space station was very active on the amateur bands.

of UoSAT-OSCAR-22, however, was to provide non-amateur radio related store-and-forward digital communications for the non-profit, humanitarian organizations VITA and SatelLife.

UoSAT-OSCAR-22's primary role in space was later modified when the satellite played a part in Amateur Radio's first "role reversal" with another spacecraft, UoSAT-OSCAR-14. One of the most unique aspects of UoSAT-OSCAR-22 was its Earth Imaging System (EIS), designed by University of Surrey doctoral student Marc Fouquet. The Earth System was designed to capture Earth images from low orbit using a charge coupled device (CCD) camera and broadcast those images to ground stations using amateur frequencies and the AX.25 packet radio "Pacsat Broadcast Protocol." After years of reliable service, UO-22 fell silent.

The year 1992 saw the launch of **KITSAT-OSCAR 23** on August 10. It was a Microsat designed by amateurs from the Korean Advanced Institute of Science and Technology (KAIST) studying at the University of Surrey in the UK. The digital store-and-forward satellite was managed by the KAIST Satellite Technology Research Center (SaTReC) of South Korea. It enjoyed several years of life before its power systems eventually failed.

The first French amateur satellite, **Arsene-OSCAR 24**, flew on May 13, 1993, aboard

The ARISS program has created an active Amateur Radio presence aboard the International Space Station.

In all, more than 100 astronauts and cosmonauts did tours of duty aboard *Mir.* The space station was finally decommissioned and allowed to reenter the atmosphere on March 23, 2001.

With the subsequent construction of the International Space Station, the American SAREX program was replaced with *ARISS* — Amateur Radio on the International Space Station. ARISS is a program that uses primarily pre-arranged Amateur Radio contacts to inspire students worldwide to pursue careers in science, technology, engineering and math.

ARISS is an international working group, consisting of delegations from 9 countries including several countries in Europe as well as Japan, Russia, Canada, and the USA. The organization is run by volunteers from the national amateur radio organizations and the international AMSAT (Radio Amateur Satellite Corporation) organizations from each country. Since ARISS is international in scope, the team coordinates locally with their respective space agency (e.g. ESA, NASA, JAXA, CSA, and the Russian Space Agency) and as an international team through ARISS working group meetings, teleconferences and through electronic mail.

In recent years, the ARISS station has been expanding with the addition of new ham gear. Also, International Space Station contacts are not limited to students. ISS crews occasionally take the time to make random 2-meter FM contacts and the ISS packet radio digipeater is often available as well.

an Ariane flight V-56A from Kourou, French Guiana. Built by the Radio Amateur Club de l'Espace, the satellite had a short, troubled life. It was originally intended to be a packet relay satellite, but the packet system was never implemented because the 2 meter transponder failed soon after launch. Arsene was then used to relay SSB and CW signals on 2.4 GHz for several months until this transponder failed as well.

There was another amateur satellite "first" in 1993. Launched alongside HealthSat-2 on the September 1993, **PoSAT-1** was Portugal's first satellite. PoSAT-1 was built at the University of Surrey in a collaborative program between a consortium of Portuguese academia and industry. The satellite carried a number of technology experiments and had the ability to function as a packet radio store-and-forward system.

PoSAT operated on amateur frequencies for several weeks in early 1994. *OSCAR News* (February 1994, p. 35) carried a letter from CT1DBS reporting that an agreement had been signed by AMSAT-PO and the PoSAT Consortium on December 6, 1993 stating, "The name of PoSat-1, when in use by the amateur radio community will be **PoSAT OSCAR 28**, OSCAR 28 or PO-28." Despite having an OSCAR designation, PoSAT-1 was never re-opened for amateur operations.

KITSAT-OSCAR 25 was launched September 26, 1993 by an Ariane rocket from Kourou, French Guiana. This satellite was essentially a twin of KITSAT-OSCAR 23, but it was designed and built entirely by the Korean Advanced Institute of Science and Technology (KAIST). KO-25 was operated from The Satellite Technology Research Center (SaTReC) in South Korea. KO-25's mission was to take CCD pictures, process numerical information, measure radiation and receive and forward messages. The Infrared Sensor Experiment (IREX) was designed to measure the characteristics of infrared sensors in space. A passive cooling structure was also devised for this experiment so ground controllers could monitor its temperature. In its Amateur Radio role, KO-25 functioned as a packet store-and-forward relay, although it is presently nonoperational.

The same rocket that took KO-25 to orbit also carried Microsats **ITAMSAT-OSCAR 26** and **AMRAD-OSCAR 27**.

ITAMSAT was the first Italian amateur satellite. It was built and operated by a small team of AMSAT members from Italy. Its mission was to store and forward amateur radio messages in the same manner as AO-16, LO-19, UO-22, KO-23 and KO-25. Unfortunately, OSCAR 26 saw little ham use.

In contrast to OSCAR 26, AMRAD-OSCAR 27 (AO-27) was highly popular. AO-27 was a secondary amateur communications payload carried aboard the EYESAT-1 experimental Microsat built by Interferometrics Incorporated of Chantilly, Virginia. The commercial side of the spacecraft's mission is the experimental monitoring of mobile industrial equipment.

AMRAD

The satellite that would eventually become OSCAR 27 undergoes final checkout.

The Amateur Radio portion of the satellite was constructed by members of AMRAD, a technically oriented, non-profit organization of radio amateurs based in the Virginia suburbs of Washington, DC. It was intended to be a platform to conduct digital satellite communications experiments.

Today AO-27 functions as an FM repeater in space, although in 2012 it suffered a major failure. Recovery efforts continue. It essentially consists of a crystal controlled FM receiver operating at 145.850 MHz and a crystal-controlled FM transmitter operating at approximately 436.795 MHz. Because of the satellite's limited power budget and a desire to maintain sufficient battery capacity for as many years as possible, the amateur transmitter on AO-27 is on for only part of the daylight portion of each orbit. AO-27 is available on daylight passes over the Northern Hemisphere.

The only amateur satellite of 1994 was **Radiosputnik 15**. It was launched on December 16. It carried a 2-meter-to-10-meter transponder and CW beacons, but their signals were so weak that it is doubtful the satellite ever became fully operational. Hams today occasionally report reception of weak CW beacons from RS-15.

On March 28, 1995 the Mexican **UNAMSAT-1** and Israeli **TechSat-1** amateur satellites were launched from Russia's Plesetsk Cosmodrome aboard a Start-1 rocket. The rocket exploded soon after liftoff, destroying both satellites.

The Japanese **Fuji-OSCAR 29**, the successor to the failing OSCAR 20, reached orbit on August 17, 1996 from the Tanegashima Space Center aboard a H-II No. 4 rocket. It was a project of the Japan Amateur Radio League. OSCAR 29 carried an SSB/CW analog linear transponder that listened on 2 meters and repeated on 70 cm. It still operates, but with diminished capacity.

The second Mexican hamsat, UNAMSAT-B, built at the National University of Mexico (UNAM)), was launched from Russia on September 5, 1996. **Mexico-OSCAR 30**, as it came to be known, was the twin of UNAMSAT-1 (see above). Unfortunately, MO-30 stopped working when its receiver failed within a few hours after reaching orbit.

Radiosputnik 16 took to the skies in February 1997 carrying 2-meter-to-10-meter transponders and a 70 cm beacon. For unknown reasons, however, the transponders were never activated. The satellite reentered the atmosphere in 1999.

An unusual satellite made a brief appearance on November 4, 1997. **Radiosputnik 17a**, also known as **Sputnik 40**, was hand launched from the Russian *Mir* space station. The little satellite was a scale model built by high school students to commemorate the 40th anniversary of the launch of Sputnik 1. RS-17a broadcast a signal for 55 days and was last heard on December 29, 1997.

Thai-Microsatellite-OSCAR 31 was launched on July 10, 1998 from the Russian Baikonur Cosmodrome into a circular sun-synchronous orbit. It carried a 9600-baud FSK digital transponder, GPS receiver and imaging subsystem. OSCAR 31 was similar to KITSAT-OSCAR 23, but included the ability to take multispectral images. At the time of this writing, it is nonoperational.

Radiosputnik 16 was a satellite shrouded in mystery. It was launched successfully in 1997, but its transponders were never activated.

Israel's **Gurwin-OSCAR 32**, otherwise known as TechSat-1b, soared into orbit on July 10, 1998. The Microsat was a project of students and scientists at the Technion Institute of Technology in Haifa, along with the Israel Amateur Radio Club. Eight years after launch, OSCAR 32 is still going strong. The satellite supports 9600-baud store-and-forward packet and a high-resolution color camera supplying stunning images of the Earth. In recent years, OSCAR 32 has also been put to use as a 9600 baud relay for the Automatic Position Reporting System (APRS) with a dedicated uplink on 145.930 MHz.

The University of Huntsville, Alabama, in conjunction

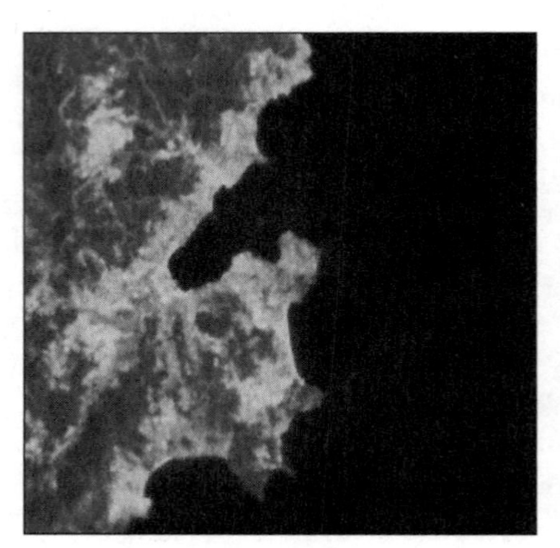

The first image transmitted from Gurwin-OSCAR 32.

with Students for the Exploration and Development of Space (SEDS) USA, designed and constructed SEDSAT-1, or **SEDSAT-OSCAR 33** as it came to be known. It reached orbit on a Delta II rocket from Vandenberg Air Force Base, California on October 24, 1998. It carried a packet store-and-forward transponder, an analog transponder and a number of experiments. It is only semi-operational at this time.

Launched October 30, 1998 from the space shuttle *Discovery*, **PANSAT-OSCAR 34** was a Microsat designed to provide a packet store-and-forward message system. Designed and built at the Naval Postgraduate School in Monterey, California, OSCAR 34 was unique among the digital satellites in that it employed direct sequence spread-spectrum communications. OSCAR 34 was only available for limited amateur access, however, and finally became silent.

Another tiny Radiosputnik, this one called **Radiosputnik-18/Sputnik-41**, was tossed out of the *Mir* space station airlock on November 10, 1998. Like its predecessor, Sputnik 40, it was a ⅓-scale replica of Sputnik 1. It transmitted a beacon signal on 2 meters along with voice greetings in English, Russian and French. The satellite reentered the atmosphere after a few months in orbit.

SUNSAT-OSCAR 35 was a Microsat built by post-graduate engineering students in the Electronic Systems Laboratory of the Department of Electrical and Electronic Engineering at the University of Stellenbosch, Matieland, South Africa. It was launched on February 23, 1999. Payloads included NASA experiments, Amateur Radio communications, a high-resolution imager, precision attitude control and school experiments. OSCAR 35 became silent in February 2001.

Yet a third satellite was hand launched from the *Mir* space station on April 16, 1999. **Radiosputnik-19/Sputnik-99** was controversial because it was originally designed to broadcast a beacon on ham frequencies that promoted the Swatch watch company of Switzerland. After substantial outcry from the ham community, the Amateur Radio element of the project was abandoned and the satellite was launched with its ham transmitter turned off.

The last Amateur Radio satellite of the 20th century was **UoSAT-OSCAR 36** built by Surrey Satellite Technology Ltd at the University of Surrey in the United Kingdom. It was launched on April 21, 1999 on a Russian rocket from the Baikonur Cosmodrome. OSCAR 36 carried a number of imaging payloads and a unique propulsion system for orbital housekeeping experiments. The S-band downlink ran at speeds up to 1Mb/s for downloading imaging data. For amateurs, it functioned as a packet store-and-forward satellite during its brief functional lifetime.

The Dawn of a New Century

A unique rocket and a complex set of amateur satellites lifted off less than a month into the 21st century. On January 27, 2000 a Minotaur rocket launched from Vandenberg Air Force Base in California. The six-story Minotaur was built from parts recycled from retired nuclear missiles. It combined the first and second stages of a decommissioned US Air Force Minuteman-2 missile with the third and fourth stages of an Orbital Sciences' commercial air-launched Pegasus rocket. The launch proved that the combo was capable of ferrying satellites to space.

OPAL-OSCAR 38 carried a 9600-baud packet radio store-and-forward system and also launched several tiny satellites, known as *picosats*.

The Minotaur carried a number of non-ham satellites to orbit, along with a group of Amateur Radio birds:

Arizona State-OSCAR 37 contained an amateur packet hardware system and a 2-meter/70-cm FM voice repeater. The satellite successfully activated, but telemetry soon confirmed that a critical problem had occurred in the power system. The solar arrays were offline and the batteries could not be recharged. As a result, OSCAR 37 died 15 hours later.

OPAL-OSCAR 38 carried a 9600-baud packet radio store-and-forward system. It is now non-operational. OSCAR 38 was designed to launch six miniature satellites, known as *picosats*. One of these picosats was an Amateur Radio project known as **StenSAT**. StenSAT was built by Amateur Radio operators in Washington, DC and was a mere 12 cubic inches in size and weighed only 8.2 ounces. It featured a single-channel FM voice repeater with uplink at 145.84 MHz and a downlink at 436.625 MHz. Although StenSAT was successfully deployed, it never became fully operational.

Weber-OSCAR 39, also known as JAWSAT, was designed to serve as a platform for deploying smaller satellites. JAWSAT stands for Joint Air Force Weber Satellite, which was a joint project between the US Air Force and Weber State University. During this mission, JAWSAT deployed OSCAR 37, OSCAR 38, OCSE — the US Air Force Research Laboratory's Optical Calibration Sphere Experiment — and FalconSat. The latter satellite was developed by US Air Force Academy cadets to study how charged particles can build up and then wreak havoc with satellites' onboard computer systems. The telemetry from JAWSAT was transmitted on ham frequencies.

Launched September 26, 2000 aboard a converted Soviet ballistic missile from the Baikonur Cosmodrome, **SaudiSat-OSCAR 41** was one of three Amateur Radio satellites on the same launch vehicle. OSCAR 41 was one of the first Saudi Arabian Microsats with Amateur Radio capability. It contained a 9600-baud store-and-forward packet system as well as an analog FM repeater. It was built by the Space Research Institute at the King Abdulaziz City for Science and Technology. Initially, the FM repeater was active, but the satellite fell silent in February 2003 and has not been heard since.

Accompanying SaudiSat-OSCAR 41 on the flight was **Malaysian-OSCAR 46** and **Saudi-OSCAR 42**. OSCAR 46 was Malaysia's first Microsat. In addition to commercial land and weather imaging payloads, it offered FM and FSK Amateur Radio communication. It was built as a collaborative effort between the Malaysian government and Surrey Satellite Technology Ltd. Opened briefly for ham use, it is now non-operational. OSCAR 42, like OSCAR 41, was a Saudi Arabian Microsat. As far as its Amateur Radio capability was concerned, it was a virtual twin of OSCAR 41. It is now off the air.

The largest, most complex Amateur Radio satellite ever constructed was launched on an Ariane 5 rocket from Kourou, French Guiana, on November 16, 2000: **AMSAT-OSCAR 40**, also known simply as AO-40. Like its Phase III predecessors, AO-40 was launched into an eccentric Molniya orbit ranging from 58,971 km at apogee and 1000 km at perigee.

OSCAR 40 offered linear transponders on several bands from VHF to microwave, a high-resolution color camera, digital transponders and scientific experiments. With its high altitude at apogee, powerful output and gain antennas, it promised hemispheric communications at signal levels never experienced before.

AMSAT

OSCAR 40 was the largest, most complex amateur satellite ever constructed.

Tragically, a protective plastic cap over a vent on the satellite's orbital insertion motor was inadvertently left in place at launch. And while precise details of exactly what happened next will forever remain sketchy, it appears that during successive firing attempts of the motor in space, pressure built up in the motor's feed lines to the point that one or more of them eventually ruptured, spewing volatile hypergolic (i.e. "no spark needed") fuel and oxidizer into the interior of the spacecraft. The resulting explosion was apparently strong enough to blow the omni-antenna-laden bottom out of the satellite. Needless to say, all radio links and beacons immediately went silent. Communications were re-established weeks later and the satellite was eventually returned to limited operation. Although a shadow of its former self, hams around the world still enjoyed considerable use of OSCAR 40 until January 2004 when it suffered a catastrophic failure of the main battery. The satellite has been silent ever since and is now considered lost.

On September 30, 2001 an Athena I rocket blasted off from the Kodiak Launch Complex on Kodiak Island, Alaska carrying three Amateur Radio satellites. **Navy-OSCAR 44** (NO-44), also known as PCSAT-1, is a 1200-baud APRS digipeater designed for use by stations using handheld or mobile transceivers. It is still operational today, but it functions with a "negative power budget," meaning it comes alive in mid-day sun on every orbit, but before the end of its next eclipse in Earth's shadow 45 minutes later, the satellite has completely depleted its battery charge. It has to be in sun long enough to get enough charge to be able to provide the peak transmit power for the packet transmitter.

COURTESY OF THE SETI LEAGUE

This is a flight model of Navy-OSCAR 44 (NO-44), also known as PCSAT-1. It is a 1200-baud APRS digipeater designed for use by stations using handheld or mobile transceivers.

The Athena rocket also deployed **Starshine-OSCAR 43**, a basketball-shaped satellite covered with

Starshine-OSCAR 43 was covered with 1500 aluminum mirrors polished by an estimated 40,000 student volunteers in the United States and 25 other countries.

1500 aluminum mirrors polished by an estimated 40,000 student volunteers in the United States and 25 other countries. Starshine's primary mission was to involve and educate school children from around the world in space and radio sciences. In addition to helping build Starshine, students were able to visually track the satellite during morning and evening passes by recording its telltale mirror flashes and reporting their observations to Project Starshine headquarters. Starshine-OSCAR 43 transmitted its telemetry on ham frequencies, but is now silent.

The last passenger on the Kodiak launch was **Navy-OSCAR 45**, otherwise known as Sapphire. Sapphire was a Microsat designed and built by students at Stanford University and Washington University-St Louis. The primary mission of Sapphire was to space-qualify two sets of "Tunneling Horizon Detector" infrared sensors designed and built by the Jet Propulsion Laboratory and Stanford University. Secondary experiments included a digital camera and voice synthesizer. Today OSCAR 45 is non-operational.

Radiosputnik 21/Kolibri had a brief, but interesting life. The tiny educational satellite was built by the Special Workshop of Space Research Institute of the Russian Academy of Sciences, Tarusa, Kaluga. Participating students in Sydney, Australia and Obninsk, Russia named it Kolibri-2000. (Kolibri means "hummingbird.")

The satellite rode the Russian *Progress M-17* cargo freighter to the International Space Station in December 2001. On March 20, 2002, as the *Progress* rocket departed, the 44-pound Kolibri was ejected into space. Kolibri dropped slowly, circled Earth 711 times and burned up in the atmosphere after about four months. It sent telemetry data and digitally recorded voice messages on the downlink frequency of 145.825 MHz.

French amateurs returned to space in 2002 with two picosats known before launch as IDEFIX CU1 and IDEFIX CU2. They flew into orbit on May 3 on Ariane-4 flight V151 from Kourou, French Guiana. The Ariane also ferried the SPOT-5 photo satellite to orbit. The satellites were later officially named **BreizhSAT-OSCAR 47** and **BreizhSAT-OSCAR 48**. The satellites remained attached to the Ariane rocket third stage and transmitted pre-recorded narrowband FM (NBFM) voice messages and digital telemetry data. Both battery-powered birds became silent two weeks later.

Radiosputnik 21/Kolibri rode a Russian *Progress* M-17 cargo freighter to the International Space Station in December 2001. On March 20, 2002, as the *Progress* rocket backed away and departed the station, the 44-pound Kolibri was ejected into space.

Two ham satellites lifted off on December 20, 2002 on a converted Soviet ballistic missile from the Baikonur Cosmodrome. **AATiS-OSCAR 49** was a German Amateur Radio payload onboard the small German scientific satellite RUBIN-2. It was designed as a store-and-broadcast system for APRS, but failed about a month after reaching orbit. Its launch companion, **Saudi-OSCAR 50**, remains operational today as an orbiting FM repeater.

It is fair to say that 2003 marked the debut of the *CubeSats*. As the name implies, CubeSats are tiny cube-shaped satellites dedicated to specific missions in low Earth orbit, some with short-design lifespans. Between 2003 and the present day, a horde of CubeSats have been placed in orbit, but only a few

Table 1.1
Currently Active Amateur Radio Satellites

As of March 2013

Satellite	*Uplink (MHz)*	*Downlink (MHz)*	*Mode*
AAUSAT-II	—	437.425 (packet)	Telemetry
AMSAT-OSCAR 7		29.502	Beacon
		145.975	Beacon
		435.106	Beacon
	145.850 - 145.950	29.400 - 29.500	SSB/CW, non-inverting
	432.125 - 432.175	145.975 - 145.925	SSB/CW, inverting
AMSAT-OSCAR 27	145.850	436.795	FM repeater
ARISS (ISS)	144.490	145.800	Crew contact, FM (Rgn. 2/3)
	145.200	145.800	Crew contact, FM (Rgn. 1)
	145.990	145.800	Packet BBS
	145.825	145.825	APRS digipeater
	—	144.490	SSTV downlink
	437.800	145.800	FM repeater
Compass-One	—	437.275 (CW)	Telemetry
	—	437.405 (packet)	Telemetry (including images)
CubeSat-OSCAR 57	—	436.8475 (CW)	Telemetry
		437.4900 (packet)	Telemetry
CubeSat-OSCAR 58	—	437.4650 (CW)	Telemetry
		437.3450 (packet)	Telemetry
CubeSat-OSCAR 66	—	437.485 (packet)	Telemetry (including images)
Cute1.7 + APDII	1267.600	437.475	9600 baud packet
Dutch-OSCAR 64	145.870 (packet)	Telemetry	
	435.530 – 435.570	145.880 – 145.920	SSB/CW, inverting
Fuji-OSCAR 29	145.900-146.00	435.800-435.900	SSB/CW, inverting
Navy-OSCAR 44	145.827	145.827	1200 baud APRS packet
Saudi-OSCAR 50	145.850	436.795	FM Repeater—67 Hz CTCSS required
VUSat-OSCAR 52	—	145.860	CW Beacon
		145.936	Carrier Beacon
	435.220 - 435.280	145.870 - 145.930	SSB/CW, inverting

A CubeSat is aptly named!

have received OSCAR designations.

CubeSat-OSCAR 55 (known as Cute-1) and **CubeSat-OSCAR 57** were launched on June 30, 2003 from Baikonur Cosmodrome aboard a Dnepr rocket. Both were University of Tokyo projects and transmitted telemetry on ham frequencies. They remain active today.

On the same day, three CubeSats with ham capability were launched from Plesetsk MSC aboard a Rockot booster. **CanX-1**, **DTUSat** and **AAU Cubesat** have since reentered the atmosphere.

RS-22, the second Radiosputnik of the 21st century, was launched on September 27, 2003 from Baikonur Cosmodrome aboard a Dnepr rocket. It was a project of Mozhaisky Military Space University and only transmits CW telemetry.

The only Amateur Radio satellite of 2004 turned out to be one of the most popular hamsats of the new century. **AMSAT-OSCAR 51**, also known as Echo, was launched on June 28, 2004 from Baikonur Cosmodrome aboard a Dnepr rocket. AO-51 contained an FM repeater with 144 MHz and 1.2 GHz uplinks and 435 MHz and 2.4 GHz downlinks. Additionally, AO-51 contained a digital subsystem that transmitted telemetry on 70 cm and provided a complete BBS that could be configured on both V and S band uplinks. There was also a 10 meter PSK uplink.

OSCAR 51 went silent in late 2011.

The Indian **VUSat-OSCAR 52** was launched on 5 May, 2005 from Sirharkota, India aboard a PSLV rocket. It has since become the most popular satellite for SSB and CW operating. The satellite listens for signals on the 70-cm band and repeats what it hears on the 2 meter band. Rather than listening and repeating on a single frequency, however, the OSCAR 52 linear transponder relays signals across an entire range of frequencies, allowing it to carry many conversations at once. See the frequency list in **Table 1.1**.

PCSAT-2 was a "satellite" that resembled a suitcase. It was installed by astronaut Soichi Noguchi on the outside of the International Space Station on August 3, 2005. PCSAT-2 completed a four-week test of its PSK-31 transponder mode and operated in APRS packet digipeater mode as well. It was returned to Earth on space shuttle mission STS-115 on September 21, 2006.

AMSAT-OSCAR 51 functions primarily as an FM repeater in space.

AMSAT-OSCAR 54, better known as *SuitSat*, is one of the most unusual Amateur Radio satellites ever placed in orbit. SuitSat was a payload installed inside a discarded Russian Orlan EVA suit that was ejected from the International Space Station on September 8, 2005. It carried a 2-meter Amateur Radio beacon transmitter. Unfortunately, there was an apparent malfunction of the transmitter (or the antenna, or both) and the beacon became very weak, copyable only by hams with well-equipped stations. The last confirmed reception of SuitSat was on Saturday February 18, 2006 by Bob King, VE6BLD.

eXpress-OSCAR 53, known as SSETI Express, was launched on October 27, 2005 from the Plesetsk Cosmodrome in Russia. Shortly after reaching orbit, SSETI entered safe mode and began to send 9600-baud and carrier pulse telemetry. OSCAR 53 apparently deployed its three research CubeSats, but fell silent thereafter.

The Japanese **CubeSat-OSCAR 56** was launched on February 21, 2006 from Kagoshima Space Center aboard a

JAXA M-V 8 rocket. It was a project of Tokyo Institute of Technology Matunaga LSS.

Another Japanese CubeSat, **HITSat-OSCAR 59** was launched on September 22, 2006 from Kagoshima Space Center aboard a JAXA M-V rocket. It features a 1200-baud packet radio bulletin board system. OSCAR 59 reentered the atmosphere in June 2008.

Navy-OSCAR 61 was carried aboard space shuttle mission STS-116. It functioned as an APRS relay until it reentered the atmosphere on December 25, 2007.

Ten satellites reached orbit April 28, 2008 aboard an Indian PSLV-C9 rocket launched from the Satish Dhawan Space Center. The primary payloads were India's CARTOSAT-2A and IMS-1 satellites. In addition to the NLS-5 and RUBIN-8 satellites, the rocket carried six CubeSat research satellites, all of which communicate using Amateur Radio frequencies:

The **SEEDS** satellite was designed and built by students at Japan's Nihon University. SEEDS downloads telemetry in Morse code and 1200-baud FM AFSK packet radio at 437.485 MHz. The satellite also has Slow-Scan TV (SSTV) capability.

AAUSAT-II was the creation of a student team at Aalborg University in Denmark. It downlinks scientific telemetry at 437.425 MHz using 1200 or 9600-baud packet.

Can-X2 was a product of students at the University of Toronto Institute for Aerospace Studies, Space Flight Laboratory (UTIAS/SFL). Can-X2 downlinks telemetry at 437.478 MHz using 4 kbps GFSK, but the downlink is active only when the satellite is within range of the Toronto ground station.

Compass-One was designed and built by students at Aachen University of Applied

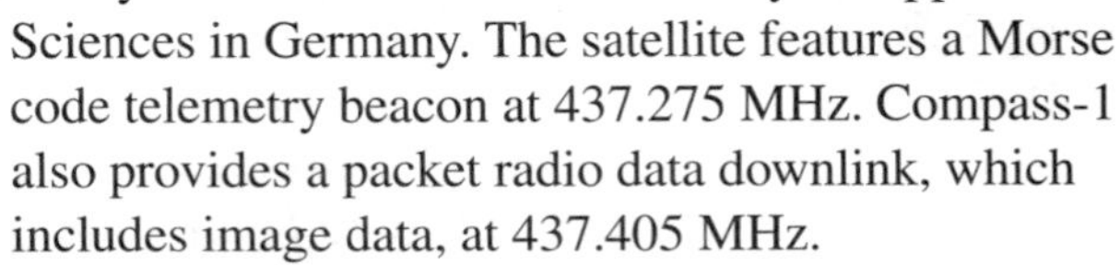

Sciences in Germany. The satellite features a Morse code telemetry beacon at 437.275 MHz. Compass-1 also provides a packet radio data downlink, which includes image data, at 437.405 MHz.

Cute 1.7 + APDII is a satellite created by students at the Tokyo Institute of Technology. This satellite not only provides telemetry, it also offers a 9600-baud packet store-and-forward message relay with an uplink at 1267.6 MHz and a downlink at 437.475 MHz.

Delfi-C³, later designated Dutch-OSCAR 64, was designed and built by students at Delft University of Technology in the Netherlands. It includes an SSB/CW linear transponder. Delfi-C3 downlinks 1200-baud packet telemetry at 145.870 MHz. The linear transponder has an uplink passband from 435.530 to 435.570 MHz and a corresponding downlink passband from 145.880 to 145.920 MHz.

Subbandila-OSCAR 67, a South African satellite was launched in late 2009 and functions as

ISS012E15666

AMSAT-OSCAR 54, better known as SuitSat, is one of the most unusual Amateur Radio satellites ever placed in orbit. SuitSat was a payload installed in a discarded Russian Orlan EVA suit that was ejected from the International Space Station on September 8, 2005.

an FM repeater. Just before the end of 2009, Hope-OSCAR 68, the first Chinese Amateur Radio satellite, reached orbit. It offered an FM repeater and a linear transponder.

The Future?

As this book was written, hams were longing for a high altitude "DX" satellite to replace OSCAR 40. The problem, however, is that it has become increasingly difficult for non-profit amateur satellite organizations to build such complex, powerful birds while relying solely on member dues and donations, thousands of hours of volunteer labor and leftover satellite parts. To make matters worse, affordable launch opportunities for large satellites have also become increasingly scarce, not to mention prohibitively expensive.

The German AMSAT organization, AMSAT-DL, is building a Phase III satellite similar to OSCAR 13 that is presently known as Phase 3 Express. Their goal is to put Phase 3 Express into orbit within a few years.

One intriguing possibility is a Phase IV — *geosynchronous* — amateur satellite created in collaboration with the commercial satellite giant Intelsat. In late 2007 AMSAT-NA announced that it was engaging in discussions with Intelsat about including an Amateur Radio multi-transponder module aboard a future communications satellite. The module would physically reside inside the Intelsat spacecraft and use a common power source. If successful, this collaboration would be a tremendous boon to the amateur community, providing continuous 24/7, hemisphere-wide communications.

In the meantime, you can count on a series of smaller low-Earth orbiting spacecraft to be launched in the future. Not only are they less expensive and easier to launch, they are often funded by colleges and universities.

Satellite Orbits And Tracking

If you've ever been to an arena sports event, you may remember the sellers who roved through the stadium, hawking colorful "program" booklets that contained the complete rosters of each team, player statistics, photographs and more. You could hear them shouting over the public address system, "Get your program! You can't tell who the players are without a program!" Strange as it may sound, the sports program has a parallel in the satellite world.

When this book was written, there were no amateur satellites traveling in geostationary orbits. A satellite in a geostationary orbit appears to be stationary in space from our perspective here on Earth. It remains fixed at a single point in the sky 24 hours a day. There is never any doubt about where it is located. You simply aim your antenna at the bird and communicate. Home satellite TV systems are good examples of this concept. The rooftop parabolic dish antennas never move—they don't need to. Their target is always in the same place.

Amateur satellites, however, are in orbits close to the Earth, or in oblong, elliptical orbits that take them far into space (beyond where geostationary birds reside) before bringing them back toward Earth for a close, slingshot pass. Since the Earth and the satellites are traveling at different speeds, amateur birds do not remain at fixed points in the sky. Instead, amateur satellites rise above the horizon, soar to a certain altitude (elevation) and then set below the horizon once again. Depending on the nature of the orbit, a satellite may be above the horizon for hours, or only for a few minutes. The satellite may appear several times each day, but each

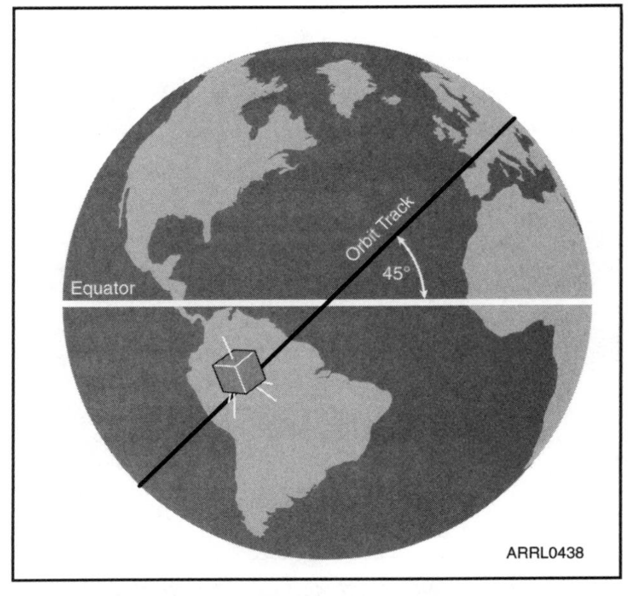

Figure 2.1 — An *inclined* orbit is one that is inclined with respect to the Earth's equator. In this example, the satellite's orbit is inclined at 45° to the Equator.

pass will be at a different maximum elevation and will follow a different track across the heavens. To add to the confusion, a satellite may not appear at the same time each day, although it will follow predictable arrival patterns when plotted over days or weeks.

To enjoy an Amateur Radio satellite you need to know where it is, when it will arrive and how it will move across the sky. In other words, to identify these "players," you do indeed need a "program." You need a basic understanding of satellite orbits and a *program* in a different sense of the word: a computer program that will take the information about a satellite's orbit and turn it into accurate predictions of when it will appear.

Of course, if you *really* want to gain an appreciation of how a satellite travels in its orbit, turn to Appendix A of this book. That's where you will find a detailed of discussion of orbital mechanics by Dr Martin Davidoff, K2UBC.

Types of Orbits

Most active amateur satellites are in various types of Low Earth Orbits (LEOs), although there are satellites planned for future launch that will travel in the elliptical orbits mentioned previously. Let's take a brief look at several of the most common orbits.

An *inclined* orbit is one that is inclined with respect to the Earth's equator. See **Figure 2.1**. A satellite that is inclined 90° would be orbiting from pole to pole; smaller inclination angles mean that the satellite is spending more time at lower latitudes. The International Space Station, for example, travels in an orbit that is inclined about 50° to the equator. Satellites that move in these orbits frequently fall into the Earth's shadow (eclipse), so they must rely on battery systems to provide power when the solar panels are not illuminated. Depending on the inclination angle, some locations on the Earth will never have good access because the satellites will rarely rise above their local horizons. This was true, for example, in the days when the US Space Shuttles carried Amateur

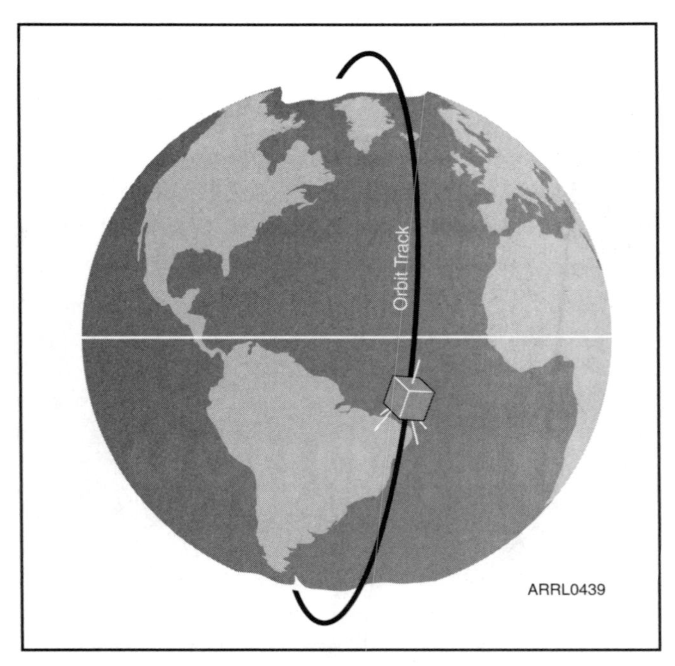

Figure 2.2 — A *sun-synchronous* orbit takes the satellite over the north and south poles. A satellite in this orbit allows every station in the world to enjoy at least one high-elevation pass per day.

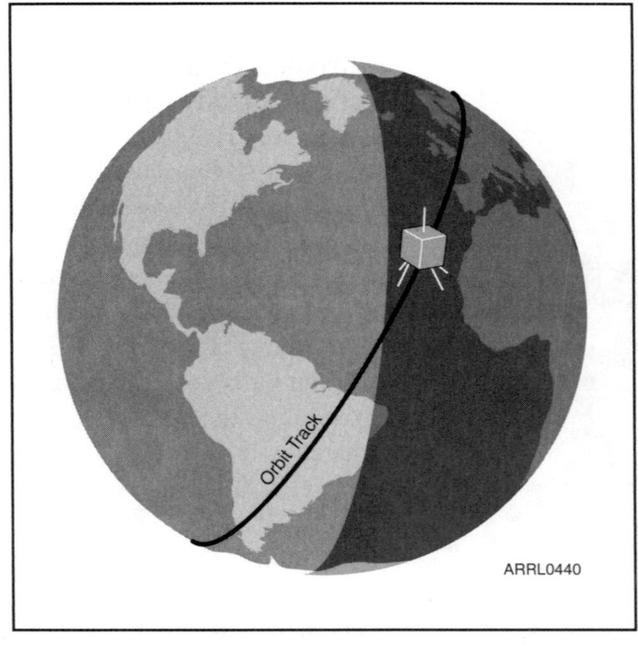

Figure 2.3 — A *dawn-to-dusk* orbit is a variation on the sun-synchronous model except that the satellite spends most of its time in sunlight and relatively little time in eclipse.

Radio operators. Shuttle orbits were usually inclined at low angles and hams living in the northern US rarely enjoyed passes that brought the shuttle to a decent elevation above their local horizon.

A *sun-synchronous* orbit takes the satellite over the north and south poles. See **Figure 2.2**. There are two advantages to a sun-synchronous orbit: (1) the satellite is available at approximately the same time of day, every day and (2) everyone, no matter where they are, will enjoy at least one high-altitude pass per day. OSCAR 51 is a good example of a satellite that travels in a sun-synchronous orbit.

A *dawn-to-dusk* orbit is a variation on the sun-synchronous model except that the satellite spends most of its time in sunlight and relatively little time in eclipse. OSCAR 27 travels in a dawn-to-dusk orbit. See **Figure 2.3**.

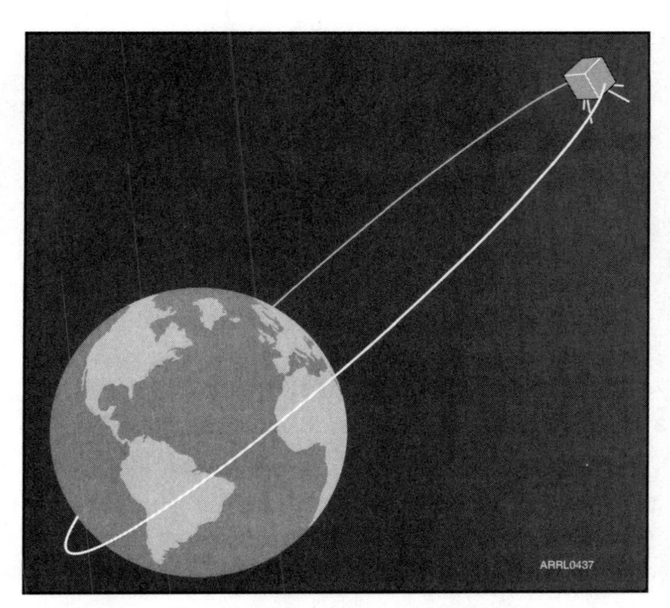

Figure 2.4 — The *Molniya* orbit is an elliptical orbit that carries the satellite far into space at its greatest distance from Earth (apogee). To observers on the ground, the satellite at apogee appears to hover for hours at a time before it plunges earthward and (often) sweeps within 1000 km at its closest approach (perigee).

The *Molniya* orbit (**Figure 2.4**) was pioneered by the former Soviet Union. It is an elliptical orbit that carries the satellite far into space at its greatest distance from Earth (apogee). To observers on the ground, the satellite at apogee appears to hover for hours at a time before it plunges earthward and sweeps to (sometimes) within 1000 km or so of the Earth at its closest approach (perigee). One great advantage of the Molniya orbit is that the satellite is capable of "seeing" an entire hemisphere of the planet while at apogee. Hams can use a Molniya satellite to enjoy long, leisurely conversations spanning thousands of kilometers here on Earth. Another great advantage of the Molniya orbit is that a single satellite can give most hams around the world such "leisurely access" to it at least *part* of the time. By contrast, it takes three geostationary (and interlinked) satellites to provide continuous worldwide coverage. When this book was being written, there were no active Molniya hamsats in orbit. However, that may change within a few years.

Satellite Footprints

Speaking of how much of our planet a satellite sees, it is important to understand the concept of the satellite's *footprint*. A satellite footprint can be loosely defined as the area on the Earth's surface that is "illuminated" by the satellite's antenna systems at any given time. Another way to think of a footprint is to regard it as the zone within which stations can communicate with each other through the satellite.

Unless the satellite in question is geostationary, footprints are constantly moving. Their sizes can vary considerably, depending on the altitude of the satellite. The footprint of the low-orbiting International Space Station is about 600 km in diameter. In contrast, the higher orbiting OSCAR 52 has a footprint that is nearly 1500 km across. See the example of a satellite footprint in **Figure 2.5.** The amount of time you have available to communicate depends on how long your station remains within the footprint. This time can be measured in minutes, or in the case of a satellite in a Molniya orbit, hours.

It is worthwhile to note that the size and even the shape of a footprint can also vary according to the type of antenna the satellite is using. A highly directional antenna with a narrow beamwidth will create a small footprint even though the satellite is traveling in a high-altitude orbit. This usually isn't an issue for amateur satellites, however.

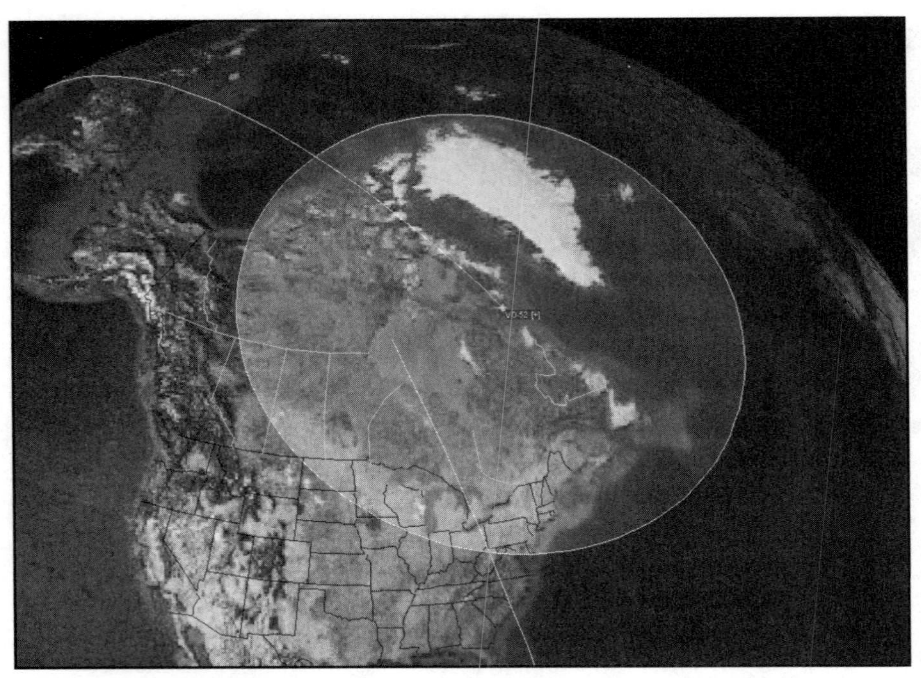

Figure 2.5 — This image shows the circular footprint of OSCAR 52 as depicted by *Nova* satellite-tracking software. The footprint indicates the area of the Earth that is visible to the satellite at any given time.

Understanding Your Place in the World

Before you can track an amateur satellite and communicate with it, you must first determine your own location with reasonable accuracy and understand your orientation to the expected path of the satellite.

Determining your location on the globe in terms of latitude and longitude coordinates is much easier today than it used be. If you own a Global Positioning System (GPS) receiver, you can use it to determine your coordinates almost instantly. You simply take the receiver outdoors (or hold it up to a window), wait for it to obtain enough signals to determine your position, and write down the resulting latitude and longitude.

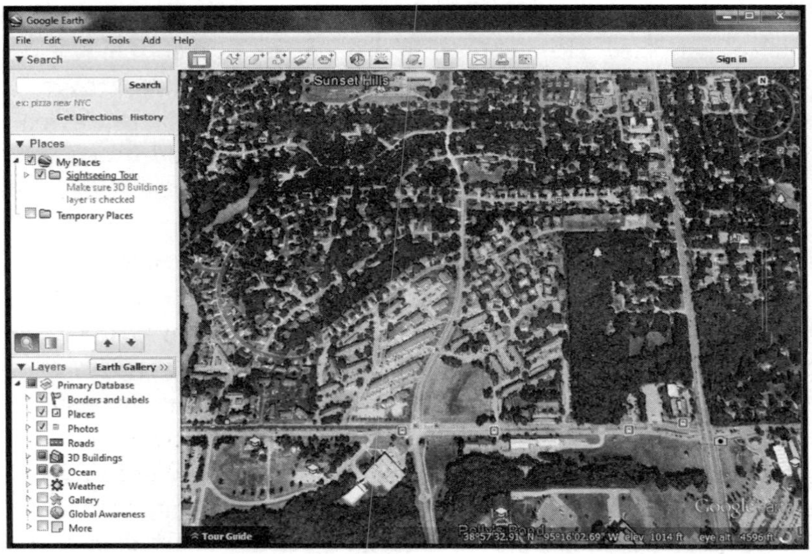

If you don't own a GPS receiver, the Internet is your next best option. There are a number of mapping programs can be downloaded.

For example, there is Google Earth at **www.google.com/earth**. Simply drag the cursor over your location and you will see your coordinates in the bottom right corner. See **Figure 2.6**.

Figure 2.6 — Google Earth.

How precise do you need to be? If you plan on using movable directional "beam" antennas for your satellite station, the more precision the better. These antennas create more focused

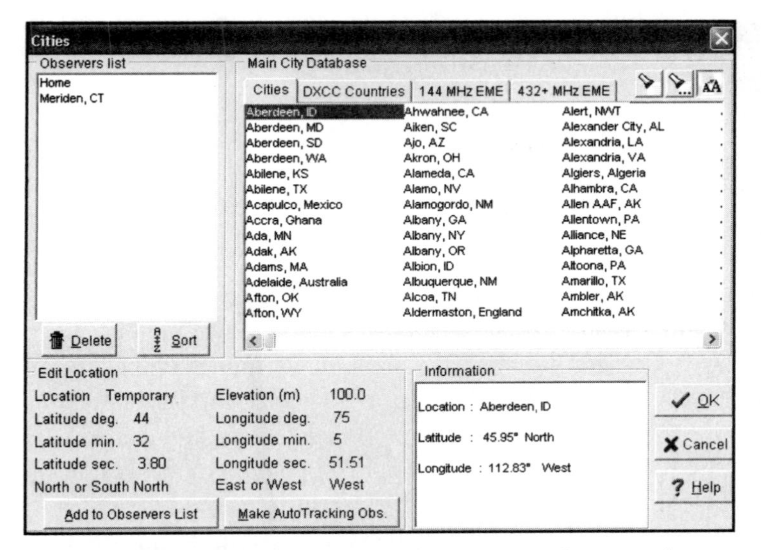

Some satellite tracking programs allow you to select your location from an extensive list of cities.

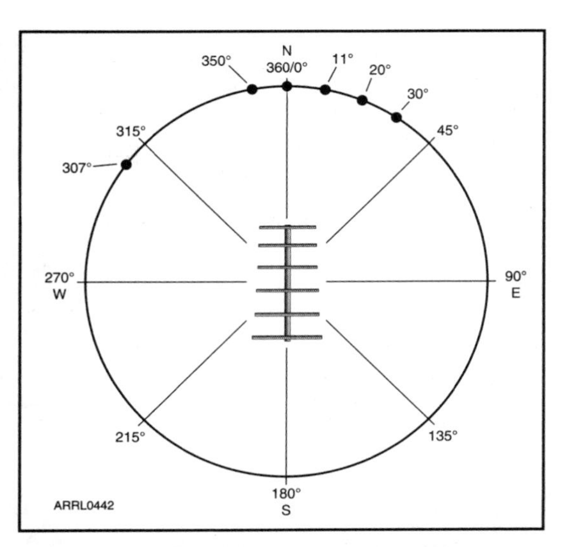

Figure 2.8 — The azimuth path of the International Space Station for our hypothetical pass.

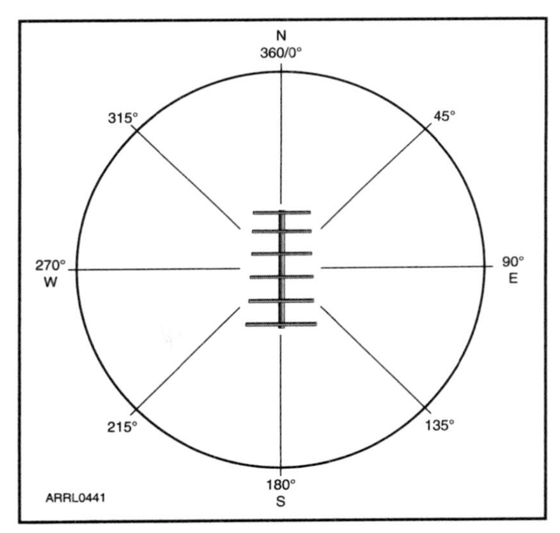

Figure 2.7 — Azimuth is the direction, in degrees referenced to true north, that an antenna must be pointed to receive a satellite signal. Imagine your station in the center of a giant compass circle that is divided in degree increments from 0 to 360. North is 0° (actually, it is also 360°), east is 90°, south is 180° and west is 270°.

(and therefore more narrow) transmitting and receiving signal patterns, so you want to be reasonably sure they are pointing in the proper direction. Your satellite tracking software will determine this direction for you, but its ability to give you accurate aiming information is highly dependent on it "knowing" where you are located in the first place.

On the other hand, if you are using omnidirectional antennas that create broad signal patterns, or directional antennas that don't move, the need for precision is less critical. In fact, the latitude and longitude of the nearest city will suffice. You can get this information from the US Geological Survey Web site at **www.usgs.gov**. Look for the link to the Geographic Names Information System.

Azimuth and Elevation

In addition to pinpointing your station location, there is the additional matter of your orientation (or your antenna orientation) to the satellites as they pass overhead. Your satellite tracking software will display a satellite's position in terms of *azimuth* and *elevation*.

Azimuth is the direction, in degrees referenced to true north, that an antenna must be pointed to receive a satellite signal. See **Figure 2.7** and imagine your station in the center of a giant compass circle that is divided in degree increments from 0 to 360. North is 0° (actually, it is also 360°), east is 90°, south is 180° and west is 270°. If your tracking software indicates that you need to point your antenna to an azimuth of 135°, you're going to point it southeast.

Let's take a look at a more detailed example. Once again, your station in **Figure 2.8** is in the center of the compass circle. According to your satellite tracking program, the International Space Station (ISS) is scheduled to rise above your local horizon at precisely 03:57:30 UTC. The program may describe the satellite's azimuth path like this:

Time	*Azimuth (degrees)*
03:57	307
03:58	350
03:59	0
04:00	11
04:01	20
04:02	30

When you plot these azimuth points on the circle in Figure 2.8, you can quickly see the horizontal path the satellite is going to take. The bird is going to rise in your northwestern sky and quickly move toward the east, finally dipping below your horizon at about 30°. If you have rotating antennas, you can now see that they'll need to be pointing northwest at the beginning of the satellite's path (or "pass" as it is sometimes called) across your section of sky, and then track around the circle from 307°, to 0° and so on until they are pointing to 30° when the satellite disappears.

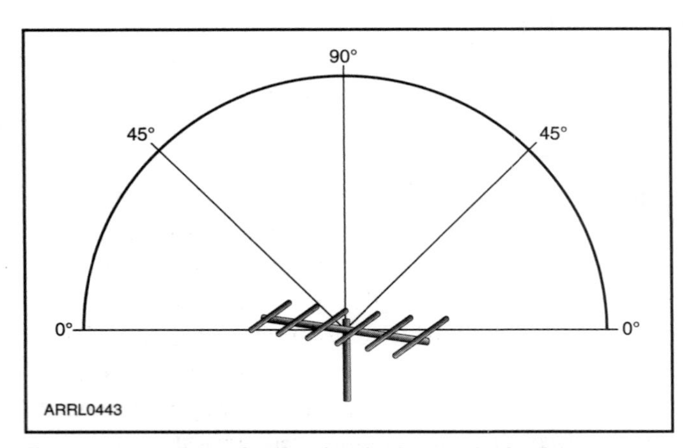

Figure 2.9 — Elevation is simply the angle, in degrees, between your station and the satellite, referenced to the Earth's surface.

Let's add another dimension to our satellite track—*elevation*. Elevation is simply the angle, in degrees, between your station and the satellite, referenced to the Earth's surface. See **Figure 2.9**. The elevation angle begins at 0° with the satellite at the horizon and increases to 90° when the satellite is directly overhead. Elevation is every bit as critical as azimuth if you are using directional antennas. Not only do your antennas need to be pointed at the satellite as it appears to move in the azimuth plane, they must also tilt up and down to track the satellite as it appears to move in the vertical plane.

Many amateur satellite stations use devices known as *az/el* (azimuth/elevation) *rotators* to move their antennas in both planes as the satellite streaks across the sky. However, az/el rotators are not strictly necessary to enjoy satellite operating. A well-designed omnidirectional antenna system can offer acceptable performance even though it doesn't move to track the satellite. You can even use directional antennas that are fixed at about 45° elevation and only turn horizontally using a conventional antenna rotator. We'll discuss station antennas in more detail later in this book.

Even if you are not using directional antennas, knowing a satellite's elevation track is important for another reason. Unless you live in Kansas or a similarly flat location, chances are you do not have a clear view to the horizon in every direction. Maybe there is a mountain, hill or building blocking the way. If you are trying to receive a microwave signal from a satellite, the RF absorption properties of trees can present serious obstacles, too. The elevations of these objects represent your true *radio horizons* in whichever direction they may lie (**Figure 2.10**). If you have a ridge to the north with an elevation of 25° above the horizon as viewed from your station, your northern radio horizon *begins* at 25° elevation. You can't communicate with a satellite in your northern sky until it rises above 25°, so you'll have to take that fact into account when you view the information provided by your satellite tracking software. Your software may tell you that the AOS (Acquisition Of Signal) time is 0200 UTC as the satellite rises in the north, but you won't be able to receive the bird until it reaches 25°.

The man who explained why sound waves (and radio waves) change as an object speeds toward us, then away: Christian Doppler.

Usually…and particularly for satellites in low Earth orbits…as the satellite's elevation angle increases, its distance from you decreases. This is a good thing since the closer the satellite, the stronger the radio signal. With that idea in mind, the higher the elevation of a satellite pass, the better, right? Well…yes and no. Remember that satellites are moving at high speeds relative to your position. As they move closer to you (move higher in elevation), the *Doppler Effect* increasingly comes into play.

The Doppler Effect

The Doppler Effect, named after scientist Christian Doppler, is the change in frequency and wavelength of a wave (radio waves, in this case) as perceived by an observer moving relative to the source of the waves. See **Figure 2.11**. Thanks to the Doppler Effect, as a satellite moves toward your location, its signal will appear to *increase* in frequency; as it moves away from you, its signal will *decrease* in frequency.

It is important to realize that the frequency of the signal that the satellite transmits *does not actually change*, regardless of what is happening at your station. To understand this, consider the following baseball analogy illustrated in **Figure 2.12**. A baseball pitcher throws one ball every second and the ball takes one second to travel the distance between the pitcher's mound and home plate. If the pitcher is stationary, the catcher will receive one ball every second. That's because the velocity of the ball and the distance between the pitcher and the catcher remain unchanged. So far, so good. This is exactly the condition present when a satellite is geostationary.

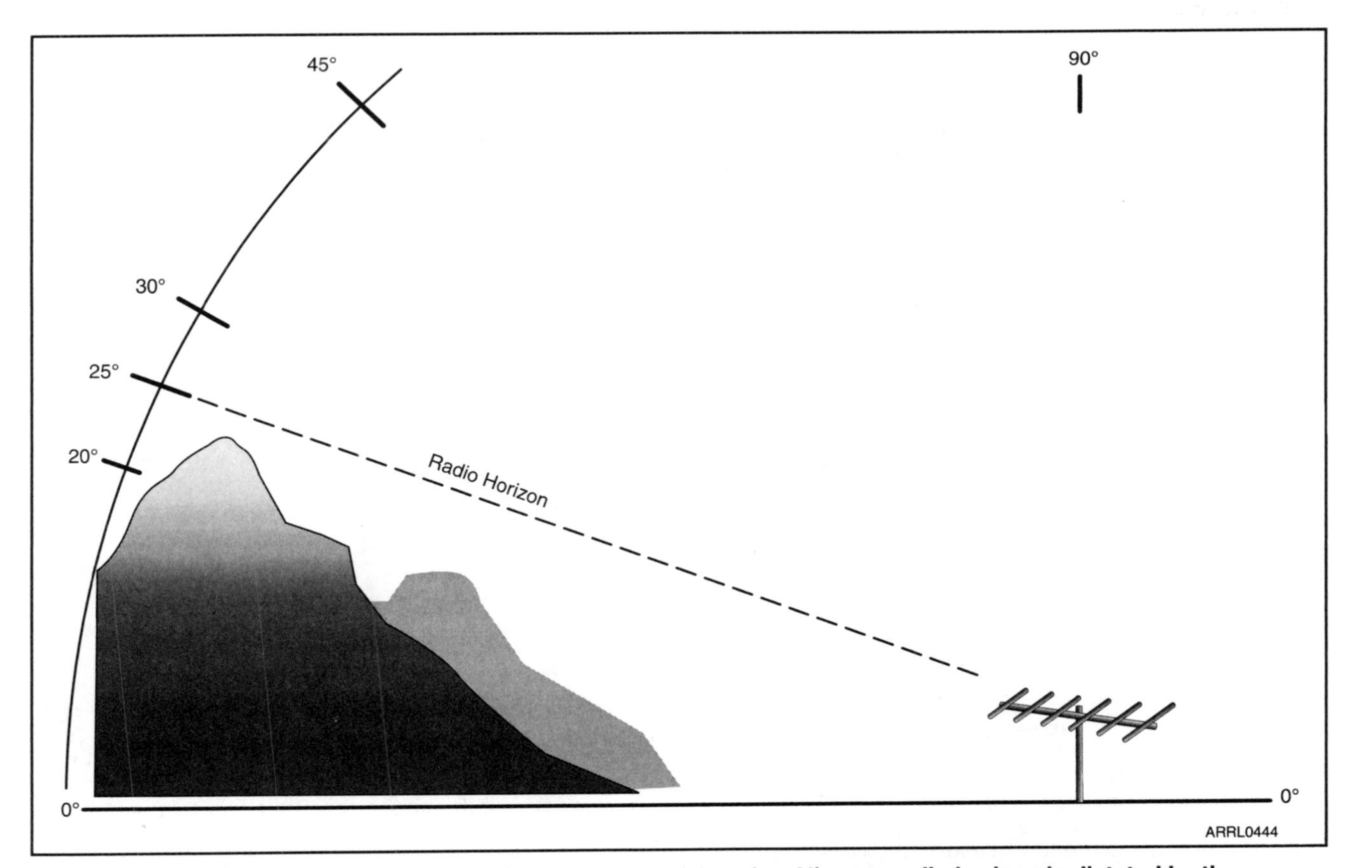

Figure 2.10 — Unless you have a clear shot to the horizon (elevation 0°), your radio horizon is dictated by the maximum elevation of any obstacles between you and the satellite.

However, if the pitcher begins moving toward the catcher, the time it takes the ball to travel between the pitcher and the catcher *decreases*. From the catcher's point of view, the pitcher may still be tossing balls at a rate of one per second, but the balls are taking less than one second to reach him.

If you imagine each ball representing the crest of a wave and the wavelength being the distance between one ball and another, the wavelength is decreasing as the pitcher moves toward the catcher. Since a shorter wavelength translates to a higher frequency, the frequency appears to increase. Conversely, as the wavelength increases (as the satellite moves away), the frequency decreases.

You probably experience the Doppler Effect almost every day. When a fire truck approaches at high speed on a nearby freeway, you hear its siren blaring at a higher pitch, shifting downward as the truck passes and speeds off into the distance. The same thing happens with satellites, except the frequency change is applied to radio waves.

At a practical level, a high elevation satellite pass can be problematic because the frequency change caused by the Doppler Effect can be considerable. This means that you'll need to constantly change your transceiver frequency to compensate. This can be quite a juggling act when you're also trying to carry on a conversation and move your antennas at the same time. That's why many satellite operators rely on computers to control their antennas or transceivers, or both. We'll

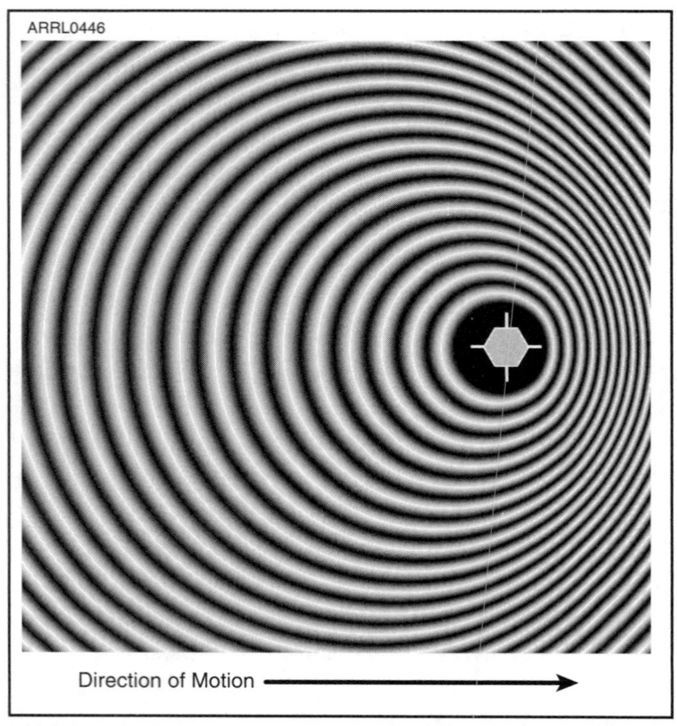

Figure 2.11 — You can think of the Doppler Effect as being caused by radio waves "crowding up" as a satellite moves toward your location. As a result, its signal will appear to *increase* in frequency as it moves toward you; as it moves away from you, its signal will *decrease* in frequency.

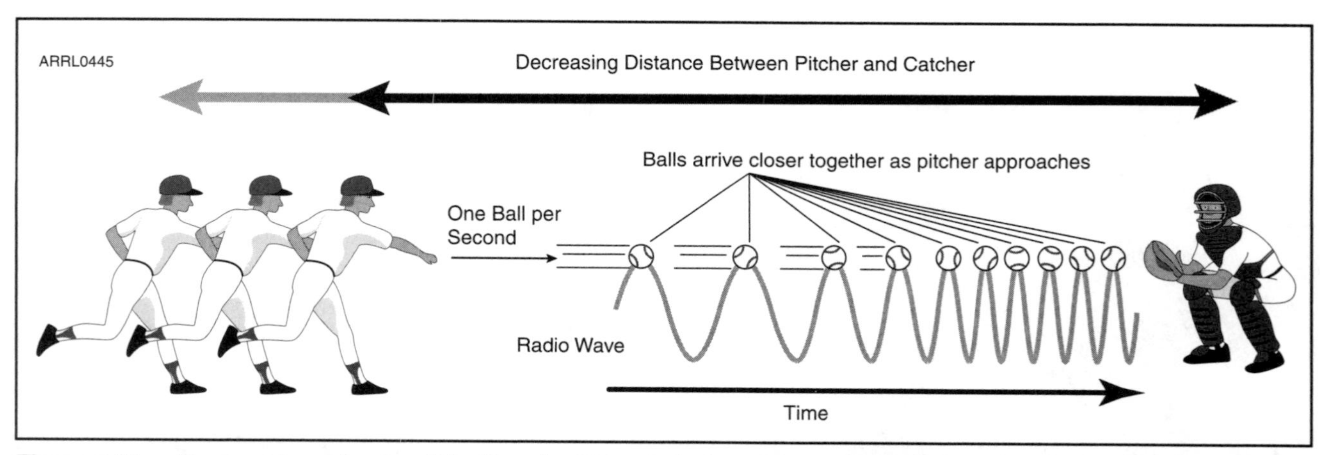

Figure 2.12 — One way to understand the Doppler Effect is by imagining a baseball pitcher who throws one ball every second. The ball takes one second to travel the distance between the pitcher's mound and home plate. If the pitcher is stationary, the catcher will receive one ball every second. That's because the velocity of the ball and the distance between the pitcher and the catcher remain unchanged. However, if the pitcher begins moving toward the catcher, the time it takes the ball to travel between the pitcher and the catcher decreases. If you imagine each ball representing the crest of a wave and the wavelength being the distance between one ball and another, the wavelength is decreasing as the pitcher moves toward the catcher. Since a shorter wavelength translates to a higher frequency, the frequency appears to increase. Conversely, as the wavelength increases, the frequency decreases.

discuss the operational aspects of coping with Doppler in a later chapter. For now, suffice it to say that while high elevation passes are best for signal strength, they present their own challenges thanks to the Doppler Effect.

Azimuth and Elevation Combined

Let's combine azimuth and elevation for a truly realistic satellite track, using our previous example with the International Space Station. We'll add the station's downlink frequency so we can see the Doppler Effect in action.

Time	*Azimuth (degrees)*	*Elevation (degrees)*	*Frequency (MHz)*
03:57	307	0	145.804
03:58	350	10	145.803
03:59	0	18	145.800
04:00	11	9	145.798
04:01	20	5	145.797
04:02	30	0	145.795

In this example, the International Space Station rises to an elevation of 18° at 03:59 UTC before sinking back down to the horizon at 04:02 UTC. This is considered a low-elevation pass. If you have objects in your northern sky that rise above 18° elevation, you won't be able to communicate with the space station during this pass. The space station is transmitting at 145.800 MHz, but you'll notice that the frequency change caused by the Doppler Effect is minimal because the distance between you and the space station doesn't change dramatically. Remember: It is the relative motion between you and the satellite that increases the effect. Less relative motion means less Doppler.

Now we'll modify our example, making it a high-elevation pass.

Time	*Azimuth (degrees)*	*Elevation (degrees)*	*Frequency (MHz)*
03:57	307	0	145.810
03:58	350	10	145.808
03:59	0	25	145.806
04:00	11	40	145.804
04:01	20	65	145.802
04:02	30	80	145.800
04:03	36	60	145.798
04:04	41	45	145.796
04:05	50	29	145.794
04:06	55	15	145.792
04:07	59	0	145.790

This pass of the ISS is also plotted graphically in **Figure 2.13**. In this illustration, we combine the azimuth and elevation, creating a "radar screen" display with your station in the center.

There are several interesting things to note in this example. Did you notice that this high-elevation pass (topping out at 80° at 04:02 UTC) had a longer overall duration than the previous low-elevation pass? The low-elevation pass lasted only 5 minutes; this pass was a full 10 minutes in length. Obviously, when an object is tracking to a high elevation

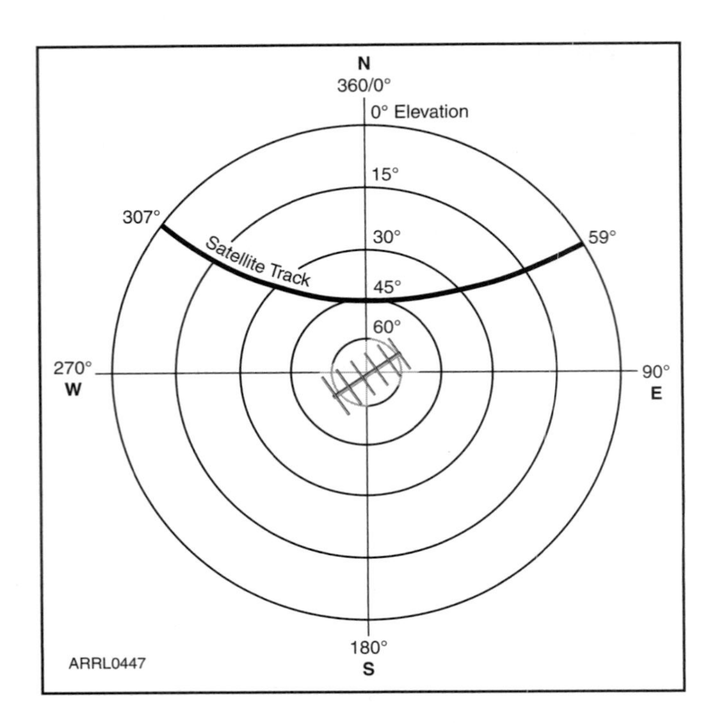

Figure 2.13 — In this illustration, we combine the azimuth and elevation, creating a "radar screen" display of a satellite pass with your station in the center. The concentric rings represent elevation; the outer ring represents both zero degrees of elevation as well as azimuth.

in the sky (almost directly overhead in this example), it is in view for a longer period.

Did you also notice what the Doppler Effect did to the space station's downlink signal frequency? It started out at 145.810 MHz, shifted down to 145.800 MHz at maximum elevation, and then continued downward until it reached 145.790 MHz as the station slipped below the horizon. That's a 20-kHz frequency shift throughout the pass!

Satellite Tracking Software

You'll find many satellite software programs have been written for *Windows*, *Mac* and *Linux* operating systems. Several popular applications are listed in **Table 2.1**. When computers were first employed to track amateur satellites, they provided only the most basic, essential information: when will the satellite be available (AOS, acquisition of signal), how high will the satellite rise in the sky and when the satellite is due to set below your horizon (LOS, loss of signal). Today we tend to ask a great deal more of our tracking programs. Modern applications still provide the basic information, but they usually offer many more features such as:

★ The spacecraft's operating schedule, including which transponders and beacons are on.

★ Predicted frequency offset (Doppler shift) on the link frequencies.

★ The orientation of the spacecraft's antennas with respect to your ground station and the distance between your ground station and the satellite.

★ Which regions of the Earth have access to the spacecraft; that is, who's in QSO range?

★ Whether the satellite is in sunlight or being eclipsed by the Earth. Some spacecraft only operate when in sunlight.

★ When the next opportunity to cover a selected terrestrial path (mutual window) will occur.

★ Changing data can often be updated at various intervals such as once per minute…or even once per second.

A number of applications do even more. Some will control antenna rotators, automatically keeping directional antennas aimed at the target satellite. Other applications will also control the radio to automatically compensate for frequency changes caused by Doppler shifting.

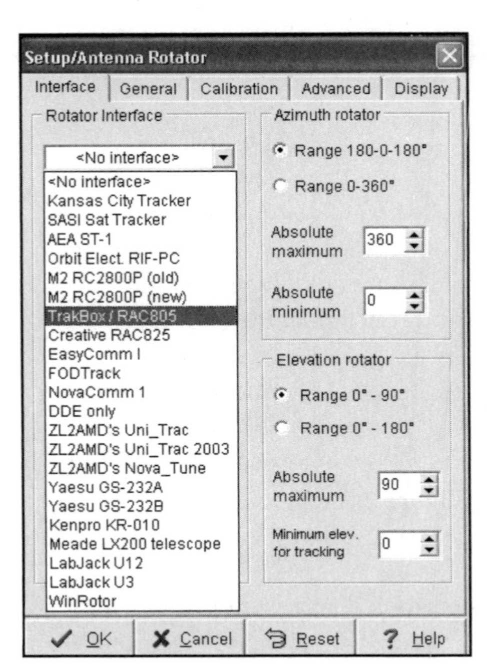

The antenna rotator control configuration screen in *Nova*.

Table 2.1
A Sampling of Satellite Tracking Software

Name	***Source***	***Operating System***	***Radio Control?***	***Antenna Control?***
Nova	**www.arrl.org** (store)	*Windows*	No	Yes
SatPC32	**www.amsat.org** (store)	*Windows*	Yes	Yes
SatScape	**http://satscape.info/satscape/**	*Windows*	Yes	Yes
MacDoppler	**www.dogparksoftware.com/MacDoppler.html**	Mac OS	Yes	Yes
Ham Radio Deluxe	**www.hrdsoftwarellc.com**	*Windows*	Yes	No
Predict	**www.qsl.net/kd2bd/predict.html**	*Linux*	No	Yes
WinOrbit	**www.sat-net.com/winorbit/**	*Windows*	No	No

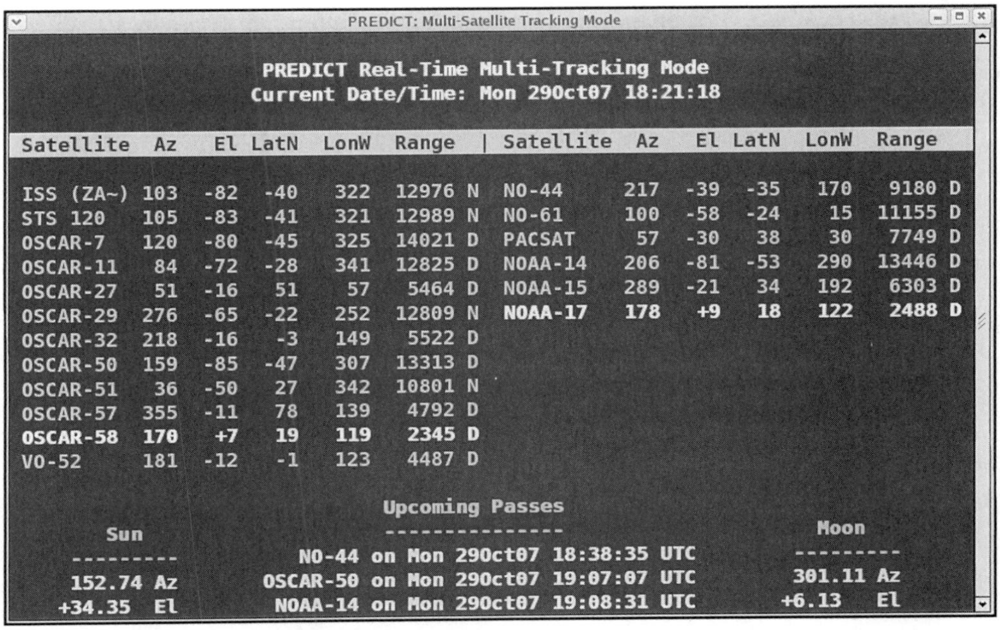

```
PREDICT: Multi-Satellite Tracking Mode

                PREDICT Real-Time Multi-Tracking Mode
               Current Date/Time: Mon 29Oct07 18:21:18

Satellite   Az   El LatN  LonW  Range  | Satellite   Az   El LatN  LonW  Range

ISS (ZA~) 103  -82  -40   322  12976 N  NO-44       217  -39  -35   170   9180 D
STS 120   105  -83  -41   321  12989 N  NO-61       100  -58  -24    15  11155 D
OSCAR-7   120  -80  -45   325  14021 D  PACSAT       57  -30   38    30   7749 D
OSCAR-11   84  -72  -28   341  12825 D  NOAA-14     206  -81  -53   290  13446 D
OSCAR-27   51  -16   51    57   5464 D  NOAA-15     289  -21   34   192   6303 D
OSCAR-29  276  -65  -22   252  12809 N  NOAA-17     178   +9   18   122   2488 D
OSCAR-32  218  -16   -3   149   5522 D
OSCAR-50  159  -85  -47   307  13313 D
OSCAR-51   36  -50   27   342  10801 N
OSCAR-57  355  -11   78   139   4792 D
OSCAR-58  170   +7   19   119   2345 D
VO-52     181  -12   -1   123   4487 D

                          Upcoming Passes
      Sun                 --------------                    Moon
   ---------      NO-44 on Mon 29Oct07 18:38:35 UTC      ---------
   152.74 Az   OSCAR-50 on Mon 29Oct07 19:07:07 UTC      301.11 Az
   +34.35 El    NOAA-14 on Mon 29Oct07 19:08:31 UTC      +6.13   El
```

Predict **is satellite tracking software for** ***Linux.***

Adding additional spacecraft to the scenario suggests more questions: which satellites are currently in range, how long will each be accessible, will any new spacecraft be coming into range in the near future and so on. Obviously there is a great deal of information of potential interest. Programmers developing tracking software often find that the real challenge is not solving the underlying physics problems, but deciding what information to include and how to present it in a useful format. This is especially true since users have different interests, levels of expertise and needs. Some prefer to see the information in a graphical format, such as a map showing real-time positions for all satellites of interest. Others may prefer tabular data such as a listing of the times a particular spacecraft will be in range over the next several days.

There are also several Internet sites where you can do your tracking online. This eliminates all the hassles associated with acquiring and installing software. The currently available online tracking sites are not as powerful or flexible as the software you can install on your PC, however. One interesting site of this type is maintained by AMSAT-NA and you'll find it at **www.amsat.org/amsat-new/tools/predict/**.

Getting Started With Software

There are so many different types of satellites software, and they change so frequently, it would be foolhardy to attempt to give you detailed operational descriptions in any book. The book would be obsolete a month after it came off the press!

Even so, there are a number of aspects of satellite-tracking software that rarely change.

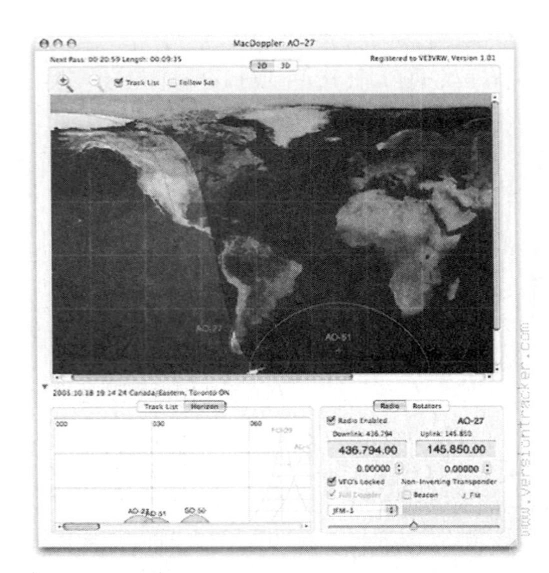

If you own a Mac, you can track amateur satellites with *MacDoppler*.

For example, we spent some time discussing how to determine your location with sufficient accuracy to be useful for satellite tracking. The next step is to get that information into your chosen program.

Most programs will ask you to enter your station location as part of the initial setup process. Some applications use the term "observer" to mean "station location," but the terms are synonymous for the sake of our discussion. Sophisticated programs will go as far as to provide you with a list of cities that you can select to quickly enter your location. (Yes, the location of a nearby city is adequate for most applications.) Other programs will ask you to enter your latitude and longitude coordinates manually.

When entering latitude, longitude (and other angles), make sure you know whether the computer expects degree-minute or decimal-degree notation. Following the notation used by the on-screen prompt usually works. Also make sure you understand the units and sign conventions being used. For example, longitudes may be specified in negative number for locations west of Greenwich (0° longitude). Latitudes in the southern hemisphere may also require a minus sign. Fractional parts of a degree will have very little effect on tracking data so in most cases you can just ignore it.

Dates can also cause considerable trouble. Does the day or month appear first? Can November be abbreviated Nov or must you enter 11? The number is almost always required. Must you write 2010 or will 10 suffice? Should the parts be separated by colons, dashes or slashes? The list goes on and on. Once again, the prompt is your most important clue. For example, if the prompt reads "Enter date (DD:MM:YY)" and you want to enter Feb. 9, 2010 follow the format of the prompt as precisely as possible and write 09:02:10.

When entering numbers, commas should never be used. For example, if a semi-major axis of 20,243.51 km must be entered, type 20243.51 with the comma and units omitted. It takes a little time to get used to the quirks of each software package, but you'll soon find yourself responding automatically.

Once you have your coordinates entered, you're still not quite done. The software now "knows" its location, but it doesn't know the locations of the satellites you wish to track. The only way the software can calculate the positions of satellites is if it has a recent set of *orbital elements*.

The popular *Sat32PC* software.

Orbital Elements

Orbital elements are a set of six numbers that completely describe the orbit of a satellite at a specific time. Although scientists may occasionally use different groups of six quantities, radio amateurs nearly always use the six known as Keplerian Orbital Elements, or simply *Keps*. (Kepler, you may recall, discovered some interesting things about planetary motion back in the 17th century!)

These orbital elements are derived from very precise observations of each satellite's orbital motion. Using precision radar and highly sensitive optical observation techniques, the North American Aerospace Defense Command (NORAD) keeps a very accurate catalog of almost everything in Earth orbit. Periodically, they issue the unclassified portions of this information to the National Aeronautics and Space Administration (NASA) for release to the general public. The information is listed by individual catalog number of each satellite and contains numeric data that describes, in a mathematical way, how NORAD observed the satellite moving around the Earth at a very precise location in space at a very precise moment in the past.

Without getting into the complex details of orbital mechanics (or Kepler's laws!) suffice it to say that your software simply uses the orbital element information NASA publishes that describe where a particular satellite was "then" to solve the orbital math and make a prediction (either graphically or in tabular format) of where that satellite ought to be "now". The "now" part of the prediction is based on the local time and station location information you've also been asked to load into your software.

Orbital elements are frequently distributed with additional numerical data (which may or may not be used by a software tracking program) and are commonly available in two forms (see **Tables 2.2** and **2-3**).

Let's use the easier-to-understand AMSAT format to break down the meaning, line by line.

The first two entries identify the spacecraft. The first line is an informal ***satellite name***. The second entry, ***Catalog Number***, is a formal ID assigned by NASA.

Table 2-2
Typical NASA Two-Line Elements

```
AO-27
1 22825U 93061C   08024.00479406 -.00000064  00000-0 -86594-5 0  8811
2 22825 098.3635 349.6253 0008378 336.4256 023.6532 14.29228459747030
```

Table 3-3
An Example of AMSAT Verbose Elements

Satellite:	AO-51
Catalog number:	28375
Epoch time:	08024.16334624
Element set:	14
Inclination:	098.0868 deg
RA of node:	056.7785 deg
Eccentricity:	0.0083024
Arg of perigee:	232.8417 deg
Mean anomaly:	126.5166 deg
Mean motion:	14.40594707 rev/day
Decay rate:	7.0e-08 rev/day^2
Epoch rev:	18752
Checksum:	310

The next entry, ***Epoch Time***, specifies the time the orbital elements were computed. The number consists of two parts, the part to the left of the decimal point that describes the year and day, and the part to the right of the decimal point that describes the (very precise) time of day. For example, 96325 .465598 refers to 1996, day 325, time of day .465598.

The next entry, ***Element Set***, is a reference used to identify the source of the information. For example, 199 indicates element set number 199 issued by AMSAT. This information is optional.

The next six entries are the six key orbital elements.

Inclination describes the orientation of the satellite's orbital plane with respect to the equatorial plane of the Earth. Recall from earlier in this chapter that the higher a satellite's orbital inclination, the more time the bird spends away from the Equator.

RAAN, Right Ascension of Ascending Node, specifies the orientation of the satellite's orbital plane with respect to fixed stars.

Eccentricity refers to the shape of the orbital ellipse. You may recall our earlier discussion of elliptical Molniya orbits. These orbits are highly eccentric. The value of the eccentricity element also yields some rough information as to the shape of the orbit the satellite is following. The closer this number is to "0", the more circular the orbit of the satellite tends to be. Conversely, an eccentricity value approaching "1", indicates the satellite is following a more elliptically shaped (possibly a Molniya) orbital path. For example, many Molniya orbit satellites have eccentricities in the .6 to .7 range.

Argument of Perigee describes where the perigee of the satellite is located in the satellite orbital plane. Recall that a satellite's perigee is its closest approach to the Earth. When the argument of perigee is between 180° and 360° the perigee will be over the Southern Hemisphere. Apogee—a satellite's most distant point from the Earth—will therefore occur above the Northern Hemisphere.

Mean Anomaly locates the satellite in the orbital plane at the epoch. All programs use the astronomical convention for mean anomaly (MA) units. The mean anomaly is 0° at perigee and 180° at apogee. Values between 0° and 180° indicate that the satellite is headed up toward apogee. Values between 180° and 360° indicate that the satellite is headed down toward perigee. More about this later.

Mean Motion specifies the number of revolutions the satellite makes each day. This element indirectly provides information about the size of the elliptical orbit.

Decay Rate is a parameter used in sophisticated tracking models to take into account how the frictional drag produced by the Earth's atmosphere affects a satellite's orbit. It may also be referred to as rate of change of mean motion, first derivative of mean motion, or drag factor. Although decay rate is an important parameter in scientific studies of the Earth's atmosphere and when observing satellites that are about to reenter, it has very little effect on day-to-day tracking of most Amateur Radio satellites. If your program asks for drag factor, enter the number provided. If the element set does not contain this information enter zero—you shouldn't discern any difference in predictions. You usually have a choice of entering this number using either decimal form or scientific notation. For example, the number –0.00000039 (decimal form) can be entered as –3.9e–7 (scientific notation). The e–7 stands for 10 to the minus seventh power (or 10 exponent –7). In practical terms e–7 just means move the decimal in the preceding number 7 places to the left. If this is totally confusing, just remember that in most situations entering zero will work fine.

Epoch revolution is just another term for the expression "Orbit Number" that we discussed earlier. The number provided here does not affect tracking data, so don't worry if different element sets provide different numbers for the same day and time.

The ***Checksum*** is a number constructed by the data transmitting station and used by the receiving station to check for certain types of transmission errors in data files. It does not bear any relationship to a satellite's orbit.

In the "old days" of satellite-tracking software you had to enter the orbital elements by hand. This was a tedious and risky process. If you entered an element number incorrectly, you would generate wildly inaccurate predictions.

Today, thankfully, most satellite-tracking programs have greatly streamlined the process. One method of entering orbital elements is to grab the latest set from the AMSAT-NA Web site at **www.amsat.org** (look under "Keps" in the main menu). You can download the element set as a text file and then tell your satellite-tracking program to read the file and create the database. Another excellent site is **CelesTrak** at **celestrak.com**. Your program will probably be able to read either the AMSAT or NASA formats.

If you're fortunate to own sophisticated tracking software such as *Nova*, and you have access to the Internet, the program will reach into cyberspace, download and process its Keps automatically. All it takes is a single click of your mouse button. Some programs can even be configured to download the latest Keps on a regular basis without any prompting from you.

The Need for Fresh Keps

So how often do you need to grab a set of fresh orbital elements? The answer depends on several factors including the satellite orbit, the location of the ground station, the directional patterns of the ground station antennas, whether one is interested in short DX windows at AOS/LOS, if automatic Doppler compensation is being used and so on. With all these factors affecting the situation, there can't be a single simple answer, but looking at some relevant information and considering a number of typical situations can provide some helpful guidelines. *Note that the age of a set of elements is measured from the epoch time, not from when they're received.*

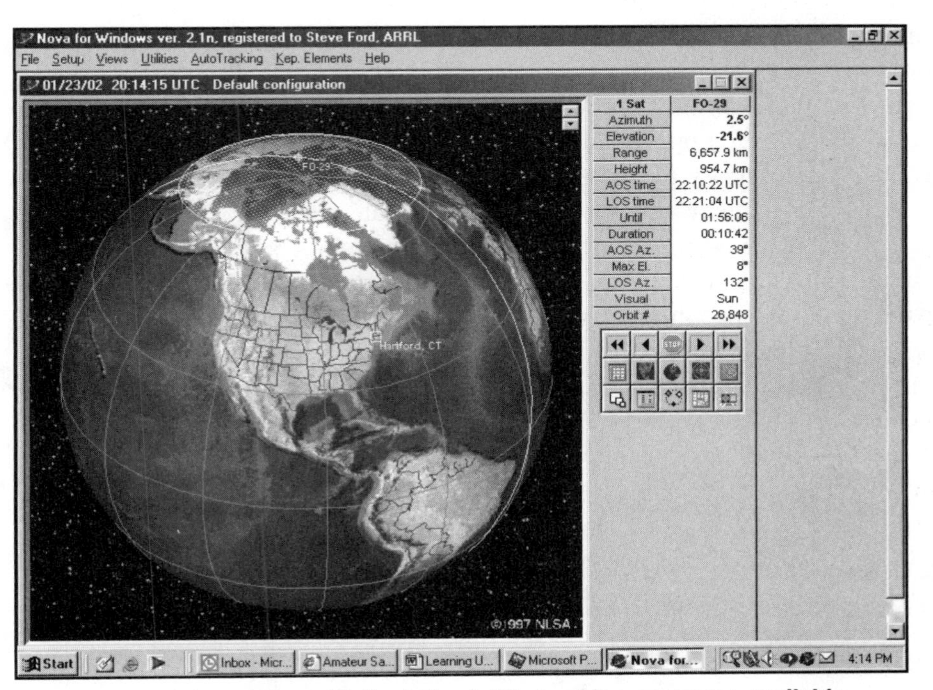

***Nova* is one of the most sophisticated satellite tracking programs available.**

The mathematical algorithms used by government agencies who distribute orbital elements are designed to provide very good predictions for periods less than 10 days. One result is that the long term effects of small periodic and sporadic perturbations due to atmospheric drag, magnetic storms and other factors can either be over- or understated. In the early days of the OSCAR program (mid 1970s) amateurs produced their own smoothed orbital elements for LEO spacecraft that averaged-out these small perturbations. In many cases, the amateur-produced elements gave better long term results than today's high precision values. For example, with OSCARs 6, 7 and 8 and early RS spacecraft, orbital elements produced in the amateur community were used to generate "Orbit Calendars" that were accurate to within a couple of minutes over a 12-month period. With Phase III satellites the situation is similar. When OSCAR 13 was still operating, G3RUH would periodically produce a set of smoothed orbital elements for the command team. These elements provided excellent predictions for up to a year.

Since the elements currently distributed by AMSAT and other groups are generally from government sources (optimized for short term predictions) our discussion will focus on them. A ground station using either a low gain beam or an omnidirectional antenna working a linear transponder on a satellite in a low altitude circular orbit, will generally find that an accurate set of orbital elements will provide good results for three to six months if the satellite orbital altitude is above 1000 km. For satellites with orbital altitudes in the range 600 to 800 km, updating every second month should be sufficient. For satellites in orbital altitudes below 600 km (such as the International Space Station [ISS]), daily updating is often required. In particular, the ISS does a lot of maneuvering so its orbit changes frequently. What's more, because the ISS orbit is already relatively low (and it is a big orbiting object!) upper atmospheric drag has a much more pronounced effect on it than on smaller satellites at higher altitudes. Therefore, obtaining more frequent orbital element updates is always prudent when tracking the ISS. This becomes particularly true when the Space Station is doing a lot of maneuvering in conjunction with periodic Space Shuttle visits. Of course, you may find it desirable to update more frequently if you're using a very high gain narrow beamwidth antenna, if you're especially interested in mutual DX windows lasting fractions of a minute, or if you use computer software to compensate for Doppler frequency shifts.

For hamsats operating in Molniya orbits, the elements should be updated at least two or three times a year. If operation around perigee is important, however, updating every month or two may be necessary. Of course, all these values are just suggestions and the details of your particular situation may warrant different values.

Aside from orbital maneuvers, which pertain mostly to the International Space Station, the main cause for change in the orbital elements of low altitude satellites is atmospheric drag. When the Sun is inactive the average status of the atmosphere can be well predicted and drag can be taken into account. However, when the sun is active, atmospheric composition changes radically over a short period of time, making it impossible to take drag into account. As a result, we're placed in a rather poor position. Drag effects can be accurately incorporated in a tracking model only when they're small and relatively unimportant. When they're large, we have no reliable way of modeling them. When using the suggested time intervals for updating orbital elements, keep in mind that you might want to shorten the intervals near sunspot maxima and lengthen them near sunspot minima.

A Satellite's "Phase"

Your software will probably be able to tell you more about a satellite than simply

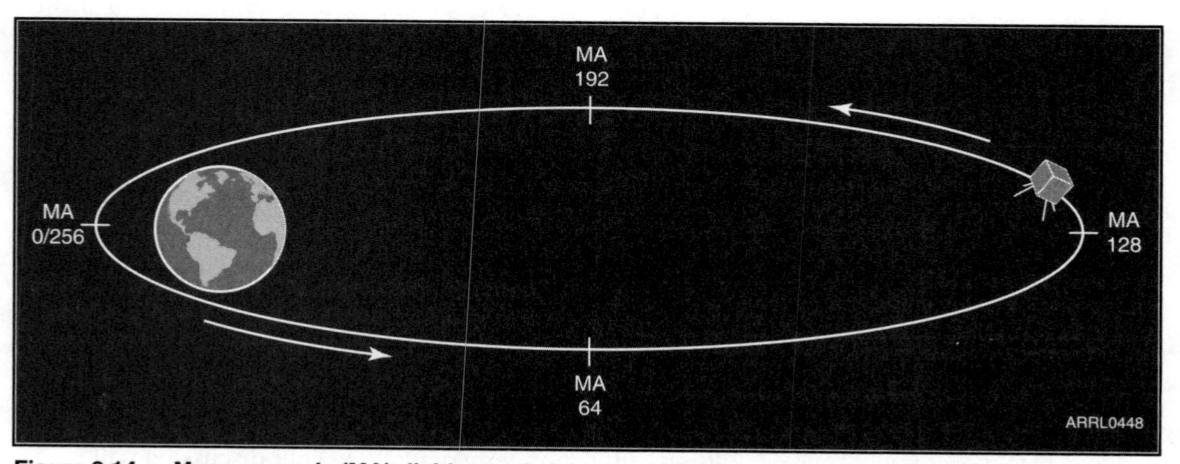

Figure 2.14 — Mean anomaly (MA) divides each orbit into 256 segments of equal time duration. Radio amateurs refer to these as "Mean Anomaly" or "Phase" units. The duration of each segment is the satellite's period divided by 256. For example, a mean anomaly unit for OSCAR 13 was roughly 2.68 minutes. At MA 0 (beginning of orbit) and MA 256 (end of orbit) the satellite is at perigee (its lowest point). At MA 128 (halfway through the orbit) the satellite is at apogee (or it's highest point).

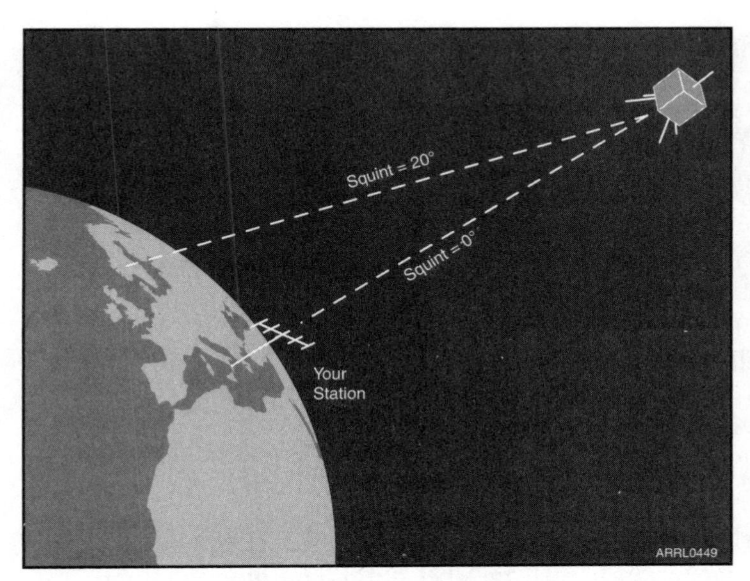

Figure 2.15 — The squint angle describes how the directive antennas on a satellite (such as Phase III satellites) are pointed with respect to your ground station. Squint angles can vary between 0° and 180°. A squint angle of 0° means the satellite antennas are pointed directly at you, which, in turn, means good link performance can usually be expected. When the squint angle is above 20° signal levels begin to drop.

where it is located or when it will arrive. Some satellites, especially the big multi-transponder birds that travel in Molniya orbits, use operating schedules to determine which transponders and antennas are active at any given time. This "Phase" information is determined by the satellite's *mean anomaly*, or *MA*.

This isn't as complicated as it sounds. The expression "anomaly" is just a fancy term for *angle*. Astronomers have traditionally divided orbits into 360 mean-anomaly units, each containing an equal time segment. Because of the architecture of common microprocessors, it was much more efficient to design the computers controlling spacecraft to divide each orbit into 256 segments of equal time duration. Radio amateurs refer to these as mean anomaly or phase units. The duration of each segment is the satellite's period divided by 256. For example, a mean anomaly unit for OSCAR 13 was roughly 2.68 minutes. At MA 0 (beginning of orbit) and MA 256 (end of orbit) the satellite is at perigee (its lowest point). At MA 128 (halfway through the orbit) the satellite is at apogee (its high point). See **Figure 2.14**.

Because radio amateurs and astronomers use the term mean anomaly in a slightly different way, there's sometimes a question as to which system is being used. Any confusion is minor and usually easily resolved. Most OSCAR telemetry with real-time MA values and schedules use the 256 system. The term "phase," and the fact that no numbers larger than 256 ever appear, are significant hints. Computer tracking programs designed for non-radio amateur audiences generally use the traditional astronomical notation. It's easy to determine when this is the case because the mean anomaly column will contain entries between 257 and 360.

If a satellite is using an MA-based schedule, the schedule will be posted on the AMSAT-NA Web site at **www.amsat.org**. (When this book was written there were no Amateur Radio satellites using MA scheduling, but this may change as currently planned satellites reach orbit.) Depending on its sophistication, your tracking software may allow you to enter this schedule. The software will then automatically tell you which operating schedule is in effect for any given time. This comes in very handy when you're about to sit down for an afternoon of satellite operating.

Schedules are generally modified every few months when satellite orientation is adjusted to compensate for changes in the Sun angle on the spacecraft. A typical schedule used for OSCAR 13, with its corresponding uplink (transmitting) and downlink (receiving) frequency band requirements, looked like this:

Off: from MA 0 until MA 49
Mode U/V (uplink on 70 cm/downlink on 2 meters): on from MA 50 until MA 128
Mode U/S (uplink on 70 cm/downlink on 2.4 GHz): on from MA 129 until MA 159

Mode U/V (uplink on 70 cm/downlink on 2 meters): on from MA 160 until MA 255

If you wanted to operate Mode U/S, you needed to be at the radio when the satellite was between MA129 and MA159 in its orbit.

Squint Angle

Another dilemma your software may help resolve is the squint angle. The squint angle describes how the directive antennas on a satellite (such as Phase III satellites) are pointed with respect to your ground station. Squint angle can vary between 0° and 180°. A squint angle of 0° means the satellite antennas are pointed directly at you and that generally indicates that good link performance can be expected (**Figure 2.15**). When the squint angle is above 20° signal level begins to drop and a disruptive amplitude flutter called spin modulation on uplinks and downlinks may become apparent.

Programs that include algorithms to calculate squint angle require information about the orientation or attitude of the satellite. This information is generally available from sources that provide the basic orbital elements and on telemetry sent directly from the satellite of interest. The parameters needed are labeled **Bahn latitude** and **Bahn longitude**. They are also known as BLAT and BLON or ALAT and ALON where the prefix "A" stands for attitude.

Programs that provide squint angle information may also contain a column labeled *Predicted Signal Level*. Values are usually computed using a simple prediction model that takes into account satellite antenna pattern, squint angle and spacecraft range. For Phase III satellites the model assumes a 0 dB reference point with the satellite overhead, at apogee and pointing directly at you. At any point on the orbit the predicted level may be several dB above (+) or below (–) this reference level.

You won't need to be concerned about squint angle for most low-Earth orbiting satellites. It only becomes a factor with the high-altitude birds since they generally use directive antennas. As with MA scheduling, you'll want to know that the satellite's antennas are pointing at your location before you fire up your equipment. With the right kind of software, you'll know well ahead of time.

Satellite Communication Systems

Writer Arthur C. Clarke is often credited with popularizing the idea of orbiting radio relays. The October 1945 issue of *Wireless World* carried his article "Extra-Terrestrial Relays—Can Rocket Stations Give Worldwide Radio Coverage?" In it he proposed the possibility of a space-based radio relay station traveling in an equatorial circular orbit at a distance of approximately 42,164 km from the center of the Earth, i.e., approximately 35,787 km (22,237 miles) above mean sea level, with a period equal to the Earth's rotation on its axis. From its orbit, this *geostationary* relay would "see" a vast amount of the planet (almost an entire hemisphere). Because its speed matched the Earth's rotation, the relay would appear to remain fixed at a single point in the sky 24 hours a day. It would listen for signals directed to it and instantly relay those signals anywhere within its coverage area.

Less than 20 years later, Arthur Clarke's vision became reality. Today there are hundreds of satellites in geostationary orbit, an orbit often referred to as the *Clarke Belt* in his honor. Hundreds more travel in lower orbits, or elliptical Molniya orbits that carry them far into space at apogee, only to bring them sweeping back within a thousand kilometers or so at perigee.

SHAHIDUL ALAM

Arthur C. Clarke was among the first to propose that radio relays orbiting the Earth could provide global communications.

Some of these satellites are devoted to scientific applications, such as space exploration or weather observation. The Hubble Space Telescope is perhaps one of the best-known scientific satellites. Other satellites are dedicated to government and military activities. GPS (Global Positioning System) satellites are actually military spacecraft, although they also function in a tremendously popular and important civilian role as well.

But the great majority of satellites still conform to Arthur Clarke's original vision. They are relays, passing information from one point to another on our planet. Despite the costs and risks of placing satellites into orbit, they remain one of the most efficient communication platforms ever devised. From their vantage points in space, they employ VHF, UHF and microwave links to handle an astonishing volume of analog and digital information, orders of magnitude more than could ever be transmitted on HF frequencies. And unlike terrestrial HF communication, satellites do not require large antennas or hefty power systems. Satellite communications are also free of the vagaries of ionosphereic propagation.

In this chapter we'll discuss the types of communication systems used by Amateur Radio satellites. The same ones, albeit in a scaled-up form, are also used on commercial birds. Understanding satellite communication systems is key to getting the most use and enjoyment possible from our amateur satellite "fleet." For the sake of discussion, it is easiest to separate typical satellite communication systems into three distinct parts: beacons, command links and transponders.

Beacons

Amateur satellite beacons typically serve several functions. In the *telemetry mode* they convey information about onboard satellite systems (solar cell panel currents, temperatures at various points, storage battery condition, and so on); in the *communications mode* they can be used for store-and-forward broadcasts; in either mode they can be used for tracking, for propagation measurements and as a reference signal of known characteristics for testing ground station receiving equipment.

Amateurs have used several telemetry encoding methods over the years. From the user's point of view, each method can be characterized by the data transfer rate and the complexity of the decoding equipment required at the ground station. To a large extent, there's a trade-off between these two factors.

The earliest amateur satellites used simple CW telemetry beacons with information being sent in Morse code. OSCAR 1, for example, varied the sending speed of "HI" in CW to communicate the status of a temperature sensor. Morse code telemetry became more sophisticated in the satellites that followed. The systems employed on OSCARs 6, 7 and 8 (and on most RS spacecraft) made these satellites very valuable to educators and amateur scientists because the Morse code information was so easy to decode. All you needed was a receiver, a piece of paper and a pencil! Restricting the code to a numbers-only format, usually at either 25 or 50 numbers/minute (about 10 or 20 words/minute), enabled individuals with little or no prior training to learn to decode the contents relatively quickly. Some Amateur Radio satellites active today still use Morse code telemetry for this reason.

> ***"The earliest amateur satellites used simple CW telemetry beacons with information being sent in Morse code."***

Digital data beacons made their first appearance in the 1970s aboard satellites such as OSCAR 7. These early beacons used radioteletype (RTTY) as a means to send more complex telemetry information at higher data rates. However, a number of factors acted to displace radioteletype as a primary means for downlinking satellite data. Radioteletype had been chosen for the convenience of users in the mid 1970s, a number of whom already had RTTY receiving equipment, but careful analysis showed that RTTY was relatively inefficient when considered in terms of data rate per unit of power. With Phase III satellites and the UoSAT series it became apparent that a higher speed, more power-efficient link was required. This need arose at the same time that microcomputers were becoming commonplace at ground stations. Since the new series of spacecraft would be controlled by onboard computers, it was natural to switch to encoding techniques suitable for computer-to-computer communications. Once a ground station uses a microcomputer to capture telemetry it's only a small step to have the computer process the raw telemetry, store it, and automatically check for values that indicate developing problems, graph data over time, and so on. Both the UoSAT and Phase III satellites series used ASCII encoding, but different modulation schemes were adopted.

Designers of early low altitude UoSAT spacecraft, with their powerful beacons, selected a specialized 1200 bps system. Power efficiency was much more important with Phase III series spacecraft so a special 400 bps optimized system was developed.

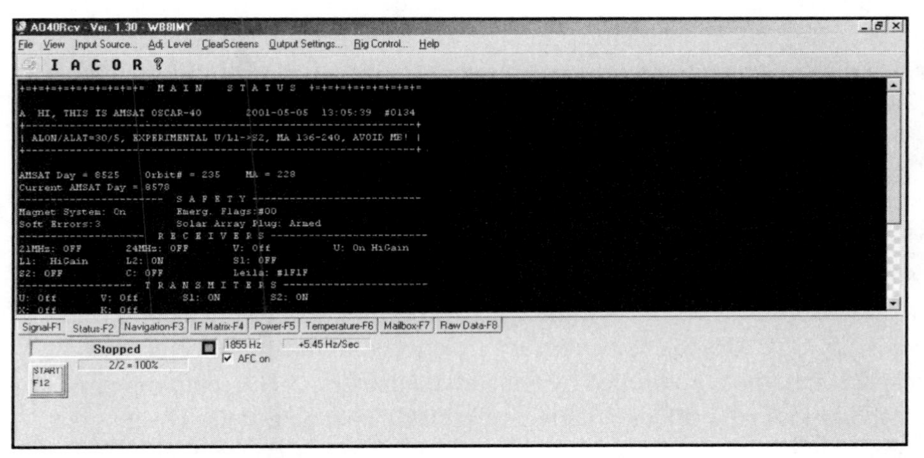

Telemetry received from OSCAR 40.

As amateur packet radio became popular in the 1980s, some satellite telemetry systems adopted the AX.25 packet radio protocol. These satellites downlinked information at either 1200 or 9600 baud. With the proliferation of packet radio terminal node controllers (TNCs), almost anyone with a computer could receive the information. DOVE-OSCAR 17 made extensive use of its packet beacon as an educational tool. A number of Amateur Radio satellites continue to use packet telemetry links to this day.

A few amateur satellites have also used digitized speech as a telemetry mode. In the digitally synthesized speech mode the telemetry is simply spoken. This produces the ultimate simplicity in ground station decoding requirements (assuming the telemetry is being spoken in a language you're familiar with). Spoken telemetry is excellent for demonstrations involving general audiences and for educational applications at lower grade levels, but the extremely low data rate (about 1 baud!) makes it unsuitable for real communications needs. Digital speech telemetry systems were used on OSCARs 9 and 11. These led to more sophisticated digital speech synthesizers capable of storing and playing back natural sounding speech that could be used for broadcasts and store-and-forward communications. Devices of this type were carried on DO-17, RS-14/AO-21, AO-27 and FO-29, although none use it today.

In either the telemetry or store-and-forward broadcast modes, a beacon with stable intensity and frequency can serve a number of functions. For example, it can be used for Doppler shift studies, propagation measurements and testing ground-based receiving equipment. In addition, stations communicating via a satellite transponder can use the satellite's beacon to adjust their uplink power properly by comparing the strength of their downlink signal to that of the beacon.

Beacon Design

Beacon power levels are chosen to provide adequate signal-to-noise ratios at well-equipped ground stations. Overkill (too much power) serves only to decrease the power available for other satellite subsystems, reduces reliability and causes potential compatibility problems with other spacecraft electronics systems. Typical power levels at

146 and 435 MHz are 40 to 100 mW on low-altitude spacecraft and 0.5 to 1.0 W on high-altitude spacecraft. As with all spacecraft subsystems, high power efficiency is essential. Phase III spacecraft have tended to use two beacons: a relatively high power unit called the *engineering beacon*, which is mainly operated when the omni antenna is used at high altitudes (during the early orbit transfer stages or in case of emergency) and the *general beacon*, which is operated continuously.

Since the telemetry system is a key diagnostic tool for monitoring the health of a spacecraft, redundant beacons, often at different frequencies, are generally flown to enhance reliability. This approach has paid off in a number of instances. Beacon failures occurred on OSCARs 5 and 6, but neither mission was seriously affected since both spacecraft carried alternate units. Beacon power output is usually one of the readings sampled by the telemetry system.

Command Links

OSCAR spacecraft are designed so that authorized volunteer ground stations with the necessary equipment can command them. The first OSCAR satellite with a command link was OSCAR 5. OSCARs 5-8 responded to a relatively limited set of commands. More recent spacecraft, those controlled by onboard computers, can accept programs via the command link. As a result, the number of possible operating states is very great. Satellite commandability is necessary for several reasons. First, it's a legal requirement of spacecraft operating in the Amateur Satellite Service. Regulations state that, should the situation arise, we must be able to turn off a malfunctioning transmitter causing harmful interference to other services. Second, it would be impossible to accomplish any of the initial orbital changes or subsequent attitude adjustments needed to keep the Phase III satellites "healthy" on orbit without also having the ability to periodically upload new command software to their onboard computers.

While the capabilities of today's sophisticated spacecraft are critically dependent on our ability to transmit software to the satellite via a command channel, the existence of a command link on even relatively simple satellites can mean the difference between an operative mission and a failure. For example, a constant stream of commands sent to OSCAR 6 turned a marginally usable spacecraft (one that was continually shutting itself down) into a reliable performer. Via the command system, ground controllers can turn off malfunctioning subsystems, adjust operating schedules to meet changing user needs, turn subsystems on/off, and so on.

Command systems are also used in the critical "housekeeping" task of activating various on-board attitude-control systems that often require periodic adjustment. Adjusting the spacecraft's attitude in space relative to the Sun helps to control its internal operating temperature and also determines how much sunlight is illuminating the spacecraft's solar panels. The latter factor directly determines how much power is available to run the satellite's various onboard systems (including the beacons and transponders) in sunlight and how much will be left over to sufficiently charge the spacecraft's batteries for use when the satellite is operating in darkness. Needless to say, the spacecraft's "whole orbit power budget" is a critical item ground controllers constantly monitor and control using the satellite's command systems.

Command stations are built and manned by dedicated volunteers. Though command frequencies, access codes and formats are considered confidential, they are available to responsible stations for projects approved by AMSAT.

Transponders

Transponders are at the heart of a satellite's ability to relay signals. There are three transponder types currently in use.

"Bent Pipe" Transponders

A bent-pipe transponder is the simplest transponder design in terms of function. It receives a signal on one frequency and simultaneously retransmits it on another. The metaphor is that of a U-shaped pipe that captures an object coming toward it and redirects the object back to its source (**Figure 3.1**).

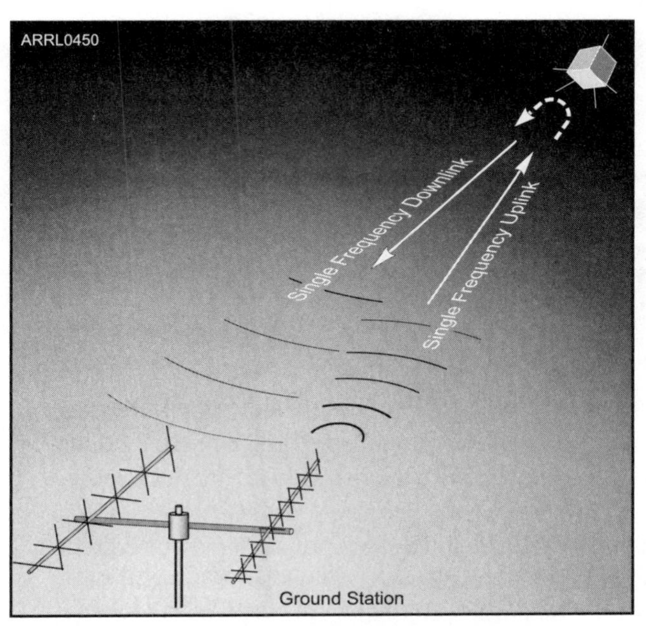

Figure 3.1 — A bent-pipe transponder gets its name from the way it functions. It takes the received uplink signal and instantly relays it on the downlink back to its source. Bent-pipe transponders are most often found on satellites that function primarily as FM repeaters.

A terrestrial FM repeater is a typical example of a bent-pipe transponder. Earthbound repeaters monitor one frequency and retransmit on another, usually (though not always) within the same band. Thanks to their elevated antenna systems, sensitive receivers and considerable output power, terrestrial repeaters can extend the coverage of mobile or handheld FM transceivers over hundreds or even thousands of square miles.

Satellite repeaters lack the output power of terrestrial repeaters because power generation in space is a difficult proposition (see the discussion of satellite power systems in Appendix B). But what satellite repeaters lack in output power they more than make up for in antenna elevation! Even something as meager as a ¼ wavelength whip antenna for the 2-meter band can become a transmitting and receiving powerhouse when it is hundreds of kilometers above the planet.

The principle advantage of a bent-pipe satellite transponder operating in the FM mode is that it is readily compatible with common Amateur Radio FM transceivers. Satellites such as Saudi-OSCAR 50 and AMRAD-OSCAR 27 can be easily worked with the same transceivers you'd otherwise use to chat through a local FM repeater. It isn't uncommon to hear mobile and even handheld portable stations through these satellites.

The principle disadvantage of a bent-pipe transponder is that it can relay only one signal at a time. Multiple signals on the satellite uplink frequency interfere with each other, usually resulting in an unintelligible cacophony on the downlink. Thanks to the "capture effect" common to FM receivers, only the strongest signal will come through clearly at any given time. With only 10 or 15 minutes available during typical low-orbit passes, an inconsiderate operator running high power can use the capture effect to monopolize the satellite, effectively shutting out all other stations whenever he or she transmits.

The solution to the bent-pipe problem is to have a satellite than can relay more than one signal at a time.

Linear Transponders

Unlike a bent-pipe transponder that can relay only one signal at a time, a linear

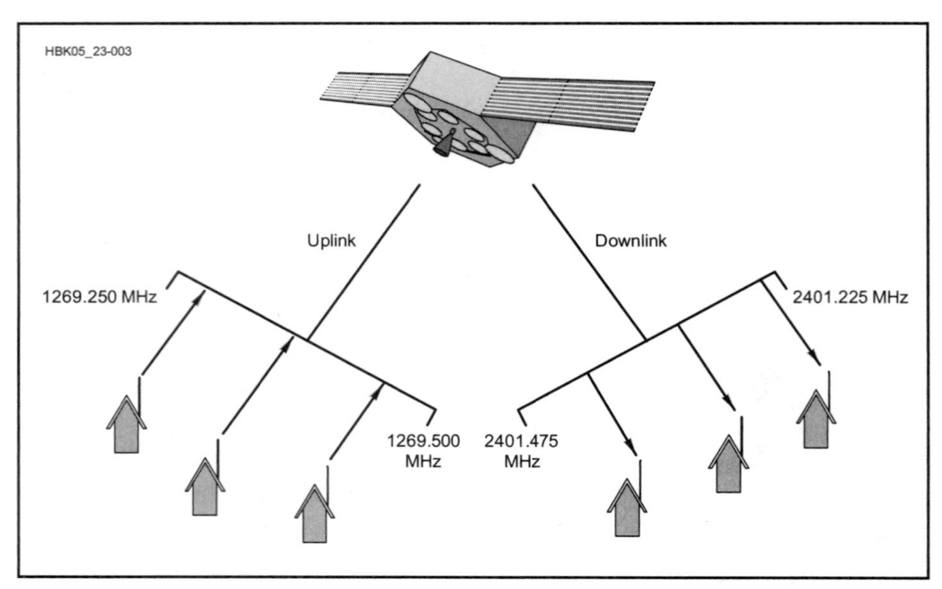

Figure 3.2 — Linear transponders repeat all signals received in an entire slice of spectrum This allows the satellite to relay many conversations simultaneously.

transponder receives signals in a narrow slice of the RF spectrum, shifts the frequency of the passband, amplifies all signals linearly, and then retransmits the entire slice. Rather than relaying only one signal, a satellite equipped with a linear transponder can relay many signals at once (**Figure 3.2**)

Total amplification is on the order of 130 dB. A linear transponder can be used with any type of signal when real time communication is desired. From the standpoint of conserving valuable spacecraft resources such as power and bandwidth, the preferred user modes are SSB and CW. Transponders are specified by first giving the approximate input frequency followed by the output frequency. For example, a 146/435-MHz transponder has an input passband centered near 146 MHz and an output passband centered near 435 MHz. The same transponder could be specified in wavelengths, as a 2-meter/70-cm unit.

However, to keep linear transponder specifications as simple as possible, the satellite community often uses so-called "Mode" designators. In the early years of amateur satellite operation, these designations were rather arbitrarily assigned and bore little

Table 3.1
Satellite Transponder Band and Mode Designators

Satellite Bands	*Common Operating Modes (Uplink/Downlink)*
10 meters (29 MHz): H	V/H (2 meters/10 meters)
2 meters (145 MHz): V	H/V (10 meters/2 meters)
70 cm (435 MHz): U	U/V (70 cm/2 meters)
23 cm (1260 MHz): L	V/U (2 meters/70 cm)
13 cm (2.4 GHz): S	U/S (70 cm/13 cm)
5 cm (5.6 GHz): C	U/L (70 cm/23 cm)
3 cm (10 GHz): X	L/S (23 cm/13 cm)
	L/X (23 cm/3 cm)
	C/X (5 cm/3 cm)

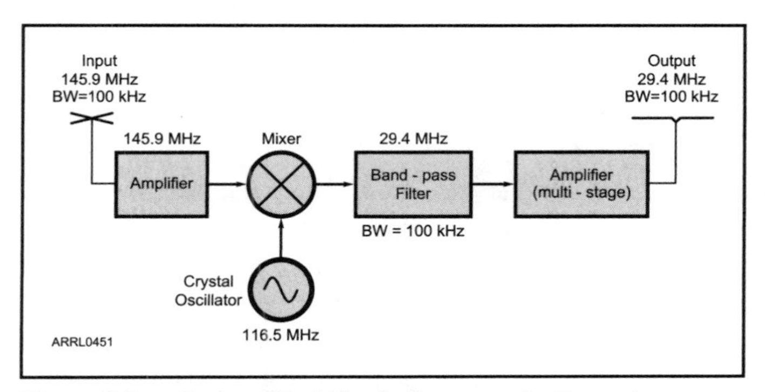

Figure 3.3 — A simplified block diagram of a linear transponder.

resemblance to the actual frequencies in use. Fortunately, the amateur satellite community has since settled on a newer series of transponder specifications that are far more intuitive. For example, in our previous example, the 2-meter band is tagged with the label "V" while the 70-cm band is labeled "U". Therefore, a transponder that listens on 2 meters and repeats on 70 cm is today usually called a Mode V/U transponder. See **Table 3.1**.

Transponder design is, in many respects, similar to receiver design. Input signals are typically on the order of 10^{-13} W and the output level can be up to several watts. A major difference, of course, is that the transponder output is at radio frequency while the receiver output is at audio frequency. A block diagram of a simple transponder is shown in **Figure 3.3**. For several reasons, flight-model transponders are more complex than the one shown.

In spacecraft applications a key characteristic of a linear amplifier is its overall efficiency (RF-output/dc-input). Once we reach power levels above a few watts the use of class A, AB or B amplifiers cannot be tolerated. The power and bandwidth of a transponder must be compatible with each other and with the mission. That is, when the transponder is fully loaded with equal-strength signals, each signal should provide an adequate signal-to-noise ratio at the ground. Selecting appropriate values accurately on a theoretical basis using only link calculations is error prone. Experience with a number of satellites, however, has provided AMSAT with a great deal of empirical data from which it's possible to extrapolate accurately to different orbits, bandwidths, power levels, frequencies and antenna characteristics.

In general, low-altitude (300 to 1600 km) satellites that use passive magnetic stabilization and omnidirectional antennas can provide reasonable downlink performance with from 1 to 10 W PEP at frequencies between 29 and 435 MHz, using a 50 to 100-kHz-wide transponder. A high-altitude (35,000 km) spin-stabilized satellite that uses modest (7 to 10 dB) gain antennas should be able to provide acceptable performance with 35 W PEP using a 300-kHz-wide transponder downlink at 146 or 435 MHz. Transponders are usually configured to be inverting in order to minimize Doppler shift. We'll discuss inverting transponders in more detail in Chapter 5.

The dynamic range problem for transponders is quite different from that for HF receivers. At first glance it may seem that the situation faced by satellite transponders is simpler. After all, an HF receiver must be designed to handle input signals differing in strength by as much as 100 dB, while a low-altitude satellite will encounter signals in its passband differing by perhaps 40 dB. Good HF receivers solve the problem by filtering out all but the desired signal before introducing significant gain. A satellite, however, has to accommodate all users simultaneously. The maximum overall gain can, therefore, be limited by the strongest signal in the passband.

Considering the state-of-the-art in transponder design and available spacecraft power budgets, an effective dynamic range of about 25 dB is about the most that can be currently obtained. In earlier satellites, the receiver AGC was normally adjusted to accommodate the loudest user. As a result, stations 25 dB weaker were not able to put a usable signal through the spacecraft even though they might have been capable of doing so if the AGC had not activated. In the ideal situation, users would adjust uplink power so that spacecraft's AGC was never activated.

The "power-hog" problem is a serious one. A single station running excessive power can effectively "swamp" a linear transponder. Far too often, an inexperienced operator may believe that by cranking up their transmit power they will improve their downlink signal strength. Instead, that increase in power only serves to depress the downlink signal levels of all other stations.

In the short-lived OSCAR 40 satellite, the designers tried an innovate approach to the power-hog problem with the addition of LEILA (*LEI*stungs *L*imit *A*nzeige [Power Limit Indicator]). LEILA's computer continuously monitored the 10.7 MHz transponder IF passband. When an uplink signal whose level exceeded a predetermined level was encountered, the computer inserted a CW message over the offender's downlink. The message indicated to the transmitting station that the transponder was being overloaded, and served as an inducement for the offending station to reduce power until the CW signal disappeared. If the overloading signal continued, or exceeded an even higher preset level, LEILA activated a notch filter tuned to the offending station's frequency. LEILA turned out to be highly effective. If you suddenly heard CW being transmitted over your signal on the downlink, you knew you had to reduce power immediately or face the consequences.

Since the transponder is the primary mission subsystem, reliability is extremely important. One way to improve system reliability is to include at least two transponders on each spacecraft; if one fails, the other would be available full time. And there are significant advantages to *not* using identical units.

Because of the large number of uplink and downlink frequencies used on the OSCAR 40 spacecraft, the design team also abandoned the old concept of a hardwired transponder with specific input and output frequency bands. In its place they introduced a design consisting of discrete receiver front ends and transmitter power amplifiers all connected to a common IF that OSCAR 40's builders called a *Matrix IF*. The Matrix IF operated at 10.7 MHz with input and output levels of –15 dBm, and it included the LEILA unit. Instructions from the ground directed the spacecraft computer to connect one or more receivers and one or more transmitters to the IF. The operating schedule was determined by an international operations committee using an extremely flexible approach that also proved helpful in accommodating various equipment failures and degraded performance as well as changes in user needs over the shortened lifetime of the spacecraft.

Digital Transponders

Digital transponders of the PACSAT or RUDAK type differ significantly from the linear transponders we've been discussing. A digital transponder demodulates the incoming signal. The data can then be stored aboard the spacecraft (PACSAT Mailbox) or used to immediately regenerate a digital downlink signal (RUDAK digipeater). The mailbox service is best suited to low altitude spacecraft. Digipeating is most effective on high altitude spacecraft. Like linear transponders, digital units are downlink limited. A key step in the design procedure is to select modulation techniques and data rates to maximize the downlink capacity. Using assumptions about the type of traffic expected one then selects appropriate uplink parameters. An analysis of both PACSAT and RUDAK suggests that, due to collisions, the uplink data capacity should be about four or five times that of the downlink.

For PACSAT Mailbox operation the designers decided to use similar data rates for the uplink and downlink and couple a single downlink with four uplinks. Fuji-OSCARs 12 and 20, and the MicroSAT series ran both links at 1200 bps. These "PACSATS" contained an FM receiver with a demodulator that accepted Manchester-encoded FSK on the uplink. To produce an appropriate uplink signal, ground stations needed FM transmitters and packet radio modems known as Terminal Node Controllers (TNCs) that were capable of generating the Manchester-encoded signal for the uplink.

The PACSAT downlink used binary phase-shift keying (BPSK) at an output of either 1.5 or 4 W. This modulation method was selected because, at a given power level and bit rate, it provides a significantly better bit error rate than other methods that were considered. One way of receiving the downlink is to use an SSB receiver and pass the audio output to a PSK demodulator. The SSB receiver is just serving as a linear downconverter in this situation. Other methods of capturing the downlink are possible but the two proven systems now operating use this approach.

> ***"A digital transponder demodulates the incoming signal. The data can then be stored aboard the spacecraft or used to immediately regenerate a digital downlink signal."***

During the heydays of the PACSATs, there were several TNCs designed specifically for this application. However, the PACSATs have gradually gone out of service and those specialized TNCs have all but disappeared from the marketplace. The remaining digital satellite transponders use 1200 bps AFSK or 9600 bps FSK data transmissions for both uplink and downlink. This means that ordinary packet radio TNCs—the kind currently used for various terrestrial applications—can be used with these digital satellites as well.

A discussion of digital transponders would not be complete without mentioning RUDAK. RUDAK is an acronym for Regenerativer Umsetzer fur Digitale Amateurfunk Kommunikation (Regenerative Transponder for Digital Amateur Communications). Early RUDAK systems took a different approach to achieving the desired ratio of uplink to downlink capacity. The one flown on OSCAR 13 used one uplink channel and one downlink channel with the data rate on the uplink (2400 bps) roughly six times that on the downlink (400 bps). The 400 bps rate on the downlink was chosen because this was the standard that had been used for downlinking Phase III telemetry since the late 1970s. Users already capturing Phase III telemetry would be able to capture RUDAK transmissions from day one. Unfortunately, the RUDAK unit on OSCAR 13 failed during launch.

A system known as RUDAK II was flown on RS-14/AO-21 and actually consisted of two units. One unit was similar to the RUDAK flown on OSCAR 13. However, the other unit, known as the RUDAK technology Experiment (or RTX), was a new experimental transponder using DSP technology. It was essentially a flying test bed for ideas being considered for future Phase III missions.

The RUDAK-U system flown on OSCAR 40 contained two CPUs, one 153.6 kbaud modem, 4 hardwired 9600 baud modems and 8 DSP modems capable of operating at speeds up to 56 kbaud. The great advantages of modern RUDAK systems are their extraordinary flexibility. Through the use of digital signal processing, the RUDAK is able to configure itself into any type of digital system desired. Had OSCAR 40 survived, the plan was to use its RUDAK system to create a highly capable digital communications platform in space.

Your Satellite Ground Station

As the term implies, a satellite "ground station" is the station on the ground. It's *your* station, the one you'll be using to communicate with amateur satellites. You have quite a few equipment and installation options to choose from, depending on which satellites you intend to access and the modes you hope to use (FM, SSB, CW or digital). In this chapter we'll attempt to make sense of this confusing array of choices.

Antenna Systems

What is true for every other aspect of Amateur Radio is also true for satellite operating: Your antenna system is the most critical component of your station. If you are wondering how to invest your equipment funds, don't cut corners when it comes to purchases for your antenna system. An expensive, full-featured transceiver will be almost worthless if it is connected to poor antennas.

And when we speak of the "antenna system," we're talking about more than just the antennas themselves. The "system" includes the feed lines that connect the antennas to your station. In the context of satellite operating, it may also encompass receive preamplifiers and antenna rotators. In addition to our discussion of antennas in this chapter, you'll find several practical satellite antenna projects in Chapter 6.

Remember that, to use most amateur satellites, you must transmit on one band (the uplink) and receive on another band (the downlink). Unless you are using a single, dual-band antenna, your satellite station will require at least two antennas. Some satellite operators have several antennas so that they can operate on various uplink/downlink band combinations as the need arises.

Antenna Directivity and Gain

All antennas, even the simplest types, exhibit directive effects, meaning that the intensity of radiation is not the same in all directions from the antenna. This property of radiating more strongly in some directions than others is called the *directivity* of the antenna.

The *gain* of an antenna is closely related to its directivity; to many hams the words are synonymous. Because directivity is based solely on the shape of the directive pattern, however, it does not take into account any power losses that may occur in an actual antenna system. Gain takes those losses into account. Gain is usually expressed in decibels, and is based on a comparison with a *standard* antenna — usually a dipole or an *isotropic radiator*. An isotropic radiator is a theoretical antenna that would, if placed in the center of an imaginary sphere, evenly illuminate that sphere with radiation. The isotropic radiator is an unambiguous standard, and for that reason it is frequently used as the comparison for gain measurements.

When the standard is an isotropic radiator in free space, gain is expressed in decibels referenced to this isotropic antenna, or *dBi*. When the standard is a dipole, *also located in free space*, gain is expressed in reference to this dipole, or *dBd*. The more the directive pattern is compressed — or focused — the greater the power gain of the antenna. This is a result of power being concentrated in some directions at the expense of others. The directive pattern, and therefore the gain, of an antenna at a given frequency is determined by the size and shape of the antenna, and by its position and orientation relative to the Earth.

When purchasing antennas for your ground station, beware of advertised gain claims. Unless the antenna has been measured at a calibrated antenna test range, or its performance "modeled" with antenna modeling software, its true gain is almost anyone's guess. This is why you will see very few antennas advertised with gain figures in ARRL's *QST* magazine. *QST* will publish gain specifications only if the antenna manufacturer has submitted proof of measurement or modeling.

Antenna Polarization

The polarity of an antenna is determined by the position of the radiating element or wire with respect to the Earth. A radiator that is parallel to the Earth radiates horizontally, while a vertical radiator radiates a vertical wave (**Figure 4.1**). These are so-called *linearly polarized* antennas. If the radiating element is slanted above the Earth, it radiates waves that have *both* vertical and horizontal components.

For terrestrial VHF+ line-of-sight communication, polarity matching is important. If one station is using a horizontally polarized antenna and the other is using a vertically polarized antenna, the mismatch can result in a large signal loss. We don't worry about polarization mismatches on HF frequencies because whenever signals are refracted through the ionosphere, as HF signals usually are, their polarities change anyway. This interesting effect is known as *Faraday Rotation.*

The problem with applying polarization concerns to spacecraft is that the orientation of a satellite's antennas relative to your ground station is constantly changing. This often results in fading when the polarities of the satellite antennas conflict with yours. As you'll read later, this problem doesn't plague satellites exclusively. Aircraft, automobiles and other moving radio platforms can suffer the same effects.

Fortunately, there is a "cure" known as *circular polarization* (CP). With CP the wavefront describes a rotational path about its central axis, either clockwise (right-hand or RHCP) or counter clockwise (left-hand or LHCP). See **Figure 4.2**.

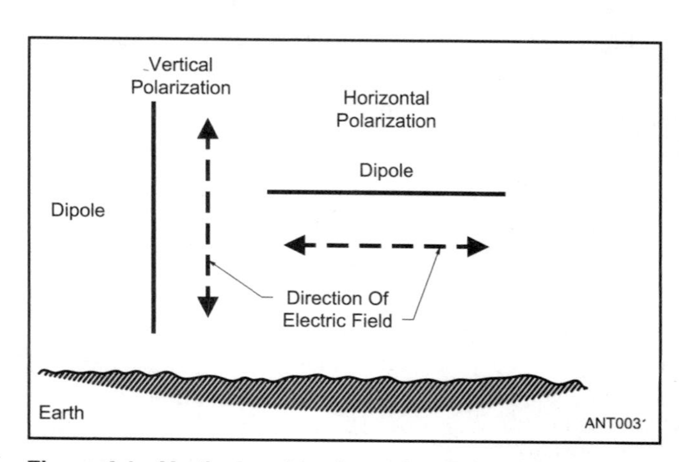

Figure 4.1—Vertical and horizontal polarization of a dipole antenna above ground. The direction of polarization is the direction of the maximum electric field with respect to the Earth.

The advantage of using circular

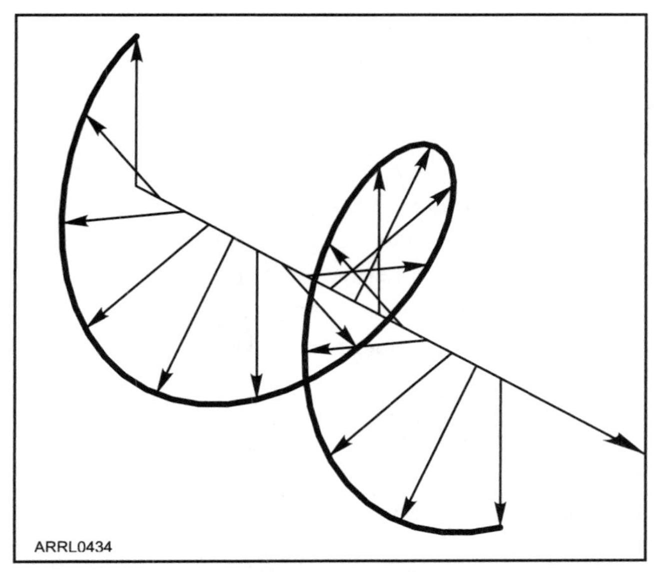

Figure 4.2—A circularly polarized wavefront describes a rotational path about its central axis, either clockwise (right-hand or RHCP) or counter clockwise (left-hand or LHCP).

polarization is that it can substantially reduce the effects of polarization conflict. Since the polarization of radio waves emitted from a CP antenna rotates through horizontal and vertical planes, the resulting pattern effectively "smoothes" the fading effects, generating consistent signals as a result.

That said, your satellite ground station antennas do *not* need to be circularly polarized to be effective! Linearly polarized antennas, either horizontal or vertical, are perfectly useful for satellite work. Some ground station antenna designs use slanted or "crossed" elements to mix the horizontal and vertical polarization components. The goal of fine-tuning your antenna polarization is to give your station an edge, something that is important when you are dealing with weak signals from deep space. But, while circular polarization of your antennas may give your station a definite advantage, it isn't an absolute requirement.

Omnidirectional Antennas

An ideal omnidirectional antenna radiates and receives signals in all directions equally. In the real world, most "omnidirectional" antennas have a certain amount of directivity. For example, a common ground plane antenna (**Figure 4.3**) is considered to be omnidirectional, but as you can see in **Figure 4.4** the radiation pattern is not uniform. Notice the deep null directly overhead. Signals will fade sharply as satellites move through this null, particularly during high-elevation passes.

Despite their low gain, omnidirectional antennas are attractive because they do not need to be aimed at their targets. This means that they don't require mechanical antenna rotators, which can add significant cost and complication to a ground station. Omni antennas are also more compact than directional antennas for the same frequency. On the other hand, their low gain makes them practical only for low-Earth orbiting satellites that have sensitive receivers and relatively powerful transmitter.

Technically speaking, any type of omnidirectional antenna can be used for a satellite ground station. Hams have enjoyed satellite operations with simple ground planes, J-poles, "big wheels" and even automobile antennas. But for best results with an omni-based antenna system, you'll want a radiation pattern that minimizes the polarization conflicts and pattern nulls. To this end, engineers have designed a number of omnidirectional antennas with these issues in mind.

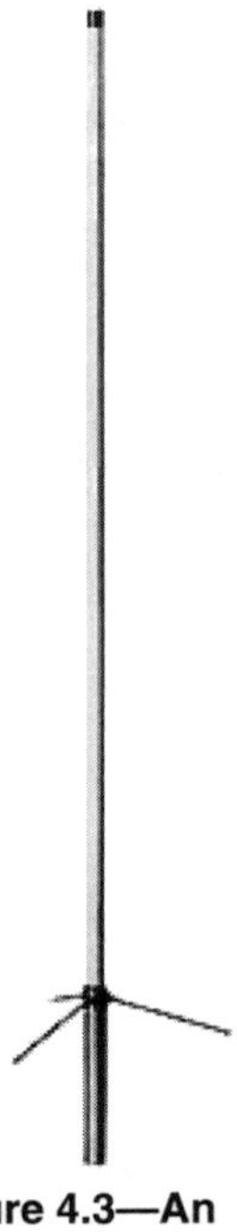

Figure 4.3—An ordinary ground plane antenna. This particular model is designed for terrestrial communications on 2 meters.

The Eggbeater Antenna

The eggbeater antenna is a popular design named after the old-fashioned kitchen utensil it resembles (**Figure 4.5**). The antenna is composed of two full-wave loops of rigid wire or metal tubing. Each of the two loops has an impedance of 100 Ω, and when coupled in parallel they offer an ideal 50 Ω impedance for coaxial feed lines. The loops are fed 90° out of phase with each other and this creates a circularly polarized pattern.

An eggbeater may also use one or more parasitic reflector elements beneath the loops to focus more of the radiation pattern upward. This effect makes it a "gain" antenna, but that gain is at the expense of low-elevation reception. Toward the horizon an eggbeater is actually linear-horizontally polarized. As the pattern rises in elevation, it becomes

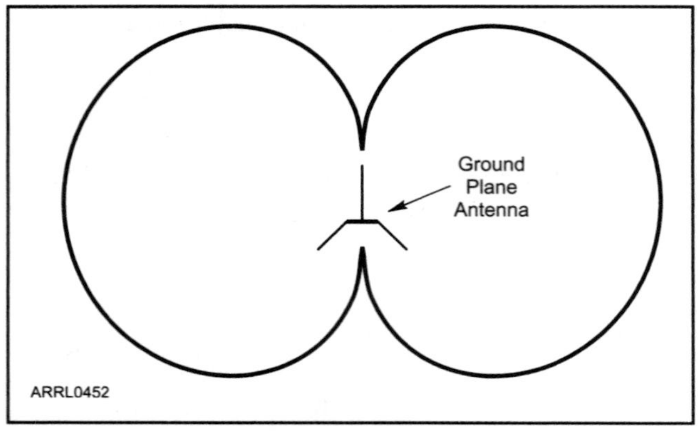

Figure 4.4—The simplified radiation pattern of a ground plane antenna. Notice the deep null directly overhead.

more and more right-hand circularly polarized. Experience has shown that eggbeaters seem to perform best when reflector elements are installed just below the loops.

Eggbeaters can be homebrewed relatively easily, but there are also a couple of commercial models available (see the advertising pages of *QST* magazine). The spherical shape of the eggbeater creates a fairly compact antenna when space is an issue, which is another reason why it is an attractive design.

The Turnstile

The basic turnstile antenna consists of two horizontal half-wave dipoles mounted at right angles to each other (like the letter "X") in the same horizontal plane with a reflector screen beneath (**Figure 4.6**). When these two antennas are excited with equal currents 90° out of phase, their typical figure-eight patterns merge to produce a nearly circular pattern.

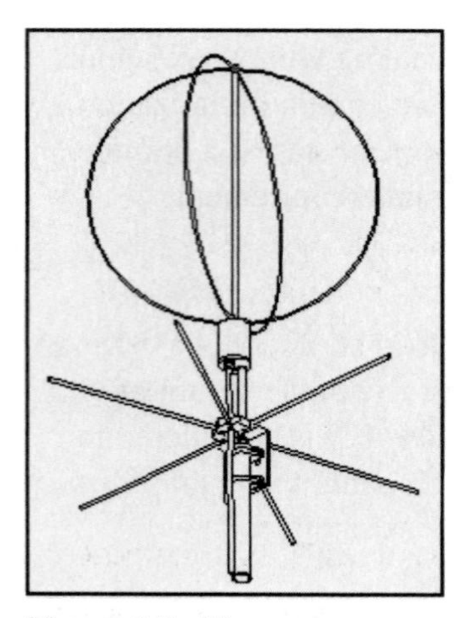

Figure 4.5—The egg-beater antenna is a popular design named after the old-fashioned kitchen utensil it resembles. The antenna is composed of two full-wave loops of rigid wire or metal tubing.

In order to get the radiation pattern in the upward direction for space communications, the turnstile antenna needs a reflector underneath. For a broad pattern it is best to maintain a distance of ⅜ wavelength at the operating frequency between the reflector and the turnstile. Homebrewed turnstile reflectors often use metal window-screen material that you can pick up at many hardware stores. (Make sure to ask for old-fashioned metal screening material, not the PVC-coated fiberglass variety.)

Like their cousins the eggbeaters, turnstiles are relatively easy to homebrew. In fact, homebrewing may be your only choice for turnstiles since they are rarely available off the shelf.

The Lindenblad Antenna

Nils Lindenblad of the Radio Corporation of America (RCA) was the inventor of the antenna that bears his name. Around 1940 he was working on antennas for the fledgling television broadcasting (TV) industry. His idea was to employ four dipoles spaced equally around a $\lambda/3$ diameter circle with each dipole canted 30° from the horizontal. The dipoles were all fed in phase and fed equal power. The spacing and tilt angles of the dipoles created the desired antenna pattern when all the signals were combined.

A few years later, George Brown and Oakley Woodward, also of RCA, were tasked with finding ways to reduce fading on ground-to-air radio links at airports. These links used linearly polarized antennas and the maneuverings of the airplanes often caused large signal dropouts if the antennas became cross-polarized.

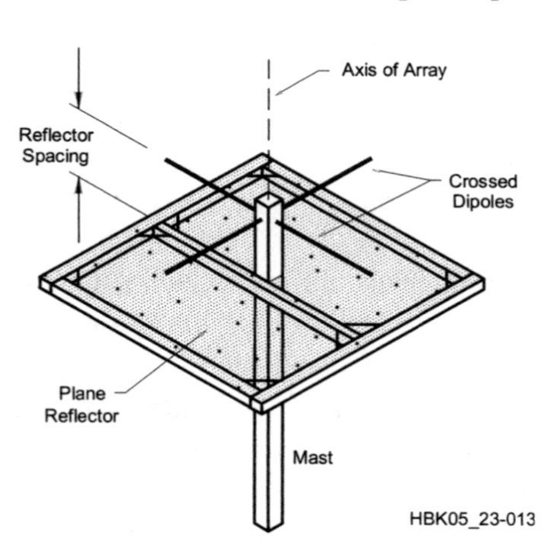

Figure 4.6—The basic turnstile antenna consists of two horizontal half-wave dipoles mounted at right angles to each other (like the letter "X") in the same horizontal plane with a reflector screen beneath.

Brown and Woodward realized that using a circularly polarized antenna at the airport could reduce or eliminate this fading so they decided to try Lindenblad's concept. Brown and Woodward designed their antenna using metal tubing for the dipole elements.

In a Lindenblad antenna, each dipole element is attached to a section of shorted open-wire-line, also made from tubing, which serves as a balun transformer. A coaxial cable runs through one side of each open-wire line to feed each dipole. The four coaxial feed cables meet at a center hub section where they are connected in parallel to provide a four-way, in-phase power-splitting function. This cable junction is connected to another section of coaxial cable that serves as an impedance matching section to get a good match to 50 Ω.

While Brown and Woodward's variation on the Lindenblad design was clever and worked well, it is quite difficult for the average ham to duplicate. Even so, many satellite operators have enjoyed success with this antenna.

The major cause of the difficulty in designing and constructing Lindenblad antennas is the need for the four-way, in-phase, power-splitting function. Since we generally want to use 50 Ω coaxial cable to feed the antenna, we have to somehow provide an impedance match from the 50 Ω unbalanced coax to the four 75 Ω balanced dipole loads. Previous designs have used combinations of folded dipoles, open-wire lines, twin-lead feeds, balun transformers and special impedance matching cables in order to try to get a good match to 50 Ω. These, in turn, increase the complexity and difficulty of the construction. In Chapter 6 you'll find a much easier Lindenblad design by Anthony Monteiro, AA2TX (who also provided some of the Lindenblad background details you've just read). See **Figure 4.7**. While certainly more elaborate than an eggbeater or turnstile, the Lindenblad creates a uniform circularly polarized pattern that is highly effective for satellite applications.

Figure 4.7—The EZ Lindenblad antenna designed by Anthony Monteiro, AA2TX.

Quadrifilar Helicoidal Antenna (QHA)

The quadrifilar helicoidal antenna (**Figure 4.8**) ranks among the best of the omnidirectional satellite antennas. It is comprised of four equal-length conductors (filars) wound in the form of a corkscrew (helix) and fed in quadrature. The result is a nearly perfect circularly polarized pattern.

QHAs can be challenging to build since the filar lengths and spacing have to be precise. (You'll find a QHA construction project in Chapter 6.) Even so, homebrewing a QHA can save you a substantial amount of money. This antenna is available off the shelf (they are favorites for maritime satellite links), but they can be costly.

Directional Antennas

As we discussed at the beginning of this chapter, gain and directivity are important antenna factors. These factors are maximized in *directional* antennas. In fact, the design goal of a directional antenna is to create a highly directional pattern along its axis (**Figure 4.9**). An ordinary flashlight is a reasonable analog for a directional antenna, although the pattern of a directional antenna isn't as well focused as a flashlight beam (antennas with parabolic reflectors come fairly close, though).

The chief advantage of a directional antenna is its considerable directivity and gain. When you are working with weak signals from spacecraft, you need all the gain you can get. Directional antennas are mandatory for Phase III satellites when they are at apogee, nearly 50,000 km distant. They also are excellent for low-Earth orbiting birds, providing strong, consistent signals that omnidirectional antennas can rarely match.

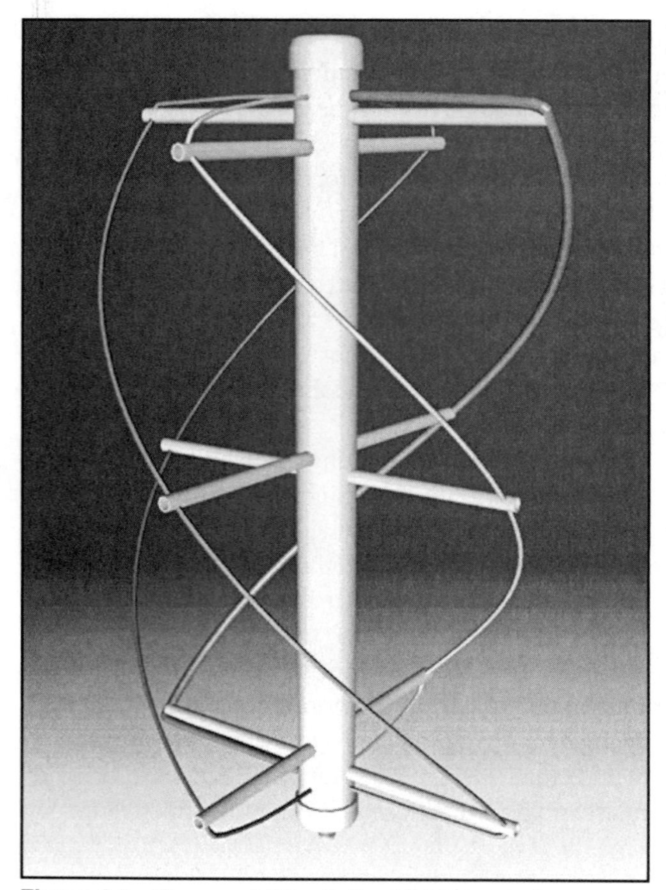

Figure 4.8—The quadrifilar helicoidal antenna is comprised of four equal-length conductors (filars) wound in the form of a corkscrew (helix) and fed in quadrature.

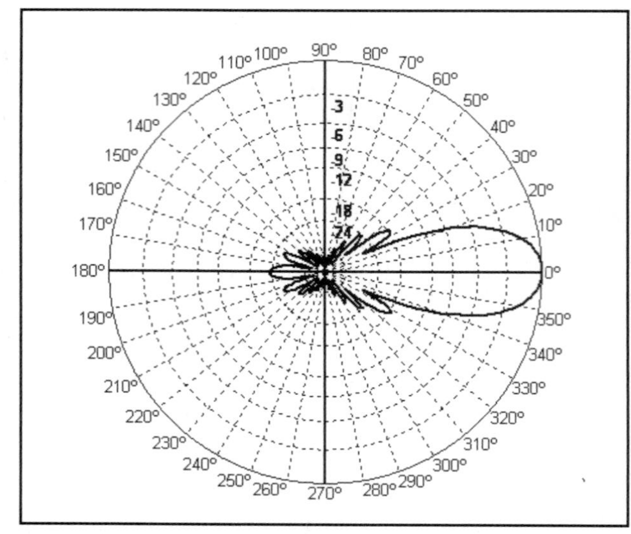

Figure 4.9—The design goal of a directional antenna is to create a highly directional pattern along its axis. This example shows the radiation pattern of a high-gain Yagi antenna.

But, a major disadvantage of a directional antenna is *also* its directivity! To achieve best results with a directional antenna, you must find a way to point it at the satellite you wish to work. This entails pointing it by hand, or by using an electric motorized device known as an *antenna rotator*. If you were to simply leave a directional antenna fixed in one place, you would enjoy good signals only during the brief moments when satellites passed through the antenna's pattern. A rotator adds significant cost to a ground station and installing one isn't a trivial exercise. On the other hand, there is a way to reduce rotator cost, which we'll discuss later.

The satellite antennas at ARRL Headquarters station W1AW include three Yagis. The two outside antennas are crossed Yagis for 2 meters and 70 cm.

If you can afford a directional antenna and rotator system, you'll never regret the investment. When properly installed, it is vastly superior to any omnidirectional antenna you are likely to encounter. Let's look at several of the designs available to you.

Yagi Antennas

The Yagi antenna, originally known as the Yagi-Uda antenna after its two inventors, has been a ham favorite since the post World War II years. This directional antenna consists of a dipole (the *driven element*) and several closely coupled parasitic elements (usually a reflector and one or more directors). See **Figure 4.10**. The reflector element, directly behind the dipole, is about 5% longer than the dipole. The director elements are placed ahead of the dipole. Thanks to phase cancellation by the reflector and reinforcement by the directors, the net effect is to create a directive radiation pattern. Yagis are directional along the axis perpendicular to the dipole in the plane of the elements, from the reflector through the driven element and out via the director(s).

The more directors a Yagi has, the higher its directivity and effective gain. That is why you will often see long Yagi antennas used for terrestrial VHF+ operations. It would seem that you'd want long Yagi designs for satellite applications as well, and this is generally true. However, keep in mind that the greater the directivity, the more focused (narrow) the antenna pattern. This translates to an antenna that must be aimed at a satellite with a fair degree of accuracy. This can be a challenge when your target is a rapidly moving low-Earth-orbiting bird.

The Yagi antennas you normally see are *linear* designs that are mounted in either horizontal or vertically polarized configurations. These same antennas can be successfully used for satellite work just as they are. However, there are ways you can optimize them to increase their effective performance.

The dipole elements of Yagi antennas radiate linearly polarized signals, but remember that the polarization direction really depends on the orientation of the antenna to the Earth. If two Yagi antennas are mounted on the same support boom, arranged for horizontal and vertical polarization, and combined with the correct phase difference (90°), a circularly polarized wave results. Because the electric fields of the antennas are identical in magnitude, the power from the transmitter will be equally divided between the two fields. Another way of looking at this is to consider the power as being divided between the two antennas; hence the gain of each is decreased by 3 dB when taken alone in the plane of its orientation. This design is known as a *crossed Yagi.*

A 90° phase shift must exist between the two antennas and the simplest way to obtain this shift is to use two feed lines. One feed-line section is ¼ wavelength longer than the other, as shown in

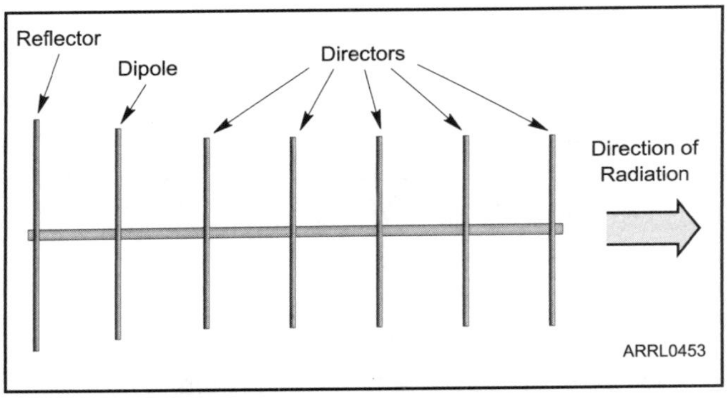

Reflector
Dipole
Directors
Direction of Radiation
ARRL0453

Figure 4.10—A simplified diagram of a 7-element Yagi antenna.

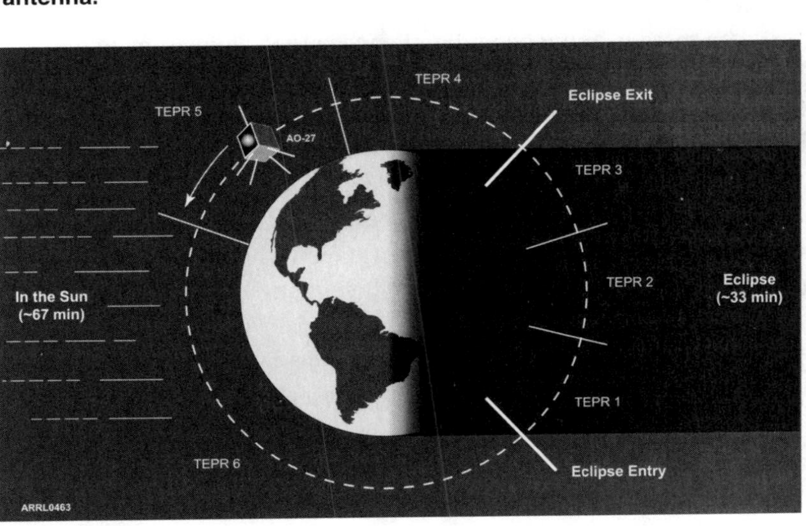

Figure 4.11—Evolution of the circularly polarized Yagi. The simplest form of crossed Yagi, A, is made to radiate circularly by feeding the two driven elements 90° out of phase. Antenna B has the driven elements fed in phase, but has the elements of one bay mounted ¼ wavelength forward from those of the other. Antenna C offers elliptical (circular) polarization using separate booms. The elements in one set are mounted perpendicular to those of the other and are spaced ¼ wavelength forward from those of the other.

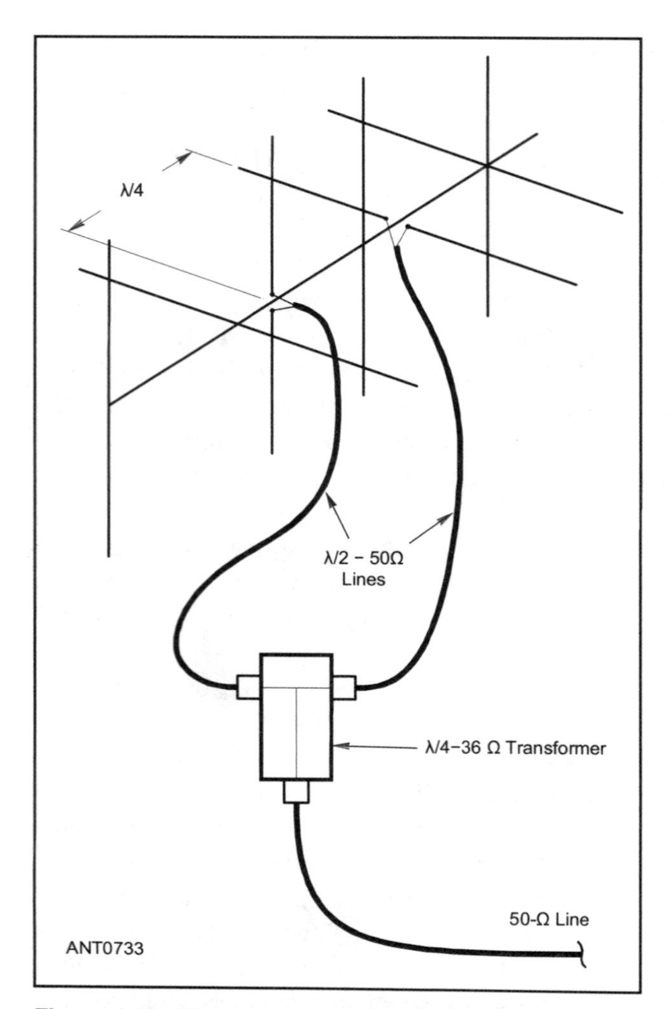

Figure 4.12—The advantage of equal-length feed lines is that identical load impedances will be presented to the common feeder, as shown here. This is a fixed circularity "sense" feed.

Figure 4.13—Offset crossed-Yagi circularly polarized antenna-phasing harness with switchable polarization.

Figure 4.11A. These separate feed lines are then paralleled to a common transmission line to the station. However, therein lies one of the headaches of this system. Assuming negligible coupling between the crossed antennas, the impedance presented to the common transmission line by the parallel combination is one half that of either section alone. (This is not true when there is mutual coupling between the antennas, as in phased arrays.)

Another method to obtain circular polarization is to use equal-length feed lines and place one antenna ¼ wavelength ahead of the other. This offset pair of Yagi-crossed antennas is shown in Figure 4.11B. The advantage of equal-length feed lines is that identical load impedances will be presented to the common feeder, as shown in **Figure 4.12**, which shows a fixed circularity "sense" feed. To obtain a switchable sense feed with the offset Yagi pair, you can use a connection like that of **Figure 4.13**, although you must compensate for the extra phase added by the relay and connectors.

Figure 4.11C diagrams a popular method of mounting two separate off-the-shelf Yagis at right angles to each other. The two Yagis may be physically offset by ¼ wavelength and fed in parallel, as shown in Figure 4.11C, or they may be mounted with no offset and fed 90° out of phase. Neither of these arrangements on two separate booms produces true circular polarization. Instead, *elliptical* polarization results from such a system. **Figure 4.14**

KD1K

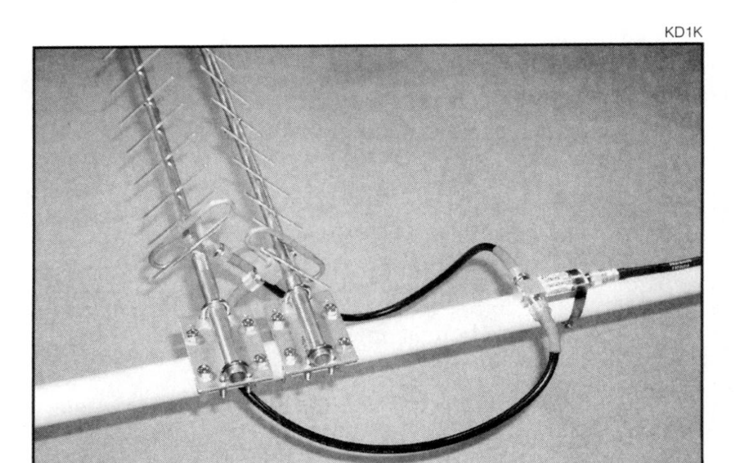

Figure 4.14—An example of offset crossed-Yagi circularly polarized antennas with fixed polarization. This example shows a pair of M² 23CM22EZA antennas for L band (1269 MHz) mounted on an elevation boom.

Figure 4.17—This antenna stack includes a loop Yagi at the very top.

is a photo of this type of mounting of Yagis on two booms for elliptical operation.

So far we've been discussing single-band crossed Yagi designs to achieve circular polarization, or something close to it. If circular polarization isn't a top criterion, you may want to consider a *dual-band Yagi* that places linear Yagi antennas for two separate bands on the same support boom. While this design doesn't produce circular polarization, it partially makes up for that disadvantage in convenience and economics. The most common design combines antennas for 2 meters and 70 cm on the same boom (**Figure 4.15**). You can feed the antennas with two separate feed lines, or use single feed line and a *diplexer* (some manufacturers incorrectly call them *duplexers*) to separate the signals at the antenna, at the radio or at both locations (**Figure 4.16**). With a single 2 meter/70 cm Yagi antenna you can enjoy nearly all available amateur satellites. That single-purchase aspect makes them attractive for budget-conscious hams.

Another type of Yagi that you may encounter is the *loop Yagi* (**Figure 4.17**). In this design, the individual elements are bent into loops and mounted on a common boom. Despite their

Figure 4.15—Joe Bottiglieri, AA1GW, uses a dual-band Yagi to work a low-Earth orbiting FM repeater satellite with a handheld transceiver.

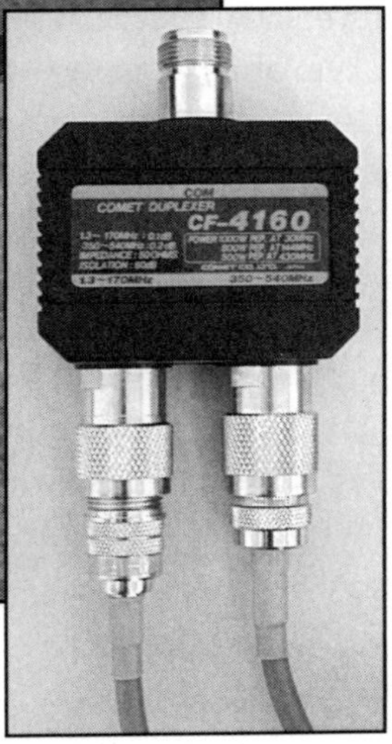

Figure 4.16—A diplexer made by the Comet Corporation. Diplexers can be used to separate or combine signals on different bands.

appearance, loop Yagis are linearly polarized antennas. Their advantage is that they create a substantial amount of gain in a relatively small physical space. Because of the sizes of the loops, however, loop Yagis are most practical and common at microwave frequencies.

Helical Antennas

Another method to create a circularly polarized signal is by means of a helical antenna. The axial-mode helical antenna was introduced by Dr John Kraus, W8JK, in the 1940s. **Figure 4.18** shows an example of a microwave helical antenna. A larger helical for 70 cm is shown in **Figure 4.19**.

KD1K

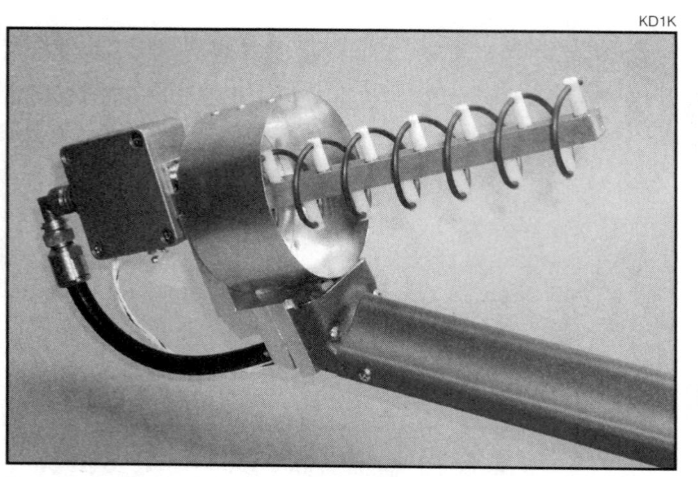

Figure 4.18—A seven-turn LHCP helical antenna for a 2.4 GHz dish feed. This helical antenna uses a cupped reflector and has a preamplifier mounted directly at the antenna feed point.

This antenna has two characteristics that make it especially interesting and useful for satellite stations. First, the helix is circularly polarized. As discussed earlier, circular polarization simply means that linearly polarized radio waves are also rotating as they travel through space. In the case of a helical antenna, this rotation is about the axis of the antenna. This can be pictured as the second hand of a watch moving at the same rate as the applied frequency, where the position of the second hand can be thought of as the instantaneous polarization of the signal. The second interesting property of the helical antenna is its predictable pattern, gain and impedance characteristics over a wide frequency range.

This is one of the few antenna designs that have both broad bandwidth and high gain. The benefit of this property is that, when used for narrow-band applications, the helical antenna is very forgiving of mechanical inaccuracies.

The helical antenna is an unusual specimen in the antenna world, in that its physical configuration gives a hint to its electrical performance. Electrically, a helix looks like a large air-wound coil with one of its ends fed against a ground plane. The ground plane often consists of a screen of 0.8 to 1.1 wavelength in diameter (or per side for a square ground plane). The circumference of the coil form must be between 0.75 and 1.33 wavelength for the antenna to radiate in the axial mode. The coil should have at least three turns to radiate in this mode. The ratio of the spacing between turns (in wavelengths), should be in the range of 0.2126 to 0.2867. The winding of the helix comes away from the cupped reflector with a counterclockwise winding direction for LHCP. A clockwise winding direction yields RHCP.

You won't find many off-the-shelf helical antennas for amateur satellites, but building your own isn't overly difficult. You'll find a helical antenna project for 70 cm in Chapter 6.

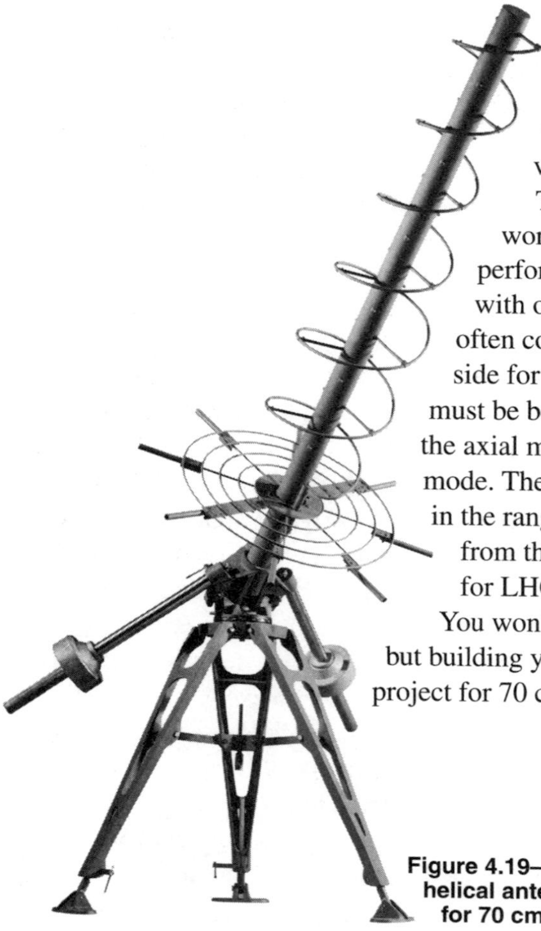

Figure 4.19—A helical antenna for 70 cm.

Antennas in the Attic?

If you are living in a setting that doesn't allow outside antennas, you may want to look at your attic. Even apartments and condos often have attics accessible to the top-floor residents. Depending on the height of the attic, you may be able to install directional antennas and even antenna rotators. If the attic is small, consider omnidirectional antennas.

The most serious problem with attic installations is the signal attenuation caused by roofing materials. Wood roofs with slate or asphalt shingles will pass VHF and UHF signals, but with considerable loss. Microwave signals will likely be attenuated to the point of being unusable. When the roof is wet, or ice and snow covered, both VHF and UHF signals will be further degraded. Metal roofs, of course, are deadly to all radio signals.

For effective attic operation with low-orbiting satellites, receive preamplifiers are highly recommended. You may also find that you need to use more RF uplink power than you would with an outdoor antenna.

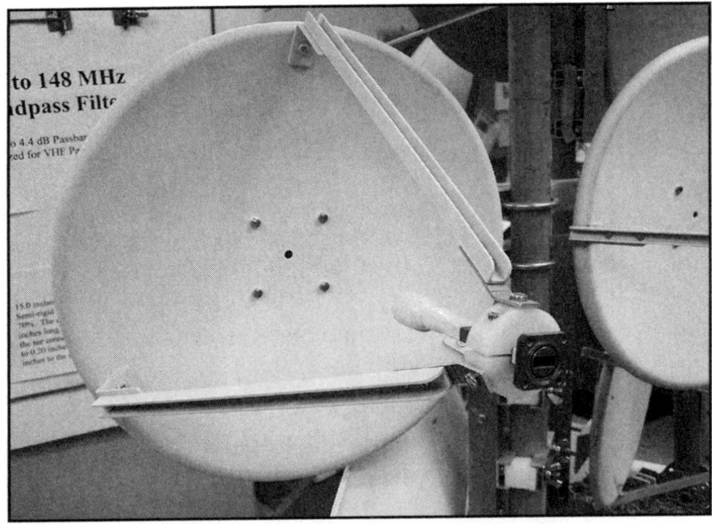

Figure 4.20—As you can see, the feed source is directly in front, making this a center-fed dish.

Parabolic Reflectors

A number of modern ham satellites now include microwave transponders and this has created a great deal of interest in effective microwave antennas. From the ground station "point of view," antennas for the microwave bands are not only small, they pack a high amount of gain into their compact packages. Among the best high-performance antenna designs for microwave use are those that employ parabolic reflectors (so-called parabolic "dishes") to concentrate the transmitted and received energy

Like a bulb in a flashlight, a parabolic antenna must have a *feed source* — the radiating and receiving part of the antenna — "looking" into the surface of the dish. Some dishes are designed so that the feed source is mounted directly in front of the dish. This is referred to as a *center-fed dish* (**Figure 4.20**). Other dishes are designed so that the feed source is off to one side, referred to as an *off-center-fed dish*, or just offset fed dish, as shown in **Figure 4.21.** (You can probably see some of these in your neighborhood doing duty as satellite TV antennas.) The offset-fed dish may be considered a side section of a center-fed dish.

The center-fed dish experiences some signal degradation due to blockage by the feed system, but this is usually an insignificantly small amount. The offset-fed dish is initially more difficult to aim, since the direction of reception is not the center axis, as it is for center-fed dishes.

The dish's parabola can be designed so the *focus point* — the point where the feed source must be — is closer to the surface of the dish, referred to a *short-focal-length* dish, or further away from the dish's surface, referred to as a *long-focal-length* dish. To determine the exact focal length (F), measure the diameter of the dish (D) and the depth of the dish (d).

$$F = \frac{D}{16d}$$

The focal length divided by the diameter of the dish gives the *focal ratio*, commonly shown as f/D. Center-fed dishes usually have short-focal ratios in the range of f/D = 0.3 to 0.45. Offset-fed dishes usually have longer focal lengths, with f/D = 0.45 to 0.80. If you attach two small mirrors to the outer front surface of a dish and then point the dish at the Sun, you can easily find the focus point of the dish. This is where you want to place the feed source.

To invoke the flashlight analogy again, the feed source should evenly illuminate the

Figure 4.21—This is a good example of an offset-fed dish. Note the feed source at the bottom edge of the photo, pointing into the center of the parabola. Also note that this is a DirectTV satellite TV antenna that has been "repurposed" for Amateur Radio satellites!

Figure 4.22—The 2.4 GHz satellite downlink antenna at ARRL Headquarters station W1AW is a so-called barbeque dish antenna originally designed for the Multichannel Multipoint Distribution Service (MMDS). These inexpensive antennas can be assembled and installed in minutes.

entire dish, and none of the feed energy should fall outside the dish's reflecting surface. However, *no* feed system is perfect in illuminating a dish. These so-called "spillover losses" affect the gain by either under-illuminating or over-illuminating the dish. Typical dish efficiency is 50%. That's 3 dB of lost gain.

A great feed system for one dish can be a real lemon on another. For example, if you can get your hands on a surplus offset-fed satellite TV dish, you may find a helical radiator is the best feed source. Designs range anywhere from 2 to 6 turns. The two-turn helices are used for very short focal-length dishes in the f/D = 0.3 region, and the 6-turn helices are used with typically longer-focal-length (f/D ~ 0.6), offset-fed dishes. Generally speaking, helix feeds work poorly on the short-focal length dishes, but really perform well on the longer-focal length, offset-fed dishes.

Parabolic antennas can be made from cast-off dishes (commercial or military surplus), including satellite TV dishes that you can "repurpose" in the finest Amateur Radio tradition. You will also find ham dealers selling so-called *barbecue* dishes that were originally designed for the Multichannel Multipoint Distribution Service, or "wireless cable TV" (see **Figure 4.22**). MMDS antennas come with built-in feed sources at the focal points, so all you have to do is connect the feed line and go. These are among the most popular dish antennas for amateur satellite enthusiats.

Taking the homebrew idea a step further, in Chapter 6 you'll discover that you can

LMR-400 coaxial cable with a Type N connector.

even craft a *corner reflector* antenna (a variation on the parabolic concept) from cardboard boxes covered in aluminum foil!

Feed Lines

At the beginning of this chapter we stated that even an expensive transceiver is nearly worthless if you connect it to a poor antenna. Expanding upon that idea, otherwise excellent antennas and transceivers can be rendered worthless if you link them with poor feed lines (also referred to as *transmission lines*). Typically found in the form of coaxial cable, the feed line is the critical pipeline between the radio and the antenna. It is responsible for delivering RF power to your antenna and received signals back to your radio.

The chief issue with feed lines is *loss*. Every feed line has some degree of loss. If you insert 100 W of RF at the station input, you will always have something less than 100 W at the antenna. The same concept works in reverse — the received signal at the antenna will always be somewhat weaker by the time it reaches your radio. Feed line loss increases with frequency and length. It also increases when the antenna impedance is mismatched to the feed line, resulting in an elevated Standing Wave Ratio (SWR).

Although feed line loss can never be eliminated, it *can* be reduced to acceptable levels by...

★ Choosing the lowest-loss feed line for the application
★ Keeping the feed line as short as possible
★ Adjusting the antenna for the lowest possible SWR

In **Table 4.1** you'll find an extensive list of commercial feed lines and their characteristics. Most of these cables are designed to present a characteristic *impedance* (*Nom.* Z_o in the table, third column from the left) of 50 Ω, which is what we typically encounter in Amateur Radio applications.

Of particular interest for this discussion are the figures in the four right-hand columns under the heading *Matched Loss*. These figures represent the total loss (in decibels [dB]) for 100 feet of feed line matched to an antenna. By "matched" we mean that the antenna has been adjusted to present the same impedance as the feed line at the point where the feed line connects to it. The resulting SWR is a perfect 1:1. Notice that the loss columns are labeled in four frequency steps: 1, 10, 100 and 1000 MHz.

Let's look at the Belden 8240 brand of RG-58 coaxial cable and assume that we're using 100 feet of it to feed a 2 meter/70 cm Yagi antenna combo. At 100 MHz (close enough to 2 meters for this discussion), the matched loss is 3.8 dB. Since we're talking about loss, 3 dB represents a halving of power, so you will stand to lose *more than half* your RF in this feed line between the radio and the Yagi. *Ouch!* At 70 cm it will be even worse, possibly a *quadruple* loss or more. It's obvious that this cable would be a poor choice in this application!

How about RG-213 (Belden 8267)? Now the loss at 2 meters drops to 1.9 dB. That's acceptable at 2 meters, but at 70 cm we're still talking about something on the order of 3 dB. Not good.

Now consider the TMS (Times Microwave Systems) LMR400 brand of RG-8 coax. At 2 meters the loss is only 1.3 dB, rising to 4.1 dB at 1000 MHz. This is an acceptable loss at 2 meters and at 70 cm it is an equally acceptable 2 dB. If the feed line turned out to be less than 100 feet long, the losses would be lower still.

When considering microwave signals, loss becomes excessive among the types of cables hams are likely to purchase. Consider LDF6-50A Heliax in Table 4.1. The matched loss at 1000 MHz is outstanding, but this cable is very expensive and difficult

Table 4.1

Nominal Characteristics of Commonly Used Transmission Lines

RG or Type	Part Number	Nom. Z_0 Ω	VF %	Cap. pF/ft	Cent. Cond. AWG	Diel. Type	Shield Type	Jacket Matl	OD inches	Max V (RMS)	Matched Loss (dB/100') 1 MHz	10	100	1000
RG-6	Belden 1694A	75	82	16.2	#18 Solid BC	FPE	FC	P1	0.275	600	0.2	.7	1.8	5.9
RG-6	Belden 8215	75	66	20.5	#21 Solid CCS	PE	D	PE	0.332	2700	0.4	0.8	2.7	9.8
RG-8	Belden 7810A	50	86	23.0	#10 Solid BC	FPE	FC	PE	0.405	600	0.1	0.4	1.2	4.0
RG-8	TMS LMR400	50	85	23.9	#10 Solid CCA	FPE	FC	PE	0.405	600	0.1	0.4	1.3	4.1
RG-8	Belden 9913	50	84	24.6	#10 Solid BC	ASPE	FC	P1	0.405	600	0.1	0.4	1.3	4.5
RG-8	CXP1318FX	50	84	24.0	#10 Flex BC	FPE	FC	P2N	0.405	600	0.1	0.4	1.3	4.5
RG-8	Belden 9913F7	50	83	24.6	#11 Flex BC	FPE	FC	P1	0.405	600	0.2	0.6	1.5	4.8
RG-8	Belden 9914	50	82	24.8	#10 Solid BC	FPE	FC	P1	0.405	600	0.2	0.5	1.5	4.8
RG-8	TMS LMR400UF	50	85	23.9	#10 Flex BC	FPE	FC	PE	0.405	600	0.1	0.4	1.4	4.9
RG-8	DRF-BF	50	84	24.5	#9.5 Flex BC	FPE	FC	PE	0.405	600	0.1	0.5	1.6	5.2
RG-8	WM CQ106	50	84	24.5	#9.5 Flex BC	FPE	FC	P2N	0.405	600	0.2	0.6	1.8	5.3
RG-8	CXP008	50	78	26.0	#13 Flex BC	FPE	S	P1	0.405	600	0.1	0.5	1.8	7.1
RG-8	Belden 8237	52	66	29.5	#13 Flex BC	PE	S	P1	0.405	3700	0.2	0.6	1.9	7.4
RG-8X	Belden 7808A	50	86	23.5	#15 Solid BC	FPE	FC	PE	0.240	600	0.2	0.7	2.3	7.4
RG-8X	TMS LMR240	50	84	24.2	#15 Solid BC	FPE	FC	PE	0.242	300	0.2	0.8	2.5	8.0
RG-8X	WM CQ118	50	82	25.0	#16 Flex BC	FPE	FC	P2N	0.242	300	0.3	0.9	2.8	8.4
RG-8X	TMS LMR240UF	50	84	24.2	#15 Flex BC	FPE	FC	PE	0.242	300	0.2	0.8	2.8	9.6
RG-8X	Belden 9258	50	82	24.8	#16 Flex BC	FPE	S	P1	0.242	600	0.3	0.9	3.1	11.2
RG-8X	CXP08XB	50	80	25.3	#16 Flex BC	FPE	S	P1	0.242	300	0.3	0.9	3.1	14.0
RG-9	Belden 8242	51	66	30.0	#13 Flex SPC	PE	SCBC	P2N	0.420	5000	0.2	0.6	2.1	8.2
RG-11	Belden 8213	75	84	16.1	#14 Solid BC	FPE	S	PE	0.405	600	0.2	0.4	1.3	5.2
RG-11	Belden 8238	75	66	20.5	#18 Flex TC	PE	S	P1	0.405	600	0.2	0.7	2.0	7.1
RG-58	Belden 7807A	50	85	23.7	#18 Solid BC	FPE	FC	PE	0.195	300	0.3	1.0	3.0	9.7
RG-58	TMS LMR200	50	83	24.5	#17 Solid BC	FPE	FC	PE	0.195	300	0.3	1.0	3.2	10.5
RG-58	WM CQ124	52	66	28.5	#20 Solid BC	PE	S	PE	0.195	1400	0.4	1.3	4.3	14.3
RG-58	Belden 8240	52	66	28.5	#20 Solid BC	PE	S	P1	0.193	1900	0.3	1.1	3.8	14.5
RG-58A	Belden 8219	53	73	26.5	#20 Flex TC	FPE	S	P1	0.195	300	0.4	1.3	4.5	18.1
RG-58C	Belden 8262	50	66	30.8	#20 Flex TC	PE	S	P2N	0.195	1400	0.4	1.4	4.9	21.5
RG-58A	Belden 8259	50	66	30.8	#20 Flex TC	PE	S	P1	0.192	1900	0.4	1.5	5.4	22.8
RG-59	Belden 1426A	75	83	16.3	#20 Solid BC	FPE	S	P1	0.242	300	0.3	0.9	2.6	8.5
RG-59	CXP 0815	75	82	16.2	#20 Solid BC	FPE	S	P1	0.232	300	0.5	0.9	2.2	9.1
RG-59	Belden 8212	75	78	17.3	#20 Solid CCS	FPE	S	P1	0.242	300	0.6	1.0	3.0	10.9
RG-59	Belden 8241	75	66	20.4	#23 Solid CCS	PE	S	P1	0.242	1700	0.6	1.1	3.4	12.0
RG-62A	Belden 9269	93	84	13.5	#22 Solid CCS	ASPE	S	P1	0.240	750	0.3	0.9	2.7	8.7
RG-62B	Belden 8255	93	84	13.5	#24 Flex CCS	ASPE	S	P2N	0.242	750	0.3	0.9	2.9	11.0
RG-63B	Belden 9857	125	84	9.7	#22 Solid CCS	ASPE	S	P2N	0.405	750	0.2	0.5	1.5	5.8
RG-142	CXP 183242	50	69.5	29.4	#19 Solid SCCS	TFE	D	FEP	0.195	1900	0.3	1.1	3.8	12.8
RG-142B	Belden 83242	50	69.5	29.0	#19 Solid SCCS	TFE	D	TFE	0.195	1400	0.3	1.1	3.9	13.5
RG-174	Belden 7805R	50	73.5	26.2	#25 Solid BC	FPE	FC	P1	0.110	300	0.6	2.0	6.5	21.3
RG-174	Belden 8216	50	66	30.8	#26 Flex CCS	PE	S	P1	0.110	1100	1.9	3.3	8.4	34.0
RG-213	Belden 8267	50	66	30.8	#13 Flex BC	PE	S	P2N	0.405	3700	0.2	0.6	1.9	8.0
RG-213	CXP213	50	66	30.8	#13 Flex BC	PE	S	P2N	0.405	600	0.2	0.6	2.0	8.2
RG-214	Belden 8268	50	66	30.8	#13 Flex SPC	PE	D	P2N	0.425	3700	0.2	0.6	1.9	8.0
RG-216	Belden 9850	75	66	20.5	#18 Flex TC	PE	D	P2N	0.425	3700	0.2	0.7	2.0	7.1
RG-217	WM CQ217F	50	66	30.8	#10 Flex BC	PE	D	PE	0.545	7000	0.1	0.4	1.4	5.2
RG-217	M17/78-RG217	50	66	30.8	#10 Solid BC	PE	D	P2N	0.545	7000	0.1	0.4	1.4	5.2
RG-218	M17/79-RG218	50	66	29.5	#4.5 Solid BC	PE	S	P2N	0.870	11000	0.1	0.2	0.8	3.4
RG-223	Belden 9273	50	66	30.8	#19 Solid SPC	PE	D	P2N	0.212	1400	0.4	1.2	4.1	14.5
RG-303	Belden 84303	50	69.5	29.0	#18 Solid SCCS	TFE	S	TFE	0.170	1400	0.3	1.1	3.9	13.5
RG-316	CXP TJ1316	50	69.5	29.4	#26 Flex BC	TFE	S	FEP	0.098	1200	1.2	2.7	8.0	26.1
RG-316	Belden 84316	50	69.5	29.0	#26 Flex SCCS	TFE	S	FEP	0.096	900	1.2	2.7	8.3	29.0
RG-393	M17/127-RG393	50	69.5	29.4	#12 Flex SPC	TFE	D	FEP	0.390	5000	0.2	0.5	1.7	6.1
RG-400	M17/128-RG400	50	69.5	29.4	#20 Flex SPC	TFE	D	FEP	0.195	1400	0.4	1.1	3.9	13.2
LMR500	TMS LMR500UF	50	85	23.9	#7 Flex BC	FPE	FC	PE	0.500	2500	0.1	0.4	1.2	4.0
LMR500	TMS LMR500	50	85	23.9	#7 Solid CCA	FPE	FC	PE	0.500	2500	0.1	0.3	0.9	3.3
LMR600	TMS LMR600	50	86	23.4	#5.5 Solid CCA	FPE	FC	PE	0.590	4000	0.1	0.2	0.8	2.7
LMR600	TMS LMR600UF	50	86	23.4	#5.5 Flex BC	FPE	FC	PE	0.590	4000	0.1	0.2	0.8	2.7
LMR1200	TMS LMR1200	50	88	23.1	#0 Copper Tube	FPE	FC	PE	1.200	4500	0.04	0.1	0.4	1.3
Hardline														
1/2"	CATV Hardline	50	81	25.0	#5.5 BC	FPE	SM	none	0.500	2500	0.05	0.2	0.8	3.2
1/2"	CATV Hardline	75	81	16.7	#11.5 BC	FPE	SM	none	0.500	2500	0.1	0.2	0.8	3.2
7/8"	CATV Hardline	50	81	25.0	#1 BC	FPE	SM	none	0.875	4000	0.03	0.1	0.6	2.9
7/8"	CATV Hardline	75	81	16.7	#5.5 BC	FPE	SM	none	0.875	4000	0.03	0.1	0.6	2.9
LDF4-50A	Heliax –1/2"	50	88	25.9	#5 Solid BC	FPE	CC	PE	0.630	1400	0.05	0.2	0.6	2.4
LDF5-50A	Heliax –7/8"	50	88	25.9	0.355" BC	FPE	CC	PE	1.090	2100	0.03	0.10	0.4	1.3
LDF6-50A	Heliax – 1¼"	50	88	25.9	0.516" BC	FPE	CC	PE	1.550	3200	0.02	0.08	0.3	1.1

to work with, so much so that it is impractical for most ham stations. The cost-effective solution for a microwave ground station is to operate the radio at a much lower frequency and convert to or from microwave frequencies at the antenna with devices such as *transverters* and *downconverters*, which we'll discuss later.

Antenna Rotators

One beneficial characteristic of satellite antennas is that they don't have to be installed at high altitudes. As long as the antenna can "see" as much of the unobstructed sky as possible, you can even resort to ground mounting. Signals to and from the satellite will already be traversing hundreds, thousands, or even *tens of thousands* of miles of free space. So, adding a few more feet of elevation to your ground station antenna won't perceptibly improve the strength of those signals. What such an arrangement *will* do, however, is needlessly increase the length of coax (and corresponding line losses) between your transceiver and your antenna.

Figure 4.23—A basic antenna rotator; the light-duty variety used to turn TV antennas. The wire exiting the bottom of the rotator housing is the multiconductor cable responsible for power and control.

If your antennas already have an unobstructed view of the sky from the ground, you usually *won't* need a tower or roof installation *unless* trees or buildings surround the antennas and block their upward view. Or, to put it another way, your antenna support only needs to be high enough to make sure the back end of the antenna array is far enough off the ground to prevent people from walking into it while it's pointing straight up.

What's more, and as we have already discussed, satellite antennas (particularly those of the circularly polarized variety, along with their associated rotators) tend to create a more complex antenna arrangement than that used for ordinary HF or VHF/UHF terrestrial operation. So, mounting them close to the ground makes performing any needed adjustments or repairs a whole lot simpler (and safer!)

If you plan to use directional antennas and don't wish to manipulate them by hand, you will need to install an *antenna rotator*. Rotators (some hams incorrectly refer to them as "rotors") are little more than high-torque electrical motors controlled remotely through a multiconductor cable (**Figure 4.23**). Making the correct decision as to how much capacity the rotator must have is very important to ensure trouble-free operation.

Rotator manufacturers generally provide antenna surface area ratings to help you choose a suitable model. The maximum antenna area is linked to the rotator's torque capability. Some rotator manufacturers provide additional information to help you select the right size of rotator for the antennas you plan to use. Hy-Gain provides an *Effective Moment* value. Yaesu calls theirs a *K-Factor*. Both of these ratings are torque values in foot-pounds. You can compute the Effective Moment of your antenna by multiplying the antenna turning radius by its weight. So long as the effective moment rating of the rotator is greater than or equal to the antenna value, the rotator can be expected to provide a useful service life.

There are several rotator grades available to amateurs. The lightest-duty rotator is the type typically used to turn TV antennas. These rotators will handle smaller satellite

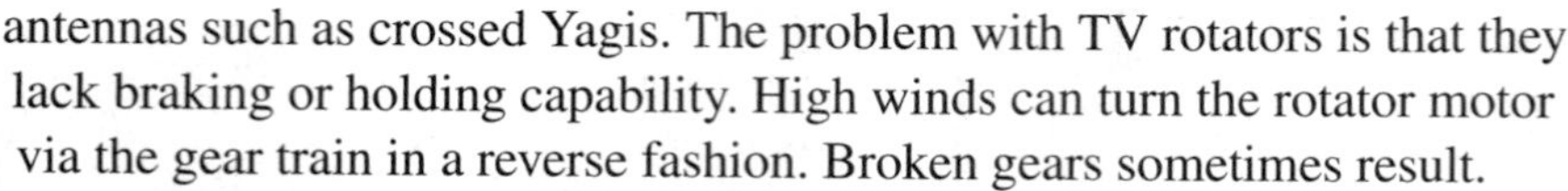

antennas such as crossed Yagis. The problem with TV rotators is that they lack braking or holding capability. High winds can turn the rotator motor via the gear train in a reverse fashion. Broken gears sometimes result.

The next grade up from the TV class of rotator usually includes a braking arrangement, whereby the antenna is held in place when power is not applied to the rotator. Generally speaking, the brake prevents gear damage on windy days. If adequate precautions are taken, this type of rotator is capable of holding and turning a stack of satellite antennas, including a parabolic dish which, by its nature, presents considerable wind loading. Also keep in mind that as rotators increase in power, they become more expensive.

Figure 4.24—The Yaesu G5500 azimuth/elevation rotator (left) and its control unit (right).

The Azimuth/Elevation Rotator — Do You Really Need One?

Perhaps the ultimate in operating convenience is the *azimuth/elevation rotator*. This rotator is capable of moving your antennas horizontally (azimuth) and vertically (elevation) at the same time. There are well-designed models available from Yaesu (**Figure 4.24**) and Alfa Radio (**Figure 4.25**). You can operate these rotators manually, or connect them to your computer for automated tracking. The downside is that az/el rotators tend to be expensive, typically on the order of $600 or more at this writing.

Figure 4.25—An az/el rotator made by Alfa Radio.

If your budget can stand the strain, az/el rotators are clearly worth the investment. On the other hand, if you're trying to shave pennies from your installation, consider using a standard rotator instead. While a traditional rotator can only move your antennas in the azimuth plane (horizontally), you can strike a compromise by installing the antennas at a permanent 25° tilt (**Figure 4.26**). Believe it or not, this configuration will allow you to work the vast majority of satellites with reasonable success. No, you won't be able to follow the satellite when it is overhead, but you'll enjoy the lion's share of every pass. Considering the fact that you'll have saved several hundred dollars in the bargain, the loss in coverage may be worthwhile.

Regardless of which type you choose, proper installation of the antenna rotator can provide many years of dependable service. Sloppy installation can cause problems such as a burned out motor, slippage, binding and even breakage of the rotator's internal gear and shaft castings or outer housing.

Most rotators are capable of accepting mast sizes of different diameters, and suitable precautions must be taken to shim an undersized mast to ensure dead-center rotation. For instance, if you decide to install your rotator on a tower, it is desirable to mount the rotator inside and as far below the top of the tower as possible. The mast absorbs the torsion developed by the antenna during high winds, as well as during starting and stopping. Some amateurs have used a long mast that stretches from the top all the way to the base of the tower. This extreme example notwithstanding, a mast length of 10 feet or more between the rotator and the antenna will add greatly to the longevity of the entire system by allowing

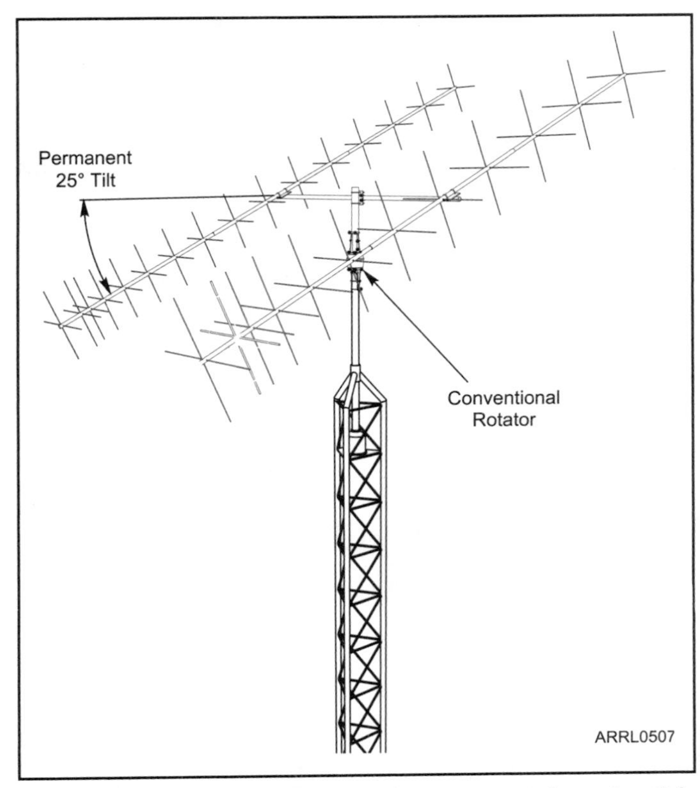

Figure 4.26—The alternative to using an expensive azimuth/elevation antenna rotator is to simply install your antennas at a 25° tilt and use a conventional rotator to move them from side to side (azimuth only). This configuration will allow access during most satellite passes.

the mast to act as a torsion shock absorber. Another benefit of mounting the rotator 10 feet or more below the antenna is that the effect of any misalignment among the rotator, mast and the top of the tower is less significant.

A tube at the top of the tower (a *sleeve bearing*) through which the mast protrudes almost completely eliminates any lateral forces on the rotator casing. All the rotator must do is support the downward weight of the antenna system and turn the array.

While the normal weight of the antenna and the mast is usually not more than a couple of hundred pounds, even with a large system, one can ease this strain on the rotator by installing a thrust bearing at the top of the tower. The bearing is then the component that holds the weight of the antenna system, and the rotator need perform only the rotating task.

Don't forget to provide a loop of coax to allow your antenna to rotate properly. Also, make sure you position the rotator loop so that it doesn't snag on anything.

A problem often encountered in amateur installations is that of misalignment between the direction indicator in the rotator control box and the heading of the antenna. With a light duty TV antenna rotator, this happens frequently when the wind blows the antenna to a different heading. With no brake, the force of the wind can move the gear train and motor of the rotator, while the indicator remains fixed. Such rotator systems have mechanical stops to prevent continuous rotation during operation, and provision is usually included to realign the indicator against the mechanical stop from inside the shack. Of course, the antenna and mechanical stop position must be oriented correctly during installation. In most cases the proper direction is true north.

Figure 4.27—The Yaesu GS-232 allows your computer and satellite tracking software to automatically control the movements of compatible azimuth/elevation antenna rotators.

In larger rotator systems with adequate brakes, indicator mis-alignment is caused by mechanical slippage in the antenna boom-to-mast hardware. Many texts suggest that the boom be pinned to the mast with a heavy duty bolt and the rotator be similarly pinned to the mast to stop slippage. But in high winds the slippage may, in fact, act as a clutch release, preventing serious damage to the rotator. On the other hand, you might not want to climb the tower and realign the system after each heavy windstorm!

As mentioned earlier, you can connect your rotator to your station computer and allow your satellite-tracking software to aim your antennas automatically (assuming your software supports rotator control). There used to be a number of commercial rotator/computer interface devices available for sale, but availability has dwindled over the years and those that remain tend to be expensive. An interface such as the Yaesu GS-232 (**Figure 4.27**) costs about $600 at the time of this writing. If you combine it with a Yaesu azimuth/elevation rotator, you will have invested $1300 total. Less expensive alternatives are now found as kits or homebrew devices. A good example is the G6LVB tracker interface at **www.g6lvb.com/Articles/LVBTracker/**. If you're willing to build it yourself, you can probably put together an LVB unit for less than $50. You'll also find an easy, inexpensive interface project in Chapter 6.

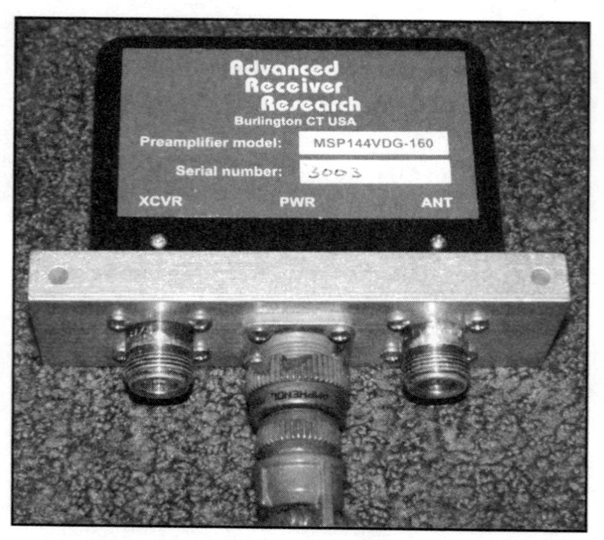

Figure 4.28—This receive preamplifier by Advanced Receiver Research gives signals a substantial boost before they travel down the coaxial cable to your radio.

Receive Preamplifiers

Signals from satellites can be exquisitely weak, which means they need as much amplification as possible to be readable. Unfortunately, there are a number of factors that may conspire to weaken your radio's ability to render a decent received signal…

★ *You're using omnidirectional antennas.* As we discussed earlier, omni antennas lack much of the signal-capturing gain of directional antennas

★ *The feed line between the antennas and the radio is long and/or contains "lossy" coax.* Remember that even with the best coax, the longer the feed line the more signal you'll lose, especially at higher frequencies.

★ *You're trying to communicate with a Phase III or IV satellite at apogee.* When a signal travels up to 50,000 km to reach your station, even the gain of a directional antenna may not be sufficient.

The way to ensure that you have a useable received signal is to install a *receive preamplifier* (**Figure 4.28**) at the antenna. This is a high-gain, low-noise amplifier with a frequency response tailored for one band only.

When shopping for a receive preamplifier, you want the most amount of gain for the least amount of noise. Every preamplifier adds some noise to the system, but you want the least additional noise possible. Preamplifier gain and noise are

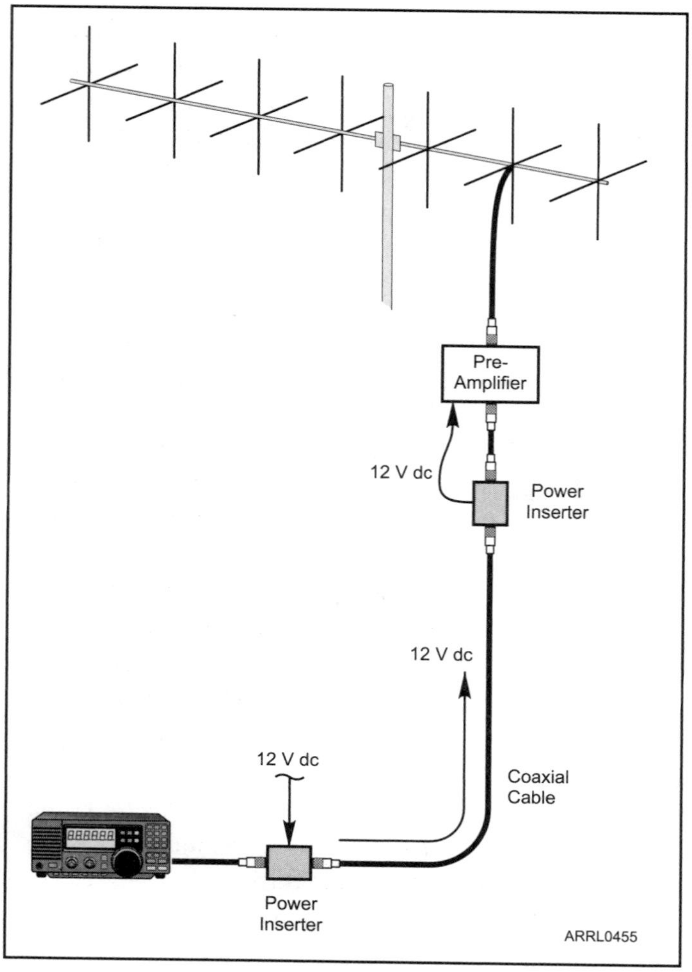

Figure 4.29—A dc power inserter acts just as its name implies — it inserts a dc voltage to the coaxial cable at the station. The inserter is designed to block dc power from going "backward" to the radio. Instead, the power flows through the coax to the antenna where another inserter picks it off and supplies it to the device (a preamplifier, in this case). Both inserters pass RF with negligible loss.

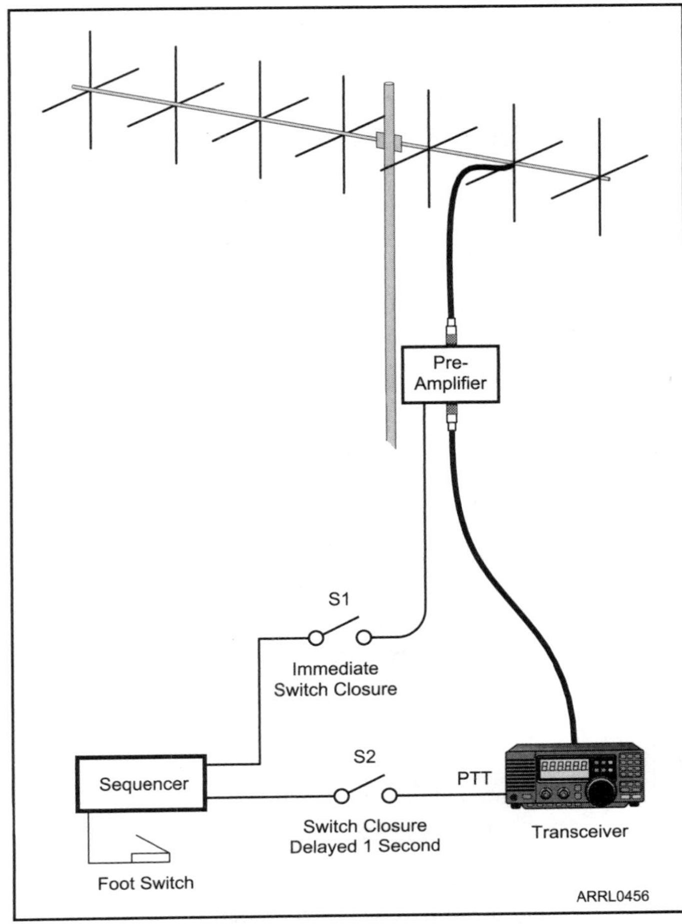

Figure 4.30—In this simplified example, the sequencer is triggered when the operator presses the foot switch. It immediately closes switch S1, which activates a "bypass" relay in the receive preamplifier at the antenna, effectively removing it from the feed line circuit. One second later, the sequencer closes switch S2, which is connected to the transceiver PTT (Push To Talk) line, keying the radio and applying RF power.

The DX Engineering TVSU-1 Time-Variable Sequence Unit (www.dxengineering.com).

specified in decibels (dB). A well-designed UHF preamplifier, for example, may have gain on the order of 15 to 25 dB and a *Noise Figure* (NF) of 0.5 to 2 dB (less is better).

If your antennas are outdoors, look for preamplifiers that are "mast mountable." These preamplifiers are housed in weatherproof enclosures.

If your preamplifiers are mast mounted, you will need to devise a means to supply dc power to the preamplifier. This can be as simple as routing a two-conductor power cable to the device. Alternatively, preamplifiers can be powered by dc sent up the feed line itself. Some transceivers have the ability to insert 12 Vdc on the feed line for this purpose. If not, you can use a *dc power inserter* to inject power at the station and/or recover it nearer the antenna (see **Figure 4.29**). Some preamplifier designs include feed line power capability, so all you need is an inserter at the "station end."

If your preamplifier is going to be installed in a feed line that will also be carrying RF power from the radio, you'll need a model that includes an internal relay to temporarily switch it out of the circuit to avoid damage to the preamplifier when you're transmitting. Some preamplifiers include this relay and nothing more; it is up to you to provide the means to energize the relay before you transmit. This is accomplished through a device known as a *sequencer*. A sequencer works with your transceiver to automatically switch the preamplifier out of the feed line before the radio can begin sending RF power (see **Figure 4.30**). A less complicated alternative is to purchase a preamplifier with *RF-sensed switching*. This design incorporates a sensor that detects the presence of RF from the radio and instantly switches the preamplifier out of harm's way. Note that RF-switched preamplifiers are rated according to the power they can safely handle. If you're transmitting 150 W, you'll need an RF-switched preamplifier rated for 150 W or more.

Downconverters

A discussion of receive preamplifiers isn't complete without bringing up their cousins, the *downconverters,* also referred to as *receive converters*. As the name implies, a downconverter converts one band of frequencies "down" to another. For example, a 2-to-10-meter downconverter would convert signals in a range from 144 to 146 MHz to 28 to 30 MHz.

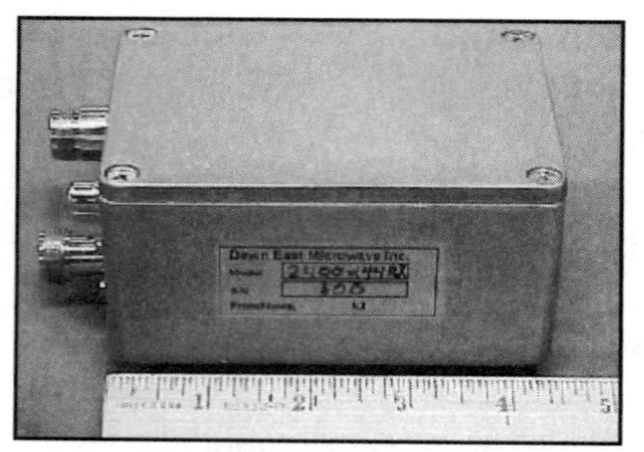

Figure 4.31—This downconverter by Downeast Microwave converts signals at 2.4 GHz to 2 meters. With its weatherproof enclosure, it is designed to be installed at the antenna.

In the days before the appearance of so-called "dc-to-daylight" transceivers, a downconverter was a popular means of receiving VHF and UHF signals on HF equipment. Today downconverters are used more often as a way to receive microwave signals. A microwave downconverter may, for example, convert a range of frequencies at 2.4 GHz to 2 meters. After stepping the microwave signals down to VHF, it becomes much easier to send those signals back to the radio over more affordable, comparatively higher-loss feed lines. A typical microwave downconverter is installed right at the antenna so that the microwave energy is immediately converted to a lower frequency before excessive loss can occur. See **Figure 4.31**.

Like receive preamplifiers, downconverters are rated by their gain and Noise Figure. Once again, you want the most gain for the least noise.

When it comes to installing a downconverter, all the same receive preamplifier issues apply. If the downconverter is installed outdoors (as it likely will be) it must be in a weatherproof enclosure. You must also supply dc power and be able to switch the downconverter out of the line if you are sending RF power to the same antenna.

ARRL0457

Recieved signals to 2-meter transceiver

144 MHz / 1.2 GHz Transverter

1.2 GHz

200 mW RF from 2-meter transceiver

Figure 4.32—In this example, the transverter is taking 1.2 GHz received signals and converting them to 2 meters. When it's time to transmit, the transverter takes RF at 2 meters and converts it to 1.2 GHz.

Transverters

Transverters are related to downconverters in that they not only convert received signals, they convert transmitted signals as well. You can use a transverter to generate, say, a 1.2 GHz uplink signal when powered by RF energy from a 2-meter transceiver. The same transverter can convert 1.2 GHz downlink signals to 2 meters as well (**Figure 4.32**).

Like receive preamplifiers and downconverters, transverters are best installed at the antenna. Therefore, once again, you must deal with the problem of supplying dc power. Transmit/receive switching in most transverters is accomplished through the use of an internal switch that is keyed through a sequencer. Some models provide automatic RF-sense switching.

When working with transverters the primary issue is supplying a safe level of RF power to the input. If your transceiver pumps out 50 W of power at 2 meters, for example, this is way too much RF for most transverters to handle. Unless the transverter has a built-in RF power attenuator, it is designed to deal with RF power levels on the order of *milliwatts* (typically 200 to 300 mW). If you are lucky enough to own a transceiver that

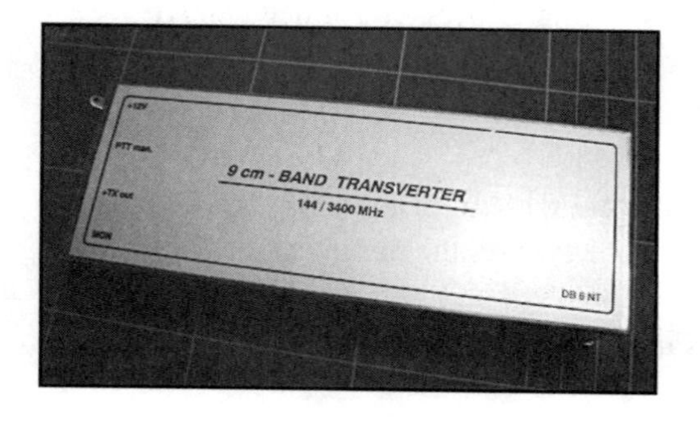

This DB6NT transverter converts 3.4 GHz signals to 2 meters, and vice versa.

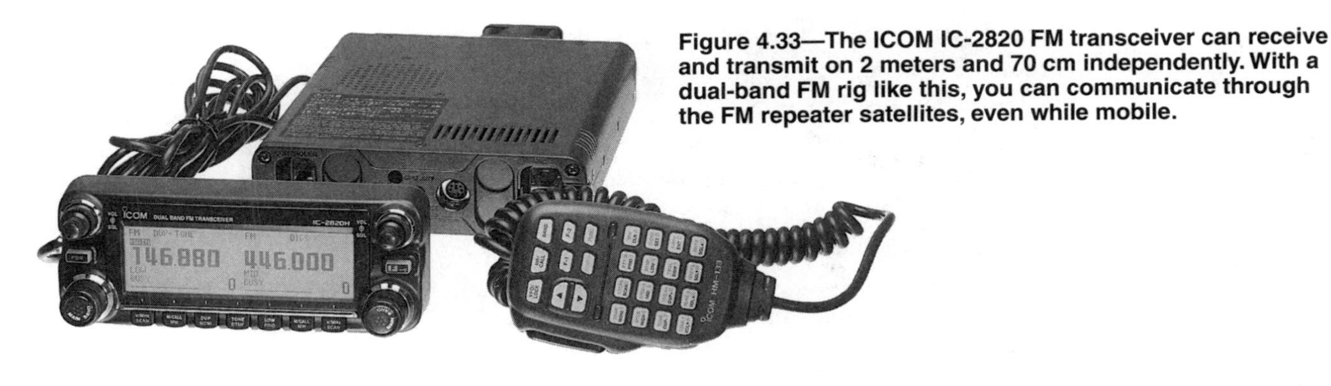

Figure 4.33—The ICOM IC-2820 FM transceiver can receive and transmit on 2 meters and 70 cm independently. With a dual-band FM rig like this, you can communicate through the FM repeater satellites, even while mobile.

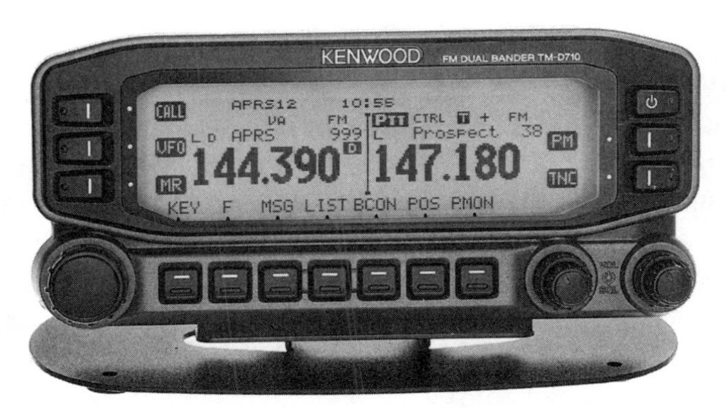

Figure 4.34—The Kenwood TM-D710 is a dual-band FM transceiver with a built in packet radio Terminal Node Controller (TNC) for digital communication.

AMSAT

Figure 4.35—The hand-held "Arrow" gain antenna is popular for FM repeater satellite operations.

features a transmit transverter port, you can obtain your milliwatt power levels there. If not, you'll need to add an attenuator at the transverter input to reduce the RF output of your radio to milliwatt levels.

Transceivers

There are many amateur transceivers that cover the VHF and UHF bands. In fact, the dc-to-daylight radios mentioned earlier in this chapter offer all the HF bands as well. The problem with choosing a transceiver for satellite operating is picking a model that will allow you to enjoy your chosen birds at a price you can afford.

FM Repeater and Digital Satellites

Almost any dual-band (2 meter/70 cm) FM transceiver will be adequate for operating FM repeater satellites such as OSCAR 51, and for digital operating with the International Space Station along with a number of other low-Earth orbiting birds as well (**Figure 4.33**).

Most modern dual-band rigs offer a "high power" output setting between 30 and 50 W. That's more than enough power to put a solid signal into a satellite with a directional antenna. It is also sufficient for omnidirectional antennas, including mobile antennas. (Yes, you can make contacts through FM repeater satellites while you drive!)

If you are considering digital operation, make sure to choose an FM transceiver that offers a data port. This will make it much easier to connect an external radio modem, such as a packet radio Terminal Node Controller (TNC). There are even a few radios with TNCs already built in (**Figure 4.34**). Also, make sure the transceiver is rated to handle 1200 and 9600 baud data signals. Nearly all FM transceivers can work with 1200 baud, but 9600 baud is another matter.

When in doubt, check the *QST* magazine "Product Reviews." ARRL members can view "Product Reviews" online at the ARRL Web site (**www.arrl.org**). Look for the bit-error-rate (BER) test results.

But what about handheld FM transceivers? Yes, you can use a dual-band handheld transceiver to work the FM birds, *but* because a hand-held's RF output is so low (5 W or less), you will definitely need to couple it to a directional antenna to be heard consistently among the competing signals (**Figure 4.35**). What's more, at this writing, most of the newer handhelds will *not* allow so-called true (or "full") duplex operation. We will discuss what full-duplex means and why it is desirable for satellite work in a moment.

Getting started with a dual-band FM radio and a repeater satellite is a worthwhile option if your funds are limited. Prices of new FM transceivers have declined over the years. You will also find many bargains on used FM rigs at hamfest fleamarkets and Web sites such as eBay (**www.ebay.com**).

Linear Transponder (SSB/CW) Satellites

FM signals tend to be wide and, by design, FM receivers are forgiving of frequency changes. That fortunate characteristic makes it easy to compensate for Doppler frequency shifting as an FM repeater satellite zips overhead. As you'll learn in the next chapter, you can program transceiver memory channels with a few uplink/downlink frequency steps and switch from one to another during the pass.

Figure 4.36—The ICOM IC-910H is a VHF/UHF satellite transceiver.

SSB and CW signals are much narrower, however, and when you're working through a linear transponder satellite your signal is sharing the passband with several others. Not only do you need to adjust your receive (downlink) frequency almost continuously to keep the SSB voice or CW sounding "normal," you also have to stay on frequency to avoid drifting into someone else's conversation. The most effective way to do this is to listen to your own signal coming through the satellite in "real time" while you are transmitting on the uplink. This type of operation is known as *full duplex*.

When it comes to considering an SSB/CW transceiver for satellite operating, the issue of full duplex capability is potentially confusing. You will find many multimode (SSB, CW, FM) transceivers that boast a feature labeled "cross-band split" or even "cross-band duplex." Be careful, though. What you require is a radio that can transmit and receive on different bands *simultaneously*. Few amateur transceivers can manage such a trick!

Figure 4.37—The Kenwood TS-2000 offers full HF coverage, plus VHF and UHF with full-duplex satellite capability.

As this book went to press, there were only two all-mode amateur transceivers on the market that were capable of full duplex operation: The VHF/UHF ICOM IC-910H (**Figure 4.36**) and the all-band Kenwood TS-2000 (**Figure 4.37**). However, if you shop the used equipment market you'll find excellent satellite radios such as the Yaesu FT-736 (**Figure 4.38**), the ICOM IC-820 and 821, the Yaesu FT-726 and FT-847 along with the Kenwood TS-790. All of these transceivers have full-duplex capability.

If you can't find or afford a full duplex transceiver, don't give up. Another option that

Figure 4.38—The Yaesu FT-736 is a legendary satellite transceiver. Although no longer manufactured, the '736 is still available on the used market.

some hams have used successfully is to exploit computer control of their ordinary non-duplex multiband transceivers to compensate for Doppler shift while transmitting "in the blind" on the uplink. As we discussed in Chapter 2, some satellite-tracking programs can automatically change the frequency of your radio during a pass. They do this by mathematically calculating the frequency shift and tweaking your radio accordingly. This solution for frequency tracking isn't as accurate as listening to your own downlink signal in full duplex, but it can work.

Another option is to use one transceiver for the uplink and a separate receiver or transceiver for the downlink. Some amateurs have even pressed old shortwave receivers into service along with VHF or UHF downconverters. They transmit with a 2 meter or 70 cm SSB radio on the uplink while listening to their shortwave receiver/downconverter combo on the downlink.

If you are working a linear transponder satellite with an uplink at 1.2 GHz or higher, you'll likely have to take the transverter approach to getting on the air, although there are some transceivers that offer optional 1.2 GHz modules. Fortunately, satellite builders are well aware that most hams own 2 meter and 70 cm transceivers and they design new birds accordingly. Even transponders with microwave downlinks are usually configured with uplinks on 2 meters or 70 centimeters. So, if you purchase a VHF/UHF transceiver, you have half the equation already solved, so to speak. All you need to add is a downconverter to receive the microwave downlink.

VHF/UHF RF Power Amplifiers

If your chosen transceiver offers at least 50 W output on the uplink band, you won't need an RF power amplifier to bring your signal to a level that can be "heard" by a low-Earth orbiting satellite, especially if you are using directional antennas.

On the other hand, if you are using omni antennas, 100 or 150 W output may help considerably. And if your target is a Phase III or IV bird orbiting at 50,000 km, 100 W or more, along with a directional antenna, is *mandatory*. If your transceiver lacks the necessary punch for the application, the solution is an external RF power amplifier.

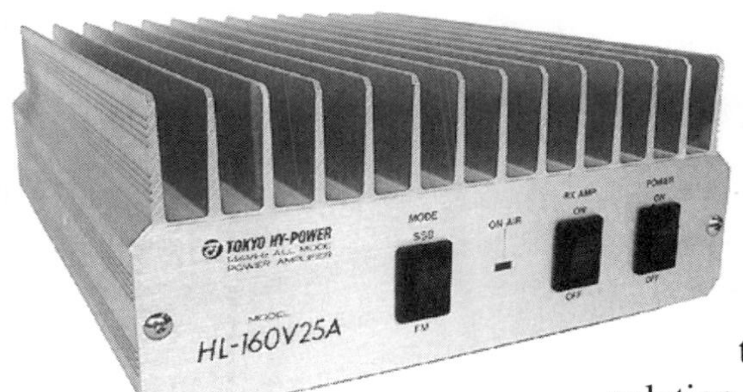

Figure 4.39—A 150W 2-meter RF power amplifier by Tokyo Hy-Power.

How much power should you buy? In most cases, a 100 or 150 W amplifier is a good choice (**Figure 4.39**). As you shop for amplifiers, take care to note the input and output specifications. How much RF at the input is necessary to produce, say, 150 W at the output? Can your radio supply that much power?

Another consideration is your dc power supply. While a 25-A 13.8 Vdc supply is perfectly adequate to run a 100-W transceiver, if you also decide to add a 100 or 150 W amplifier to your satellite station, the current demands will increase considerably. A separate power supply may be required to provide an *additional* 20 A (or more) to safely power the amplifier when both the transceiver *and* the amplifier are transmitting at the same time.

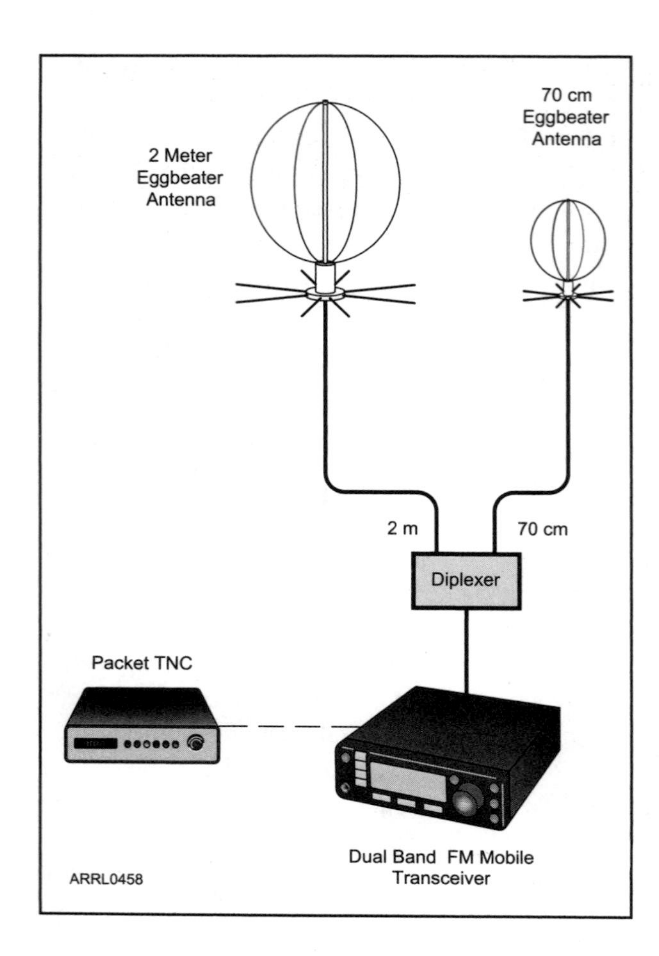

Typical Station Designs

Figure 4.40 illustrates a basic FM voice and data station for low-Earth orbiting satellites. At its core is a dual-band FM transceiver rated for 30 to 50 W output at its "high power" setting. The example uses omnidirectional antennas and it presumes that the FM transceiver has only one antenna jack, which is why a diplexer is indicated. Otherwise, you could run two separate coaxial feed lines back to the radio.

The performance of this station can be substantially enhanced by adding directional antennas or a single dual-band Yagi but, of course, that will require an antenna rotator (human or mechanical!). Although not shown in the diagram, a receive preamp on the downlink antenna is a worthwhile investment.

In **Figure 4.41** we've stepped up to a multiband multimode transceiver with full duplex capability and

Figure 4.40—A basic FM voice and data station for low-Earth orbiting satellites. At its core is a dual-band FM transceiver rated for 30 to 50 W output at its "high power" setting. This example uses omnidirectional antennas and it presumes that the FM transceiver has only one antenna jack, which is why a diplexer is indicated. A pre-amplifier at the receive antenna will provide a large performance boost.

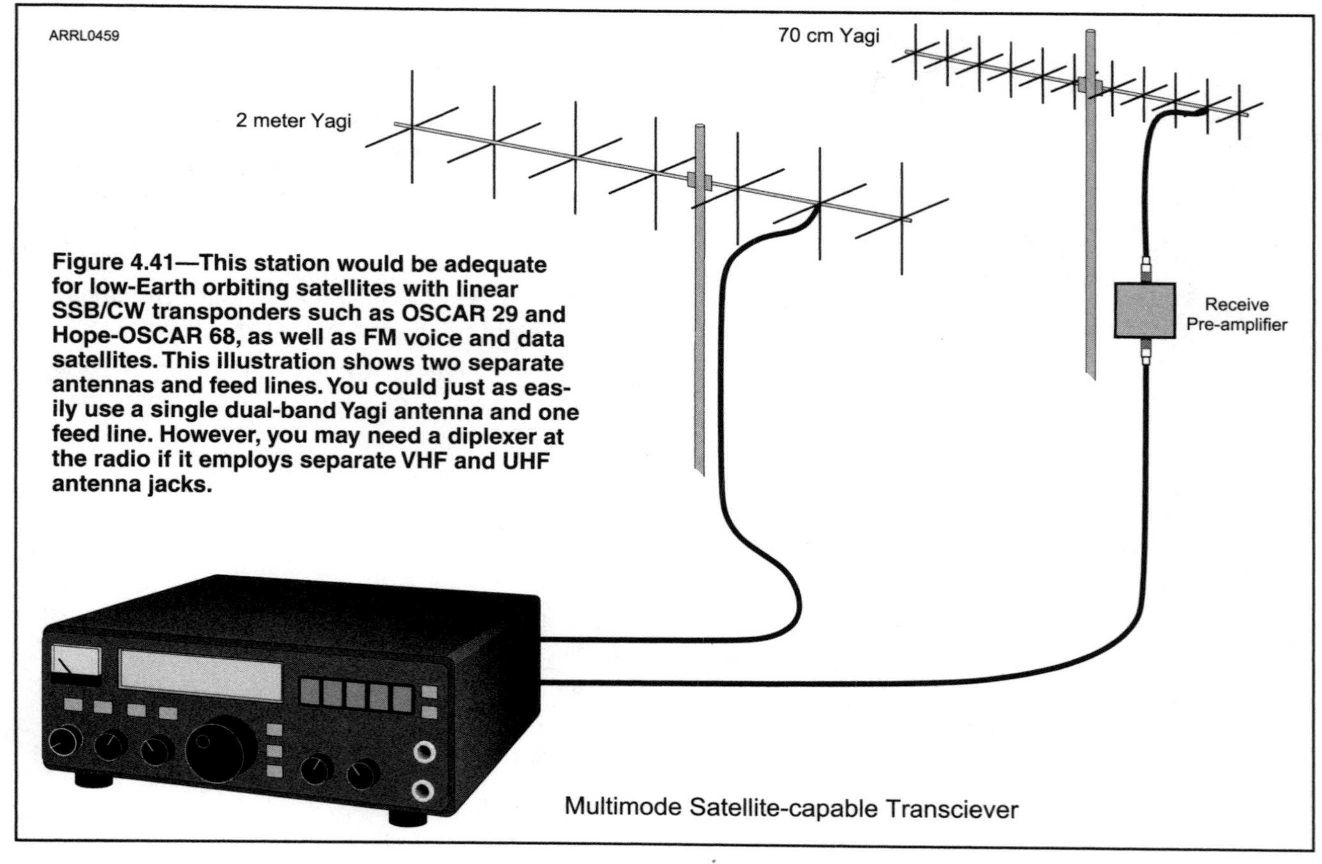

Figure 4.41—This station would be adequate for low-Earth orbiting satellites with linear SSB/CW transponders such as OSCAR 29 and Hope-OSCAR 68, as well as FM voice and data satellites. This illustration shows two separate antennas and feed lines. You could just as easily use a single dual-band Yagi antenna and one feed line. However, you may need a diplexer at the radio if it employs separate VHF and UHF antenna jacks.

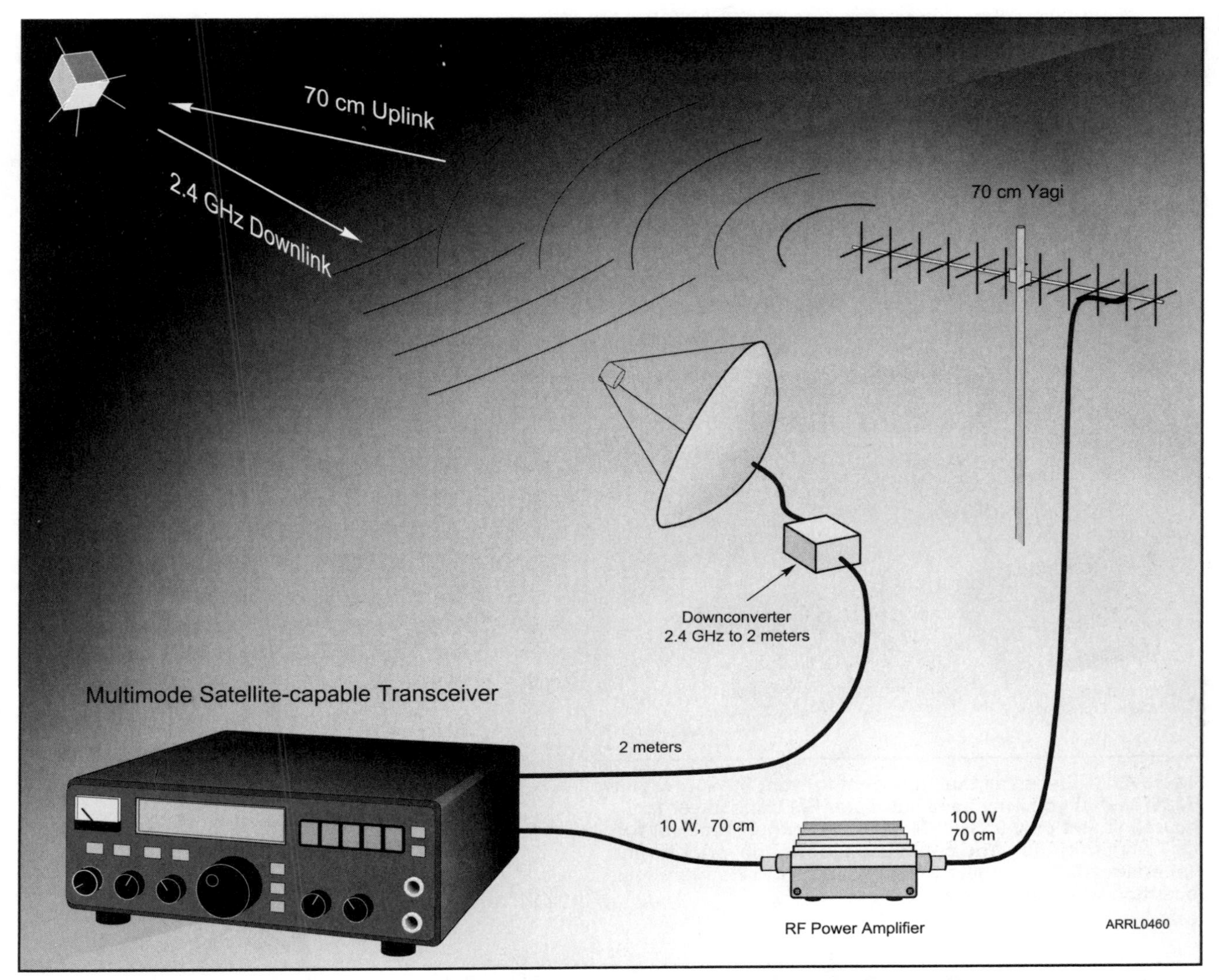

Figure 4.42—This example illustrates a station designed to work a distant Phase III or IV satellite with a microwave downlink. It assumes that the uplink is at 70 cm and the downlink is at 2.4 GHz. Note the use of the downconverter to step the 2.4 GHz signal at the antenna down to 2 meters before it is fed to the radio.

directional antennas. This station would be excellent for low-Earth orbiting satellites with linear SSB/CW transponders such as OSCAR 52, as well as FM voice and data satellites. By adding an RF power amplifier (assuming the output of the radio is less than 100 W), this station would be capable of working distant Phase III and IV satellites as well.

Note that this illustration shows two separate antennas and feed lines. You could just as easily use a single dual-band Yagi antenna and one feed line. However, you may need a diplexer at the radio if it employs separate VHF and UHF antenna jacks.

Figure 4.42 addresses at least one method to work satellites with microwave downlinks. It assumes that the uplink is at 70 cm and the downlink is at 2.4 GHz. It also assumes that the satellite in question is a high-orbiting Phase III or IV spacecraft (note the RF power amp on the uplink).

Note also the use of the downconverter to step the 2.4 GHz signal at the antenna to 2 meters before it is fed to the radio. You can actually use this same approach for any combination of bands and radios. For instance, let's say that you want to work the low-orbiting FM birds, but you have only a 2-meter FM transceiver for the uplink and a separate VHF FM "police scanner" receiver for the downlink. You could use a downconverter for the 70-cm downlink, stepping the signal down to 2 meters for reception with the scanner (**Figure 4.43**).

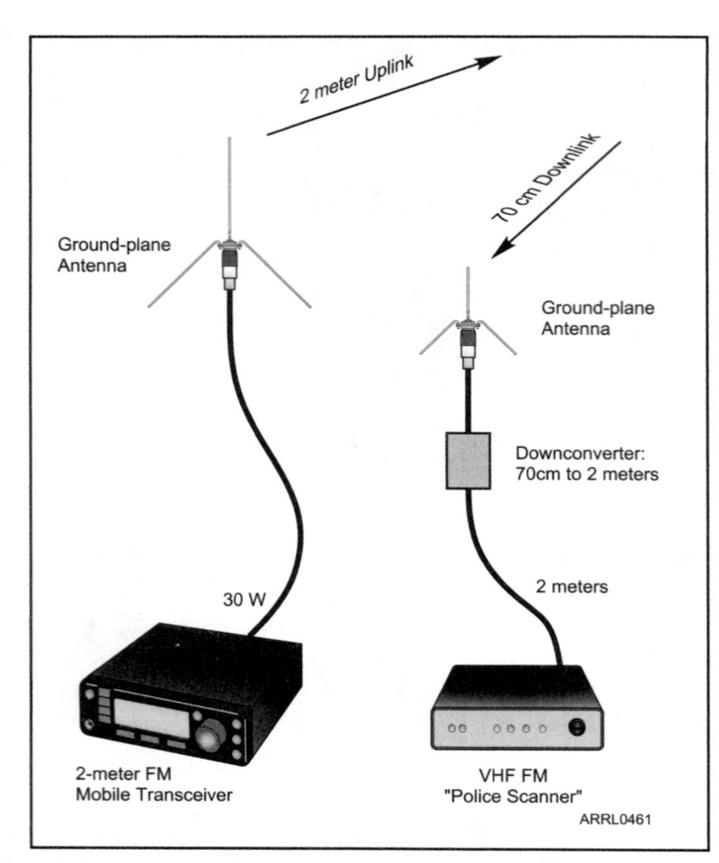

Figure 4.43—Let's say that you want to work the low-orbiting FM birds, but you only have a 2-meter FM transceiver for the uplink and only a VHF FM "police scanner" receiver for the 70-cm downlink. You could use a downconverter for the 70-cm downlink, stepping the signal down to 2 meters for reception with the scanner.

Satellite antennas don't have to be on towers or rooftops. This station for OSCAR 40 found its home on a patio in a suburban back yard. Note the barbeque grill antenna for the 2.4 GHz downlink sharing the same crossboom with the 70-cm uplink antenna.

W7ON

WD9EWK pauses to contact the OSCAR 51 FM repeater satellite from his car. He is using a small beam for the 2-meter uplink and an omnidirectional antenna on the back of the vehicle for the 70-cm downlink.

Satellite Operating

Satellite operating is unlike any other Amateur Radio activity. It is more than a matter of sitting down before your radio and making a contact. As you've learned in previous chapters, you have to know when the satellite is available, what path it will take from horizon to horizon, what uplink/downlink bands are in use and how you will deal with the Doppler Effect. All this amounts to a high-tech juggling act as you attempt to communicate with a spacecraft hurtling through the vacuum of space at many thousands of miles per hour.

But the challenge of satellite operating is part of the enjoyment. If it was as easy as making a contact on your local FM repeater, it wouldn't be nearly as fun. Veteran operators will tell you that there is nothing like the thrill of making contact through a satellite. Even in our highly technical age, the sheer wonder of what you're doing never fails to inspire.

Single-Channel "Repeater" Satellites

The single-channel repeater satellites are among the easiest birds to work, not just from an equipment standpoint, but also from an operational perspective.

As we discussed in Chapter 4, you can make contacts through the low-Earth orbiting satellites with little more than a dual band (2 meter/70 cm) FM mobile transceiver and an omnidirectional antenna—even a dual-band mobile whip will do. You can use a dual-band handheld radio as well, but don't expect success with its compact "rubber duck" antenna. Instead, you'll need something more substantial such as a dual-band Yagi. In Chapter 6 you'll find a

Waiting in the early morning fog for a Field Day satellite pass.

clever Yagi antenna by Thomas Hart, AD1B, that's specifically designed for handheld transceivers.

When this book was written, there were two FM repeater satellites in orbit. Each satellite has different characteristics and, therefore, different roads to success. Let's look at each one separately. In all likelihood, the same techniques will apply to any new single-channel birds that are launched after this book goes to press.

AMRAD-OSCAR 27

OSCAR 27 is an FM repeater bird. It operates in Mode V/U, listening on 2 meters and retransmitting on 70 cm. At the time this revised edition was printed, OSCAR 27 was inactive with recovery efforts underway. With luck, the command team will be successful.

OSCAR 27 doesn't require a CTCSS tone on the uplink, so you won't have to worry about programming one into your transceiver memory channels. Speaking of memory channels, you'll find a suggested programming configuration for OSCAR 27 in **Table 5.1**.

Perhaps the most unusual aspect of OSCAR 27 is its operating schedule. The satellite operates in Mode V/U most of the time (except when it is sending telemetry for the control operators), but because of the satellite's limited power budget the transmitter is on for only part of the daylight portion of each orbit over North America. To maintain the battery system, OSCAR 27 relies on a method known as *T*imed *E*clipse *P*ower *R*egulation (TEPR).

As Chuck Wyrick, N1UC, explains, the TEPR states describe the amount of time (in minutes) when the spacecraft enters and leaves sunlight. TEPR numbers are adjusted every few months to account for the seasonal North/South movement of the latitudes where AO-27 enters and exits sunlight.

The current software onboard AO-27 breaks the orbit into 6 different states as follows:

TEPR State 1 – Satellite enters eclipse

TEPR State 2 – Starts a programmed time interval after TEPR State 1 (during eclipse)

TEPR State 3 – Starts a programmed time interval after TEPR State 1 that's after State 2 (during eclipse)

TEPR State 4 – Satellite enters sunlight

TEPR State 5 – Starts a programmed time interval after TEPR 4 (during sunlight)

TEPR State 6 – Starts a programmed time interval after TEPR 4 that's after State 5 (during sunlight)

During a daytime pass, AO-27 will enter TEPR State 4 after coming out of eclipse and start charging its batteries. TEPR State 5 time is programmed so the satellite will stay in TEPR State 4 until the footprint reaches latitudes equal to the United States. At that time, it changes to TEPR State 5 and the transmitter turns on.

The duration of TEPR State 5 is set for the longest period that the transmitter can be left on and keep the batteries in a state that will prolong their lifetime. When this book was written, State 5 lasted about 18 minutes. If the transmitter was left on longer, the lifetime of the satellite would be shortened. The TEPR 6 period needs to be long

Table 5.1
Transceiver Memory Channels for OSCAR 27

Channel	*Time*	*Transmit (MHz)*	*Receive (MHz)*
1	AOS (start)	145.840	436.805
2	AOS+3 Minutes	145.845	436.800
3	Zenith (maximum)	145.850	436.795
4	Zenith+1 Minute	145.855	436.790
5	LOS (end)	145.860	436.785

enough to recharge the batteries before AO-27 enters eclipse again (TEPR State 1).

The time for AO-27 to stay in TEPR 6 places a limit on the number of hams in the southern latitudes that can work the satellite. See **Figure 5.1** to illustrate the following example:

AO-27 is set to charge its batteries for "x" number of minutes after entering Sunlight. This is TEPR State 4 expressed in 30-second increments. For example, TEPR 4 value of 42 equals a charging time of 21 minutes (42 × 30 seconds = 1260 seconds or 21 minutes).

TEPR State 5 is then set to the length of time that the transmitter will be on from the start of TEPR 4, but not on until the completion of TEPR 4. For example, TEPR 5 value of 78 means that the transmitter will shut off 39 minutes (78 × 30 seconds = 2340 seconds or 39 minutes) after the start of TEPR 4. Thus, the transmitter will turn on at the end of TEPR 4 and turn off at the end of TEPR 5 for a total of 18 minutes.
(78 - 42 = 36 and 36 × 30 seconds = 1080 seconds or 18 minutes).

If you find the TEPR concept difficult to follow, don't worry. The bottom line is that OSCAR 27 is only available during daylight passes.

Operating techniques for OSCAR 27 are the same as for OSCAR 51—short conversations, grid square exchanges and so on. Generally speaking, equipment requirements are the same as well. However, OSCAR 27 users have reported that directional antennas have been far more effective than omnis with this bird. That's probably because, in order to preserve its aging batteries for as long as possible, AO-27's handlers deliberately keep the power output of the satellite's downlink transmitter set very low…typically 500 mW or less.

There is the possibility that you may hear the occasional roar and buzz of digital signals on OSCAR 27. This satellite has a tradition of being a test bed for new technology

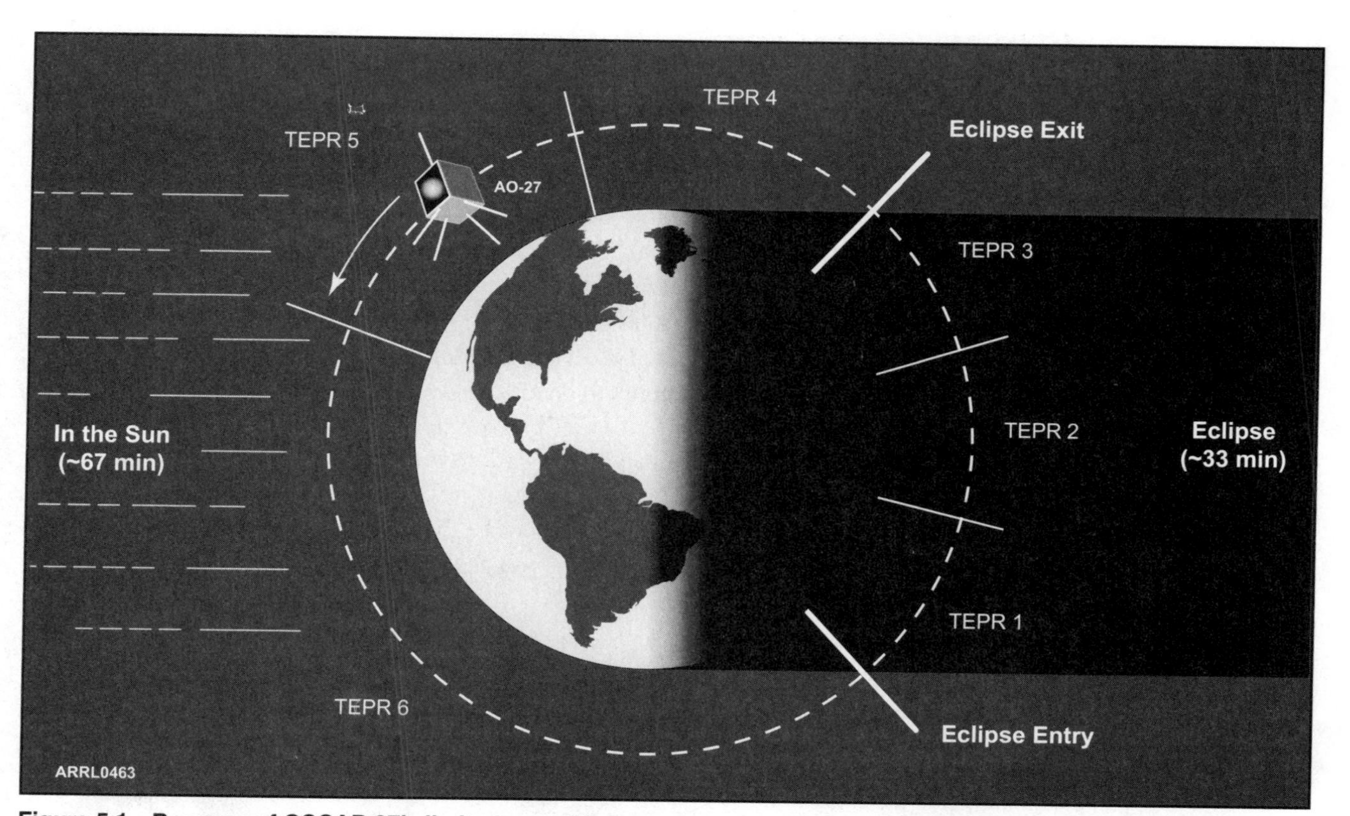

Figure 5.1—Because of OSCAR 27's limited power budget, the transmitter is on for only part of the daylight portion of each orbit over the North American hemisphere. To regulate the battery systems, OSCAR 27 relies on a method known as Timed Eclipse Power Regulation (TEPR). See text.

and with the increasing interest in digital communication, some hams have tried using OSCAR 27 for those applications. In 2007, for example, a number of amateurs used AO-27 as a test repeater for D-STAR digital communication.

OSCAR 50

Saudi-OSCAR 50 is a Mode V/U FM repeater satellite similar to OSCAR 27, but with an unusual means of controlling access. This satellite requires *two* CTCSS tones—one to "open" the satellite for use and another to pass each transmission.

Before OSCAR 50 will respond to uplink signals, it must receive a 74.4 Hz CTCSS tone for 2 seconds on the uplink. A successful transmission "arms" the bird by activating a 10-minute timer. Any transmissions that follow must carry a 67 Hz CTCSS. To reset the timer, the 74.4 Hz tone must be sent again for 2 seconds.

As OSCAR 50 rises above your local horizon, start listening for activity. If you hear voices, it obviously means that someone has already triggered the 10-minute timer. If not, this is your opportunity to open the bird and make yourself heard. **Table 5.2** lists memory channels for Doppler compensation with OSCAR 50, along with a separate channel to open the bird for use. This is where alphanumeric memory channel tags can be extremely handy. You can label one channel "SO50 OPEN" and program it with the 74.4 Hz CTCSS; the remaining channels can be programmed with the 67 Hz CTCSS. This way, you can quickly select the correct RF/CTCSS combo to trigger OSCAR 50's timer and then jump to the other channels for Doppler compensation.

International Space Station

The International Space Station has a powerful presence in the amateur satellite community thanks to the Amateur Radio on the International Space Station project (ARISS). Perhaps you've seen stories in the media about space station crews using Amateur Radio to chat with school children. These are pre-arranged contacts where hams assist schools in setting up and operating the necessary equipment.

ARISS also supports digital communication, as we'll discuss later in this chapter, and an onboard FM voice repeater. Sometimes crew members will surprise amateurs by picking up the microphone and making direct contacts.

The ARISS operating schedule is at the mercy of all the other activities aboard the station. For example, whenever a *Progress* supply vehicle is docking, the Amateur Radio station is switched off. The amateur station is also off the air during Space Shuttle visits. To stay on top of ARISS activities, your best bet is to check the AMSAT satellite status page on the Web at **www.amsat.org/**.

Table 5.2
Transceiver Memory Channels for OSCAR 50

Channel	*Time*	*Transmit (MHz)*	*Receive (MHz)*
1	AOS (start)	145.840 (67 Hz CTCSS)	436.805
2	AOS+3 Minutes	145.845	436.800
3	Zenith (maximum)	145.850	436.795
4	Zenith+1 Minute	145.855	436.790
5	LOS (end)	145.860	436.885
6	**Reset Timer**	**145.850 (74.4 Hz CTCSS)**	

Astronaut Frank Culbertson, KD5OPQ, makes a contact from the amateur station aboard the International Space Station.

As you'll see in **Table 5.3**, the FM voice repeater operates in a cross-band mode, listening at 437.800 MHz and repeating at 145.800 MHz. This is a conventional FM repeater, so a standard dual-band VHF/UHF FM transceiver will work well. As with OSCAR 50, conversations should be kept as short as possible. This is especially true for the space station because it orbits at a relatively low altitude and its "pass windows" are only about 10 minutes long.

Direct crew contacts can happen when you least expect them. If you are listening to the voice repeater output at 145.800 MHz and you suddenly hear an astronaut or cosmonaut calling CQ, quickly reconfigure your transceiver to transmit on 144.49 MHz and make a call. (It is a good idea to program a split-frequency memory channel that you can easily access if the unexpected takes place!) If the crewperson answers your call, make sure to keep the conversation very brief—name, location, etc. Remember that many other hams are trying to make contact as well.

A Russian *Progress* freighter about to dock with the International Space Station. The ARISS equipment is always turned off during docking maneuvers.

Table 5.3
Currently Active Amateur Radio Satellites

As of March 2013

Satellite	*Uplink (MHz)*	*Downlink (MHz)*	*Mode*
AAUSAT-II	—	437.425 (packet)	Telemetry
AMSAT-OSCAR 7		29.502	Beacon
		145.975	Beacon
		435.106	Beacon
	145.850 - 145.950	29.400 - 29.500	SSB/CW, non-inverting
	432.125 - 432.175	145.975 - 145.925	SSB/CW, inverting
AMSAT-OSCAR 27	145.850	436.795	FM repeater
ARISS (ISS)	144.490	145.800	Crew contact, FM (Rgn. 2/3)
	145.200	145.800	Crew contact, FM (Rgn. 1)
	145.990	145.800	Packet BBS
	145.825	145.825	APRS digipeater
	—	144.490	SSTV downlink
	437.800	145.800	FM repeater
Compass-One	—	437.275 (CW)	Telemetry
	—	437.405 (packet)	Telemetry (including images)
CubeSat-OSCAR 57	—	436.8475 (CW)	Telemetry
		437.4900 (packet)	Telemetry
CubeSat-OSCAR 58	—	437.4650 (CW)	Telemetry
		437.3450 (packet)	Telemetry
CubeSat-OSCAR 66	—	437.485 (packet)	Telemetry (including images)
Cute1.7 + APDII	1267.600	437.475	9600 baud packet
Dutch-OSCAR 64	145.870 (packet)	Telemetry	
	435.530 – 435.570	145.880 – 145.920	SSB/CW, inverting
Fuji-OSCAR 29	145.900-146.00	435.800-435.900	SSB/CW, inverting
Saudi-OSCAR 50	145.850	436.795	FM Repeater—67 Hz CTCSS required
VUSat-OSCAR 52	—	145.860	CW Beacon
		145.936	Carrier Beacon
	435.220 - 435.280	145.870 - 145.930	SSB/CW, inverting

Linear Transponder Satellites

Single-channel satellites are attractive because of the minimal ground station equipment required to work them. Their major shortcoming, however, is their inability to support more than one conversation at a time. As a single–channel satellite operator, you're under constant pressure to make short-duration contacts so that others can use the bird.

By contrast, linear transponder satellites relay an entire range of frequencies at once, not just a single channel. A linear transponder can, as a result, support many simultaneous conversations. There can still be interference issues, as we'll see later. But once you've established contact, you can chat for as long as you wish—or at least as long as the satellite is available to you. For low-Earth orbiting satellites, conversations can span 10 or 15 minutes. If the satellite in question is a high-Earth orbiter (HEO), conversations can last for *hours.*

Typical Transponders

When this book was written, there were no active amateur HEO satellites available; all linear-transponder satellites were in low orbits. Look at Table 5.3 and you'll find four linear transponder birds:

AMSAT-OSCAR 7
VuSAT-OSCAR 52
Fuji-OSCAR 29

These satellites are easy to identify because their uplinks and downlinks span frequency *ranges* in the table, not just single frequencies. Their transponder bandwidths vary from 40 kHz to 100 kHz. To put this in perspective, even a transponder with only a 40 kHz bandwidth, is capable of supporting a dozen simultaneous SSB conversations and many more CW chats.

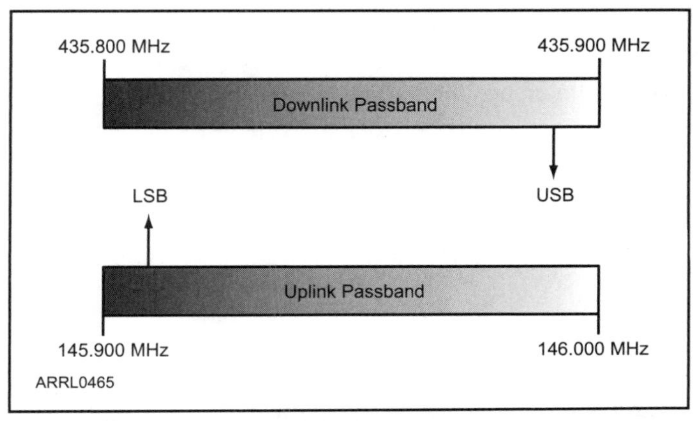

Figure 5.2—An example of a hypothetical inverting transponder. Note how the LSB signal in the lower portion of the *uplink* passband becomes a USB signal in the upper portion of the *downlink* passband.

You'll also see that some transponders are marked as *inverting* or *non-inverting*. An inverting transponder relays a mirror image of the uplink passband. This means that a *lower* sideband signal on the uplink becomes an *upper* sideband signal on the downlink. At the same time, a signal at the *high end* of the uplink passband will appear at the *low end* of the downlink passband. See the example in **Figure 5.2**. A non-inverting transponder relays the uplink signals exactly as they appear in the passband—sidebands remain unchanged and the relative position of an uplink signal in the downlink passband remains the same.

For engineering design reasons, most linear transponders are of the inverting variety.

Figure 5.3—An example of a transceiver (a Kenwood TS-2000, in this case) with dual-VFOs operating in reverse tracking mode. As the main receive VFO is adjusted, the secondary VFO (on the far right) tracks along with it.

This presents a challenge when you're on the air. As you're *increasing* your uplink frequency, for example, you have to remember that your downlink frequency is *decreasing*—"heading the other way," as it were. Fortunately, a number of transceivers designed for satellite use have *reverse VFO tracking* among their features (see **Figure 5.3**). This locks the uplink and downlink VFOs in a reverse arrangement. If you tweak the uplink VFO to increase your signal frequency by, say, 5 kHz, the downlink (receive) VFO will automatically shift downward by the same amount.

Finding Yourself...and Others

Full duplex operation is strongly recommended for linear transponder satellites. Because of the relatively narrow bandwidths of SSB and CW signals, the Doppler Effect will be more pronounced. You need to be able to hear your own signal while you are transmitting so that you can make frequent adjustments to the downlink receiver to maintain the tone (CW) or voice clarity (SSB).

It is possible to use a computer to estimate Doppler frequency shifts and apply receiver correction automatically (assuming your radio is under computer control). This is high-tech guesswork at best, though. The better, more accurate, solution is to slip on a pair of headphones and correct for Doppler by listening to your own downlink signal. (Headphones are necessary to help you avoid creating feedback through the satellite.)

Before you attempt your first conversation on a linear transponder satellite, it's best to gain some practice at receiving your own signal during a pass. For this example, let's use VuSAT-OSCAR 52, a popular bird with an inverting linear transponder. Let's also assume that you are operating SSB. **Figure 5.4** illustrates OSCAR 52's uplink and downlink passband frequencies relative to each other. If you pick 435.230 MHz, for instance, as your uplink frequency, you might expect to hear yourself somewhere in the vicinity of 145.920 MHz. For SSB, the convention is to transmit lower sideband on the uplink, which inverts to upper sideband on the downlink.

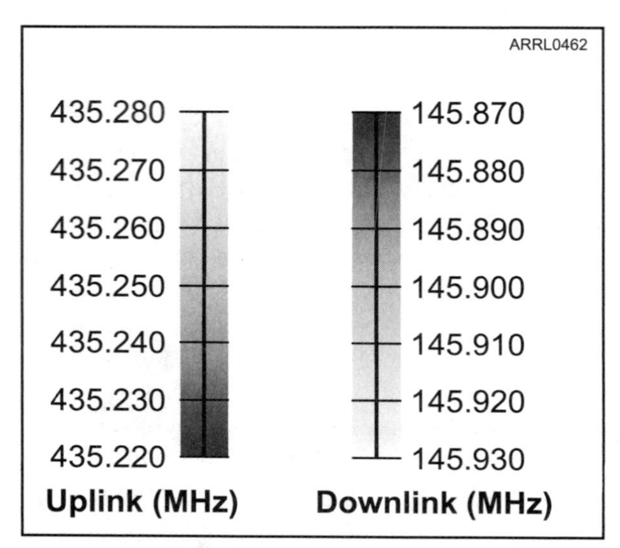

Figure 5.4—The relationship between uplink and downlink frequencies for VuSAT-OSCAR 52. Select an uplink frequency on the left and you'll see its corresponding downlink frequency on the right.

If you are using separate transmitters and receivers, set your transmitter for 435.230 MHz LSB and leave it there. If you are using a multiband satellite-capable transceiver with a VFO tracking feature, "unlock" (disable) the tracking and set the transmitter VFO for 435.230 MHz LSB.

As OSCAR 52 climbs above the horizon, start sweeping your receive VFO through the 2-meter downlink passband, listening for signals as you go. (Another technique is to listen for a satellite's beacon, if available.) As soon as you hear activity, tune your receiver to 145.920 MHz. If the frequency is clear, begin transmitting your call sign and perhaps the word "testing." As you speak, tune your downlink receiver back and forth from 145.920 to about 145.930 MHz. It may take a couple of minutes, but with luck you'll soon hear your own voice rising out of the noise. When you do, tune it in quickly until your voice sounds normal. Congratulations! You've just heard your own signal being relayed by a spacecraft!

This is a good time to experiment. Stop tuning your receiver and note how the Doppler Effect changes the sound of your voice. Practice retuning to keep your voice sounding normal. If you are using a satellite transceiver, try locking the uplink and downlink VFOs and observe how the reverse tracking affects your signal.

Calling CQ

Once you've become comfortable with finding your own signal, try calling CQ. Tune your uplink transmitter and downlink receiver to the frequencies of your choice. (If your radio has a VFO tracking feature, make sure it is unlocked.) When the satellite comes into range, start calling CQ as you listen for your voice on the downlink. Once you hear it, tweak your receiver as necessary to keep your voice sounding normal.

Don't be surprised if you suddenly hear a string of CW beeps. That's good news—it's the sound of someone who has heard your CQ and is quickly adjusting their uplink transmitter while sending a continuous series of Morse "dits" with a CW keyer. They are trying to hear their own signal and bring it to approximately the same frequency as yours. Alternatively, you may hear an off-frequency, high-pitched voice that suddenly "swoops" into your CQ. Once again, that is another station that has heard you and is preparing to answer.

> ***"Once you've become comfortable with finding your own signal, try calling CQ."***

Once the conversation is underway, all you have to do is adjust your receiver to keep your voice, and the voice of the other operator, sounding normal. If your conversational companion is operating properly, he is doing the same thing. *Do not adjust your transmit frequency*.

Answering a CQ

A full duplex transceiver with a VFO tracking feature comes in handy when you're answering someone else's CQ. Once you've set the uplink and downlink VFOs, activate the tracking function to keep them locked together. Now all you have to do is tune through the downlink passband with the receive VFO while the uplink (transmit) VFO follows you automatically. If you discover someone calling CQ, tune him in and then unlock the tracking. With your uplink VFO now operating independently, begin answering the other station as you gently adjust the *uplink* VFO to bring your signal on frequency. Once your frequency matches his (when your voice or CW tone sounds normal on the downlink), don't touch the uplink VFO again. Enjoy the conversation and compensate for Doppler by adjusting the *downlink (receive)* VFO.

If you are using separate rigs for the uplink and downlink, you'll need to tune in the station calling CQ, then make a quick estimate of the correct corresponding uplink frequency. Devising a chart like the one in Figure 5.9 will be a great help. Begin answering and adjust your uplink radio until your signal matches his on the downlink.

A Crowded Dance Floor

It is possible that your conversation will be sharing the transponder with several others at the same time. Thanks to the Doppler Effect, all the downlink signals—including yours—are drifting through the passband.

Unfortunately, you can't count on everyone playing by the same rules. Some stations will be adjusting their uplink frequencies as they speak, and that will effectively anchor them at a single frequency in the downlink passband. While everyone else is drifting, the uplink tuner is stuck in place, creating a fixed signal that another conversation may eventually drift into.

With that in mind, its best to put as much frequency "distance" as possible between you and the nearest conversation when you're about to call CQ. A 10 kHz separation would be ideal, but in a crowded transponder that won't always be possible.

How Much Power is Enough?

The issue of uplink power and linear transponders has always been controversial. Obviously, you want to use enough power to generate a listenable signal on the downlink. For low-Earth orbiting satellites, that may amount to only 30 or 50 W, depending on the type of antenna you are using. For the high-Earth orbiters, uplink power levels of 100 to 150 W are common.

Unfortunately, there is a "more-is-better" obsession among some amateurs. A listenable downlink signal is not sufficient—they want a *loud* signal. Some of these operators, for example, are working the low-Earth orbiters by using directional antennas *and* 100 W of output power or more. The result can be an effective radiated power level in excess of *1 kW!*

> ***"For the high-Earth orbiters, uplink power levels of 100 to 150 W are common."***

The net effect of such a powerful signal on a linear transponder is to swamp its receiver. All signals weaker than the high-power station will be dramatically reduced in strength on the downlink; some may disappear altogether. This is because the satellite is dedicating the lion's share of its output to relaying the loud uplink signal while starving everyone else.

The problem of "transponder hogging" is so common that the designers of the AMSAT-OSCAR 40 created *LEILA*—(*LEI*stungs *L*imit *A*nzeige [Power Limit Indicator]). As we discussed in Chapter 3, LEILA's computer continuously monitored the 10.7 MHz transponder IF passband. Whenever it encountered an uplink signal that exceeded a predetermined level, the computer inserted a CW message over the offender's downlink. When you heard the CW message, that was your cue to reduce uplink power—quickly. If you ignored the warning, LEILA activated a notch filter that effectively removed your signal. As the saying goes, if you can't play nicely, you won't be allowed to play at all. None of the satellites in orbit at the time of this writing have a LEILA system (although there is a good chance that the next HEO satellite will.)

To be a good neighbor on a linear transponder satellite, the rule of thumb is to use only the uplink power necessary to keep your signal about as strong as everyone else's. If the satellite has a telemetry beacon, another technique is to use the beacon as the standard for downlink signal strength. In other words, your downlink signal should never be louder than the beacon.

Communication Modes and Power

No doubt you've noticed that we haven't discussed using FM with linear transponder satellites. There is no rule that says that you can't transmit FM through such a bird, but doing so is considered extremely poor amateur practice. The problem isn't so much the width of an FM signal, although that is certainly a consideration, but more the fact that FM is a *100% duty cycle* communication mode. This means that an FM transmitter generates its full output during the entire transmission. If a linear transponder attempts to relay an FM signal, it must also generate full output on the downlink for as long as the ground station is transmitting. This not only reduces or eliminates all the other signals in the passband, it places a great strain on the satellite's power systems.

Unlike commercial or military satellites, hamsats are not designed with robust power supplies. That's why SSB and CW are the modes of choice for linear transponders. The output power of a ground station transmitter operating SSB or CW fluctuates considerably, resulting in an average duty cycle of 50% or less. This is well within the ability of an amateur satellite transponder and power system to accommodate.

As long as we're discussing satellite power systems, this is a good time to mention the special case of AMSAT-OSCAR 7. You may recall reading in Chapter 1 that AO-7 is a "miracle satellite" that returned to the land of the living after decades of silence. This satellite is functioning with a damaged power system. It sports Mode V/H and U/V linear transponders (see Table 5.1), but they are only operational when the satellite's solar panels receive sufficient sunlight. Mode U/V is by far the most popular, but be aware that AO-7 has a tendency to change its operating modes unpredictably, especially when it is emerging from an eclipse period.

Because of its barely functional power system, OSCAR 7 is operationally "fragile" and should be treated accordingly. Its transponders are not capable of supplying a substantial amount of output on the downlink, so make sure you use as little power as possible on the uplink. If you hear your downlink signal rapidly changing frequency or "warbling", this means that the power system is being overtaxed. Reduce your uplink output right away.

HEO Considerations

Although we don't have a high-Earth orbiting (HEO) satellite as this book goes to press, one is under construction: AMSAT-DL's *Phase 3E*. If all goes as planned, it may be in orbit within several years. Check the AMSAT Web site at **www.amsat.org** for the latest updates.

Phase 3E will feature linear transponders on several bands and the low-Earth orbiting operating techniques to access such a satellite will still apply. However, the transponders for this HEO will also most likely operate on schedules that will vary according to the orbital *phases* of the satellite. This means that you'll need to become acquainted with the phase schedules before you sit down in front of your radio. See the discussion of satellite phases in Chapter 2.

If you are working a HEO that's traveling in an elliptical orbit, you'll enjoy the bonus of reduced Doppler correction when the bird is at apogee. At this point in its orbit, the difference in relative motion between the satellite and your station is minimal, so the Doppler Effect is minimal as well. And if amateurs are ever fortunate enough to acquire a transponder on a geostationary HEO, you won't have to be concerned about Doppler at all.

Operating in full duplex with a HEO can still present an unusual challenge. If the satellite is at the apogee of an elliptical orbit, it could be as much as 50,000 km distant. At this distance, it takes a total of 3⁄10 second for your radio signal to reach the satellite on the uplink and return on the downlink. If you are listening to your own signal on the downlink, you'll definitely notice the delay. For many operators, having their own voices fed back to their headphones with a 3⁄10 second delay is disconcerting—it can even disrupt your speech, causing you to stammer or speak in an unusual cadence. The cure for this annoyance is to turn down your receive audio while you are talking. With the satellite at apogee, you won't have to be too concerned about correcting for Doppler, so it isn't critical that you listen to your own downlink signal at all times.

Digital Satellites

Before the word "Internet" became widely recognized in most households, amateur satellite operators were routinely using *packet radio* to share electronic mail, binary files and other bits of digital information. They were accomplishing this feat thanks to a number of digital relay satellites, particularly the MicroSats that hams pioneered in the early 1990s.

The rise of the Internet greatly reduced amateur packet radio activity, both on the ground and in space. Even so, there is still a portion of the amateur satellite community that enjoys digital communication and there are still satellites that support this innovative activity.

The TNC: the Heart of Packet Radio

As we discussed in Chapter 4, the TNC—*Terminal Node Controller*—is at the heart of any digital satellite ground station. It acts as the middleman between your radio and your computer. The TNC takes data from your computer, creates *AX.25* packets and then transforms the AX.25 formatted data into audio signals for transmission by the radio. Working in reverse, the TNC demodulates the received audio, changes it back into data, disassembles the AX.25 packets and sends the result to the computer.

For 1200-baud applications, TNCs create signals for transmission using audio frequency shift keying (AFSK). Twelve hundred baud packet is the most common. When creating a 1200-baud signal, a *mark* or 1 bit is represented by a frequency of 1200 Hz. A *space* or 0 bit is represented by a frequency of 2200 Hz. The transition between each successive mark or space waveform happens at a rate of 1200 baud. The frequencies of 1200 and 2200 Hz fit within the standard narrowband FM audio passband used for voice, so that AFSK is accomplished by simply generating 1200 and 2200 Hz tones and feeding them in to the microphone input of a standard FM voice transmitter.

Pure frequency shift keying is used for 9600 baud packet and this signal must be applied to dedicated 9600-baud ports on the transceiver. This requires a radio designed to pass these 9600 baud signals without distortion. Fortunately, most satellite transceivers and a number of standard FM rigs offer this capability and include a dedicated port for 9600 baud signals.

All TNCs are functionally the same. When you buy a stand-alone TNC it usually includes a cable for connecting it to the radio, but you'll have to attach the appropriate connectors for the radio you're going to use. You'll also have to furnish the cable that connects the TNC to your computer at the COM port. In most cases this is an RS-232 serial cable. Most ham TNCs have yet to migrate to USB at the time of this writing, although conversion cables and software are now commercially available to connect RS-232 serial TNC output jacks to USB computer ports. TNCs integrated into radios obviously don't require cabling to the radio itself, but you'll still need a cable between the radio's TNC port and your computer.

Note that to use the *WiSP* software mentioned later in this chapter, your TNC must be able to function in the *KISS* (Keep It Simple, Stupid!) mode. KISS passes digital information between the radio and computer in a very straightforward manner without the need for special command dialogues and so forth. Most TNCs will function in the KISS mode, but a few do not—especially those TNCs that are built into transceivers. Check ahead of time and be sure.

To test TNC-to-computer communication, any terminal program will work (Microsoft *Windows 98* and *XP* include such a program). You'll need to start that software and specify the COM port you'll be using, and set the baud and data parameters for that port. Refer to the manual for the specific program you've chosen. The baud rate of your computer must match the baud rate of your TNC. Some TNCs will automatically set their baud rate to match the computer. Other TNCs have software commands or switches for setting the baud rate. Again, you'll need to refer to your manual for specific instructions. When setting the data parameters, 8-N-1 is normally used: 8 data bits, no parity, and 1 stop bit. But like the baud rate, the computer and TNC parameters must match.

"All TNCs are functionally the same. When you buy a stand-alone TNC it usually includes a cable for connecting it to the radio, but you'll have to attach the appropriate connectors for the radio you're going to use."

Satellite APRS

The Automatic Position Reporting System, better known as *APRS*, was the brainchild of Bob Bruninga, WB4APR. In fact, APRS® is a trademark registered by WB4APR. The original concept behind APRS involved tracking moving objects, and that's still its most popular use today.

APRS stations transmit position information that is decoded at the receiving stations. Station positions are represented by symbols (called *icons*) on computer-generated maps. When a station moves and transmits a new position, the icon moves as well.

Any discussion of APRS must begin with the technology that lies at its heart: the *Global Positioning System.*

The Evolution of GPS and APRS

The Global Positioning System (GPS) is a satellite-based radionavigation system that uses 24 orbiting satellites to provide a highly accurate position finding capability anywhere on the face of the Earth anytime, day or night (**Figure 5.5**). Although GPS has become the best known electronic navigation system today, it was not the first. GPS was preceded by other well known electronic navigational aids including radio direction finders (RDF), hyperbolic systems (OMEGA, DECCA. Loran-A, and Loran-C), and the very first satellite based navigational aid, TRANSIT.

The Global Positioning System is owned and managed by the US Department of Defense. The official name of the system is NAVSTAR, which is an acronym for **NAV**igation **S**atellite **T**iming **A**nd **R**anging. To meet US requirements for a highly accurate electronic navigational system for military and intelligence communities, the Department of Defense began research and development of GPS in 1973. The United States Air Force was named as the lead agency for this multiservice program. The first GPS satellite was launched on February 22, 1978.

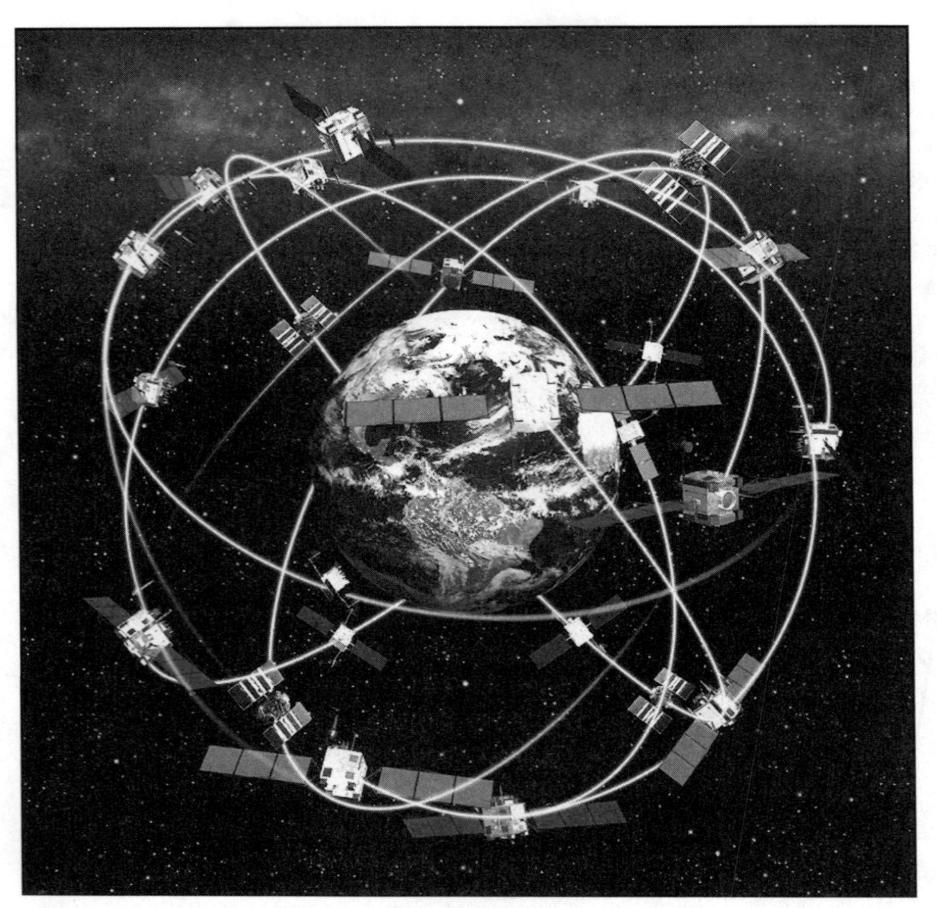

Figure 5.5—The Global Positioning System (GPS) is a satellite-based radionavigation system that uses 24 orbiting satellites to provide a highly accurate position finding capability anywhere on the face of the Earth anytime, day or night.

GPS was originally developed strictly for military use. This changed in 1983 after the downing of Korean Air Flight 007 by the Soviet Union. This tragedy occurred in part because the crew of the Korean 747 aircraft made an error in navigation which brought the aircraft over Soviet air space and the Russians shot them down. It was argued that if GPS had been available this tragedy would not have occurred. As a result, President Ronald Reagan issued an Executive Decree that certain portions of the GPS system be made available for civilian use free of charge to the entire world. The US military insisted, however, that those portions of the GPS made available for civilian use be degraded in accuracy so that the system could not be used by the enemies of the US for clandestine purposes. When the Standard Positioning Service portions of the GPS were opened up to civilian use, it came with something called Selective Availability (SA) which degraded the normal accuracy of 50 feet to 300 feet.

Even with portions of GPS now open to civilian use, there were very few GPS receivers available, and any to be found were very expensive. In 1991, during Operation Desert Storm, the use of GPS was so widespread that the military found they did not have enough GPS receivers to supply the troops. A large multi-sourced procurement by the military for GPS receivers to use in Desert Storm resulted in a tremendous spin-off of the technology into the civilian sector. This, in turn, resulted in the availability of highly capable GPS receiver equipment on the civilian market. Although GPS receivers were expensive at first, widespread acceptance of the technology and a flood of receiver equipment has resulted in a basic unit that can give position location accuracy to within 10 feet and can be purchased for less than $100.

After many studies and considerable lobbying in Congress, President Clinton ordered that SA be permanently turned off on May 2, 2000. The improvement in GPS accuracy for the civilian world since then has been considerable, and the military has found a way of locally degrading GPS accuracy for selected areas without affecting the rest of the world.

With the sudden availability of affordable GPS receivers, it wasn't long before WB4APR and others began experimenting with them. They discovered that it was possible to tap the GPS receiver's data stream and extract position information that could then be sent via amateur packet radio. At the receiving end, special software was used to decode the position information and create symbols (*icons*) on computer-generated maps. Whenever the GPS receiver moved, a new position report was transmitted. When the receiving station decoded the signal, it "moved" the map icon to the new position. APRS as we know it today was born!

Virtually all earthbound APRS activity takes place today at 144.39 MHz using 1200-baud packet TNCs and ordinary FM voice transceivers. APRS relay stations have proliferated over the years, creating an extensive network throughout the US and beyond. APRS information is also ported to the Internet where hams (and nonhams) can view the locations of amateur APRS stations at Web sites such as **www.wulfden.org/APRSQuery.shtml**.

APRS software is available for *Windows* (*UIView*: **www.ui-view.org/**), *MacOS* (*WinAPRS*: **www.winaprs.com**) and *Linux* (*Xastir*: **xastir.sourceforge.net/**). Some of these applications can load elaborate maps that show detail down to street level.

In addition to conventional packet TNCs, some APRS enthusiasts rely on dedicated APRS tracking units that, unlike TNCs, only *transmit* position data; they do not receive. Hams connect these trackers to GPS receivers and install them in vehicles along with their 2-meter FM transceivers.

If you are operating APRS from a non-moving station, you do *not* need a GPS receiver. Simply enter your latitude or longitude into the APRS software.

APRS data often include more that merely the location of a station. Other data can be included such as direction and speed, telemetry information and even short text messages.

APRS and the International Space Station

The International Space Station offers an APRS *digipeater* relay that transmits and receives on a single frequency: 145.825 MHz. This is a 1200-baud system that operates much like terrestrial APRS—you send your APRS packets to the digipeater and it quickly retransmits them on the same frequency.

The International Space Station transmitter is powerful compared to most packet satellites and it can be easily heard with little more than a standard FM voice transceiver and an omnidirectional antenna. On the other hand, the station orbits at a relatively low altitude and even its high-elevation passes last only 10 minutes at most.

The ARISS APRS digipeater is popular because you do not need specialized equipment (such as a radio that supports 9600-baud packet) to use it. In fact, to operate through the space station, conventional APRS users need only change their radio frequencies to 145.825 MHz and include **ARISS** in their TNC or software path statements. They can make the transition from terrestrial to space APRS in about 60 seconds!

If you've been using your packet TNC to communicate through Gurwin-OSCAR 32 in full duplex, make sure to switch the TNC back to normal operation by setting the *FULLDUPE* parameter to **OFF**. In addition to adding **ARISS** to your TNC or APRS software path statements, make sure your transmit intervals remain at one minute. Yes, you can be more aggressive by instructing your TNC to transmit more frequency, but this really isn't necessary and it only serves to cause more congestion on the frequency.

As the space station rises above your horizon, turn off your receiver squelch and listen for bursts of packet data. When you hear the bursts, turn up your squelch just to the point where the background noise stops. Now your TNC should begin decoding the APRS packet data and sending your packets as well.

Keep an eye on your APRS software. Within a couple of minutes you should begin to see icons on your APRS maps and even a short message or two. See **Figure 5.6**.

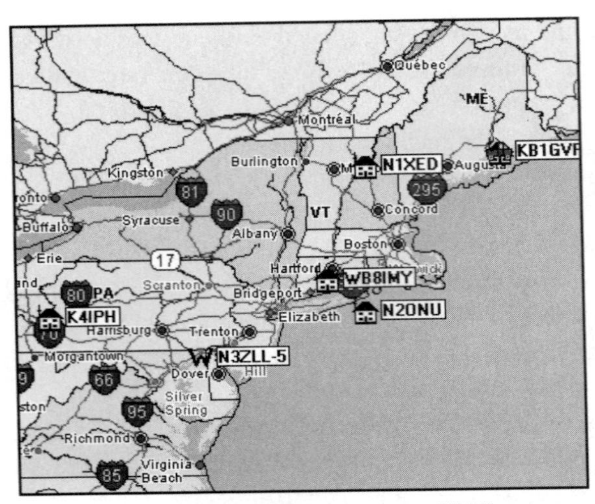

Figure 5.6—APRS icons appear on *a UI-View* map during an International Space Station pass.

The ARISS digipeater is occasionally off the air for various reasons. As discussed earlier in this chapter, the station crew may be using the equipment to make direct voice contracts, or they may enable different functions such as the FM voice repeater. Check the status of ARISS by going to the AMSAT "Satellite Status" page on the Web at **www.amsat.org/amsat-new/satellites/status.php**.

Telemetry Birds

There are several satellites you can enjoy simply by listening. In recent years a number of universities have built and launched small satellites for research purposes. Many of these satellites feature telemetry downlinks at amateur frequencies. Some of the telemetry is sent in Morse code, but most is sent as 1200-baud packet radio data.

You'll see these satellites listed in Table 5.3, such as the various "CubeSats." If you check the AMSAT-NA satellite status Web page at **www.amsat.org/**, you'll find more information about each of the satellites and, in many cases, a Web link to a site where you can download free telemetry decoding software. For example, the University of Tokyo offers free software to decode telemetry from several of their CubeSats at **www.space.t.u-tokyo.ac.jp/gs/en/application.aspx**.

To receive 1200-baud telemetry, you'll need a 1200-baud packet radio TNC and a 2 meter or 70 cm FM transceiver (or receiver). The technique is the same as monitoring APRS data from the International Space Station or NO-44, but in this case you'll have special software that will make sense of all the incoming data.

By watching the telemetry you'll get a sense of how the satellite changes from day to day. You'll see temperatures rise and fall; the ebb and flow of battery power and much more. In many instances you can observe transmitter output power and see the configuration of various satellite subsystems. The CubeSat-OSCAR 66 and Compass-One CubeSats, launched in April 2008, even include images in their downlink data streams.

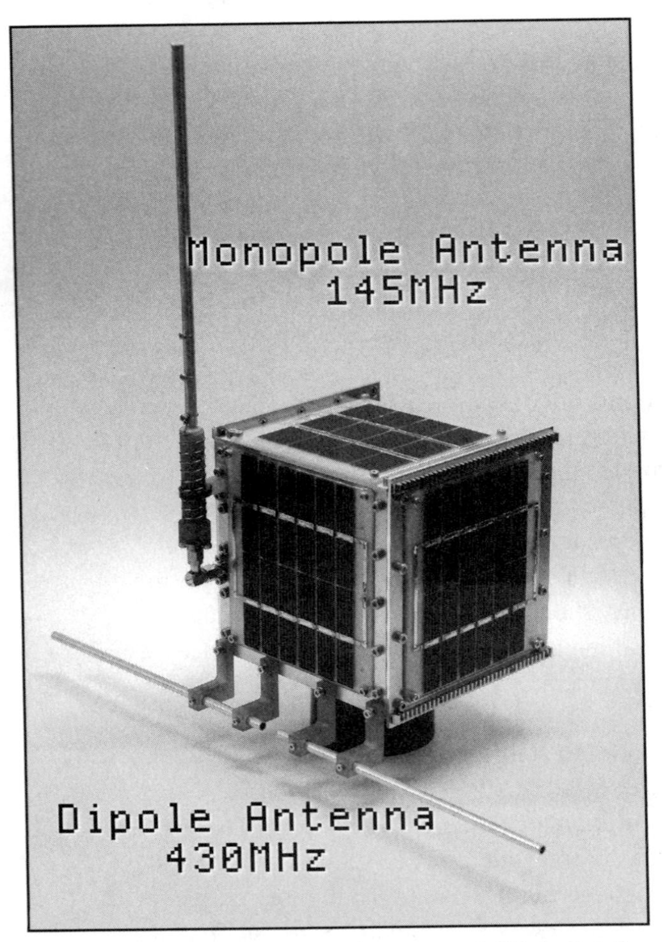

HITSAT is an example of the tiny research "CubeSats" being launched in recent years.

Amateur Radio Projects

The following projects were originally published in the *ARRL Handbook*, *ARRL Antenna Book*, *QST* magazine and *QEX* magazine. Not only are these projects informative and fun to build, you'll save a bundle of money in the process!

The W3KH Quadrifilar Helix Antenna
by Eugene Ruperto, W3KH
August 1996 *QST*

Dual Band Handy Yagi
by Thomas Hart, AD1B
May 2008 *QST*

An EZ-Lindenblad Antenna for 2 Meters
by Anthony Monteiro, AA2TX
August 2007 *QST*

Low-Profile Helix Feed for Phase 3E Satellites: System Simulation and Measurement
by Paolo Antoniazzi, IW2ACD and Marco Arecco, IK2WAQ
May/June 2006 *QEX*

A Satellite Tracker Interface
by Mark Spencer, WA8SME
September 2005 *QST*

Helix Feed for an Offset Dish Antenna
ARRL Handbook

An Affordable Az-El Positioner for Small Antennas
by Lilburn Smith, W5KQJ
June 2003 *QST*

Work OSCAR 40 with Cardboard-Box Antennas
by Anthony Monteiro, AA2TX
March 2003 *QST*

A Simple Fixed Antenna for VHF/UHF Satellite Work
by L.B. Cebik, W4RNL (SK)
August 2001 *QST*

Portable Helix for 435 MHz
ARRL Antenna Book

The W3KH Quadrifilar Helix Antenna

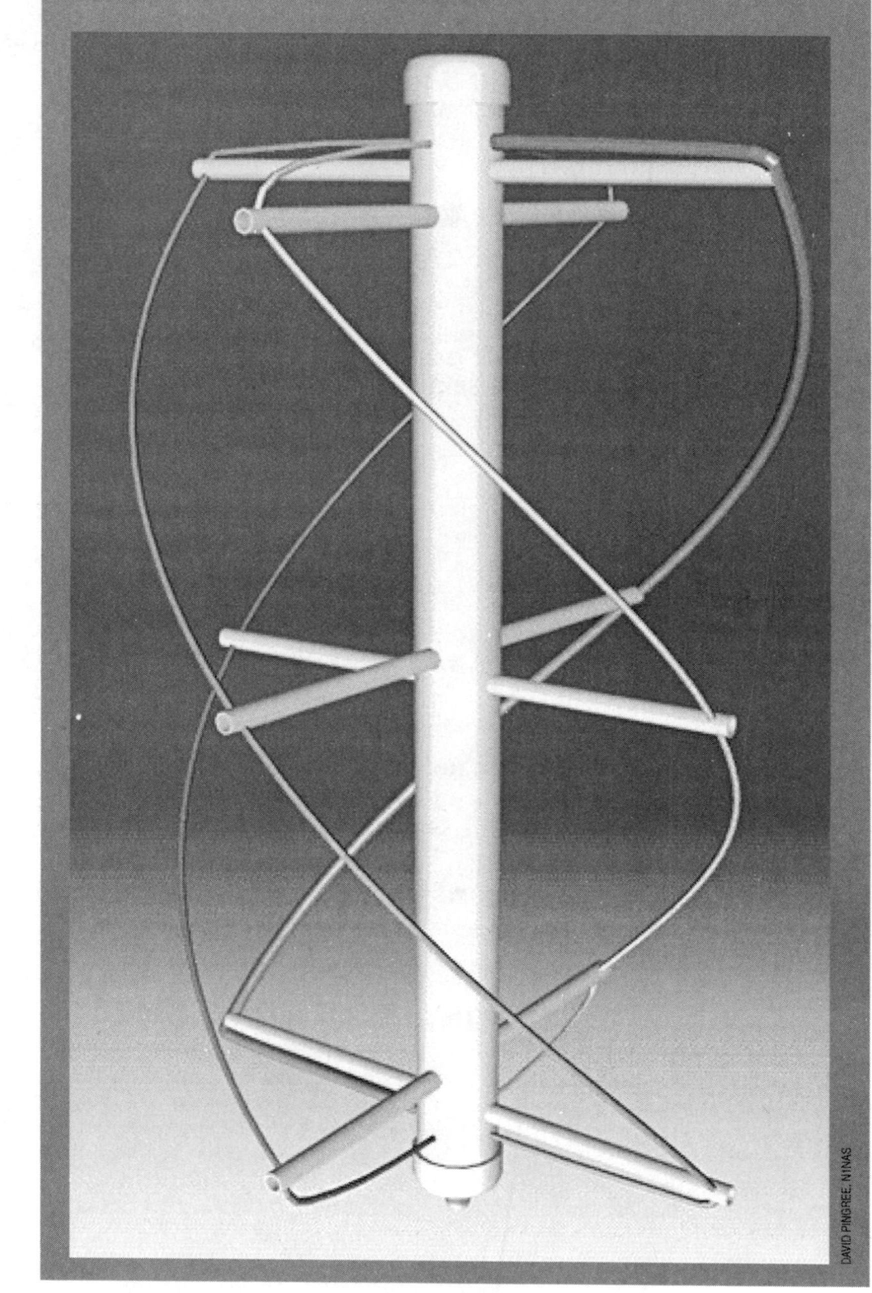

DAVID PINGREE, N1NAS

By Eugene F. Ruperto, W3KH

If your existing VHF omnidirectional antenna coverage is "just okay," this twisted 'tenna is probably just what you need!

I still remember that hollow, ghostly signal emanating from my receiver in 1957. The signal was noisy and it faded, but that was to be expected—it was coming from outer space. I couldn't help but marvel that mankind had placed this signal sender in space! They called it *Sputnik,* and it served to usher in the space race.

Little did I realize then that four decades later we would have satellites in orbit around Earth and other heavenly bodies performing all sorts of tasks. Now we tend to take satellites for granted. According to the latest information on the Amateur Radio birds, I count about 15 low-Earth-orbit (LEO) satellites for digital, experimental and communications work, and two in *Molniya*-type highly elliptical orbits (AO-10 and AO-13), with the probability of a third to be launched in early 1997.

The world has access to several VHF weather satellites in low Earth orbit. Unlike geostationary Earth-orbiting satellites (GOES), the ever-changing position of the LEOs presents a problem for the Earth station equipped with a fixed receiving antenna: signal fading caused by the orientation of the propagated wavefront. This antenna provides a solution to the problem. Although this antenna is designed primarily for use with the weather sats, it can also be used with any of the polar-orbiting satellites.

These days, technical advances and miniature solid-state devices make it relatively easy for an experimenter to acquire a weather-satellite receiver and a computer interface at an affordable price. So it was only a matter of time before I replaced my outdated weather-sat station with state-of-the-art equipment.

Yesterday

In the early '70s, I built a drum recorder that used a box with a light-tight lid. It was a clumsy affair. The box and photo equipment took up most of the 6×8-foot room in which it was housed. Next to the recorder, a 3×4-foot table supported a tube-type receiver, frequency converters, a reel-to-reel tape recorder (our data-storage medium), a 50-pound monitor oscilloscope, az/el rotator controls for the helical antennas and a multitude of other devices including the drum-driver amplifiers and homemade demodulator. This station

Figure 1—The humble beginnings of a terrific antenna.

Figure 2—The quadrifilar helix antenna with two of the four legs (filars) of one loop attached.

Figure 3—This view shows the QHA with all four legs in place. The ends of the PVC cross arms that hold the coaxial leg are notched; the wire elements pass through holes drilled in the ends of their supporting cross arms.

provided coverage of the polar-orbiting and geostationary satellites and furnished me with "tons" of data. Over time, my weather-satellite station evolved into a replica of mission control for the manned-spaceflight program! I had so much gear, it had to be housed in a shed separate from the house.

Today

Now, my entire weather-satellite station sits unobtrusively in one corner of the shack, occupying an area of less than one square foot—about the same size as my outboard DSP filter. My PC—now the display for weather-sat photos—is used for many applications, so an A/B switch allows me to toggle the PC between the printer and the weather-satellite interface.

What I needed next was a simple antenna system for unattended operation—something without rotators—something that would provide fairly good coverage, from about 20° above the horizon on an overhead pass. It was a simple request, but apparently one without a simple solution.

Background

Initially I used a VHF discone antenna with mixed results. The discone had a good low-elevation capture angle, but exhibited severe pattern nulls a few minutes after acquisition of signal and again when the satellite was nearly overhead. The fades and nulls repeated later as it approached the other horizon. About this time, Dave Bodnar, N3ENM (who got me reinterested in the antenna project), built a turnstile-reflector (T-R) array. The antenna worked fairly well but exhibited signal dropout caused by several nulls in the pattern. Dave built two more T-Rs, relocating them for comparison purposes. Unfortunately, the antennas retained their characteristic fades and nulls. Another experimenter and I built T-Rs and we experienced the same results. I suggested that we move on to the Lindenblad antenna. The Lindenblad proved to be a much better antenna for our needs than either the T-R or the discone, but still exhibited nulls and fades. Over a period of several months, I evaluated the antennas and found that by switching from one antenna to another on the downside of a fade, I could obtain a fade-free picture, but lost some data during the switching interval. Such an arrangement isn't conducive to unattended operation, so my quest for a fade-free antenna continued.

The Quadrifilar Helix Antenna

Several magazines have published articles on the construction of the quadrifilar helix antenna (QHA) originally developed by Dr. Kilgus,[1] but the articles themselves were generally reader unfriendly—some more than others. One exception is *Reflections* by Walt Maxwell, W2DU.[2] Walt had considerable experience evaluating and testing this antenna while employed as an engineer for RCA.

Part of the problem of replicating the antenna lies in its geometry. The QHA is difficult to describe and photograph. Some of the artist's renditions left me with more questions than answers, and some connections between elements as shown conflicted with previously published data. However, those who have successfully constructed the antenna say it is *the* single-antenna answer to satellite reception for the low-Earth-orbiting satellites. I agree.

Design Considerations

I had misgivings about the QHA construction because the experts implied that sophisticated equipment is necessary to adjust and test the antenna. I don't disagree with that assumption, but I *do know* that it's possible to construct a successfully performing QHA by following a cookbook approach using scaled figures from a successful QHA. These data—used as the design basis for our antennas—were published in an article describing the design of a pair of circularly polarized S-band communication-satellite antennas for the Air Force[3] and designed to be spacecraft mounted. Using this antenna as a model, we've constructed more than six QHAs, mostly for the weather-satellite frequencies and some for the polar-orbiting 2-meter and 70-centimeter satellites with excellent results—*without the need for adjustments and tuning*. Precision construction is not my forte, but by following some prescribed universal calculations, a reproducible and satisfactory antenna can be built using simple tools. The proof is in the results.

[1]Notes appear on page 6.

The ultrahigh frequencies require a high degree of constructional precision because of the antenna's small size. For instance, the antenna used for the Air Force at 2.2 GHz has a diameter of 0.92 inch and a length of 1.39 inches! Nested inside this helix is a smaller helix, 0.837 inch in diameter and 1.27 inches in length. In my opinion, construction of an antenna *that* size requires the skill of a watchmaker! On the other hand, a QHA for 137.5 MHz is 22.4 inches long and almost 15 inches in diameter. The smaller, nested helix measures 20.5 by 13.5 inches; for 2 meters, the antenna is not much smaller. Antennas of this size are not difficult to duplicate even for those of us who are "constructionally challenged" (using pre-cut pieces, I can build a QHA in *less than an hour!*).

Electrical Characteristics

A half-turn half-wavelength QHA has a theoretical gain of 5 dBi and a 3-dB beamwidth of about 115°, with a characteristic impedance of 40 Ω. The antenna consists basically of a four-element, half-turn helical antenna, with each pair of elements described as a *bifilar,* both of which are fed in phase quadrature. Several feed methods can be employed, all of which appeared to be too complicated for us with the exception of the infinite-balun design, which uses a length of coax as one of the four elements. To produce the necessary 90° phase difference between the bifilar elements, either of two methods can be used. One is to use the same size bifilars, which essentially consist of two twisted loops with their vertical axes centered and aligned, and the loops rotated so that they're 90° to each other (like an egg-beater), and using a quadrature hybrid feed. Such an antenna requires *two* feed lines, one for each of the filar pairs. The second and more practical method, in my estimation, is the self-phasing system, which uses *different-size loops:* a larger loop designed to resonate *below* the design frequency (providing an inductive reactance component) and a smaller loop to resonate *higher* than the design frequency (introducing a capacitive-reactance component), causing the current to lead in the smaller loop and lag in the larger loop. The element lengths are 0.560 λ for the larger loop, and 0.508 λ for the smaller loop. According to the range tests performed by W2DU, to achieve *optimum* circular polarization, the wire used in the construction of the bifilar elements should be 0.0088 λ in diameter. Walt indicates that in the quadrifilar mode, the fields from the individual bifilar helices combine in optimum phase to obtain unidirectional endfire gain. The currents in the two bifilars must be in quadrature phase. This 90° relationship is obtained by making their respective terminal impedances $R + jX$ and $R - jX$ where $X = R$, so that the currents in the respective helices are –45° and +45°. The critical parameter in this relationship is the terminal reactance, X, where the distributed inductance of the helical element is the primary determining factor. This assures the ±45° current relationship necessary to obtain true circular polarization in the combined fields and to obtain maximum forward radiation and minimum back lobe. Failure to achieve the optimum element diameter of 0.0088 λ results in a form of elliptical, rather than true circular polarization, and the performance may be *a few tenths of a decibel* below optimum, according to Walt's calculations. For my antenna, using #10 wire translates roughly to an element diameter of 0.0012 λ at 137.5 MHz—not ideal, but good enough.

To get a grasp of the QHA's topography, visualize the antenna as consisting of two concentric cylinders over which the helices are wound (see Figures 1 through 5). In two-dimensional space, the cylinders can be represented by two nested rectangles depicting the height and width of the cylinders. The width of the larger cylinder (or rectangle) can be represented by 0.173 λ, and the width of the smaller cylinder represented by 0.156 λ. The length of the larger cylinder or rectangle can be represented by 0.260 λ, and the length of the smaller rectangle or cylinder can be represented by 0.238 λ. Using these figures, you should be able to scale the QHA to virtually any frequency. Table 1 shows some representative antenna sizes for various frequencies, along with the universal parameters needed to arrive at these figures.

Figure 4—Another view of the QHA.

Table 1
Quadrifilar Helix Antenna Dimensions

Freq (MHz)	Wavelength λ (inches)	Small Loop Leg Size (0.508 λ)	Small Loop Diameter (0.156 λ)	Small Loop Length (0.238 λ)	Big Loop Leg Size (0.560 λ)	Big Loop Diameter (0.173 λ)	Big Loop Length (0.26 λ)
137.5	85.9	43.64	13.4	20.44	48.10	14.86	22.33
146	80.9	41.09	12.6	19.25	45.30	14.0	21.03
436	27.09	13.76	4.22	6.44	15.17	4.68	7.04

Figure 5—An end-on view of the top of the QHA prior to soldering the loops and installing the PVC cap.

Physical Construction

After several false starts using plywood circles and plastic-bucket forms to hold the helices, I opted for a simple PVC solution that not only is the simplest from a constructional standpoint, but also the best for wind loading. I use a 25-inch-long piece of schedule 40, 2-inch-diameter PVC pipe for the vertical member. The cross arms that support the helices are six pieces of $^1/_2$-inch-diameter PVC tubing: three the width of the large rectangle or cylinder, and three the width of the smaller cylinder. Two cross arms are needed for the top and bottom of each cylinder. The cross arms are oriented perpendicularly to the vertical member and parallel to each other. A third cross arm is placed midway between the two at a 90° angle. This process is repeated for the smaller cylindrical dimensions using the three smaller cross arms with the top and bottom pieces oriented 90° to the large pieces. Using $^5/_8$-inch-diameter holes in the 2-inch pipe ensures a reasonably snug fit for the $^1/_2$-inch-diameter cross pieces. Each cross arm is drilled (or notched) at its ends to accept the lengths

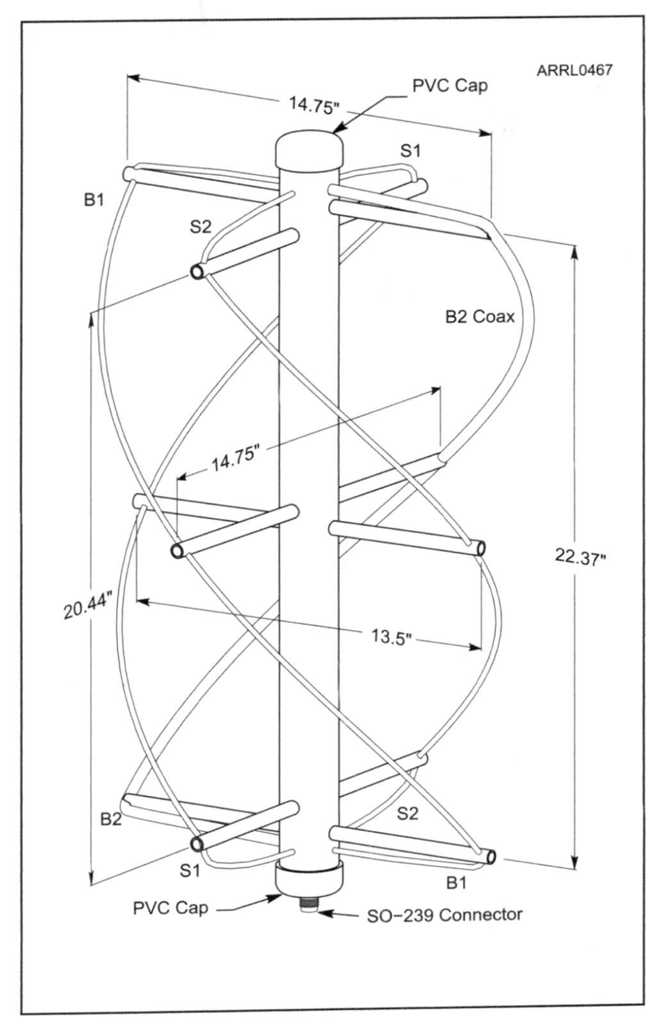

Figure 6—Drawing of the QHA identifying the individual legs; see text for an explanation.

The prototype worked better than expected and duplicates required no significant changes.

Figure 8—It's said that "The proof of the pudding is in the eating." To a weather-satellite tracker, clear, no-fade, no-noise pictures such as this one—compliments of W3KH's quadrifilar helix antenna—are delicious fare!

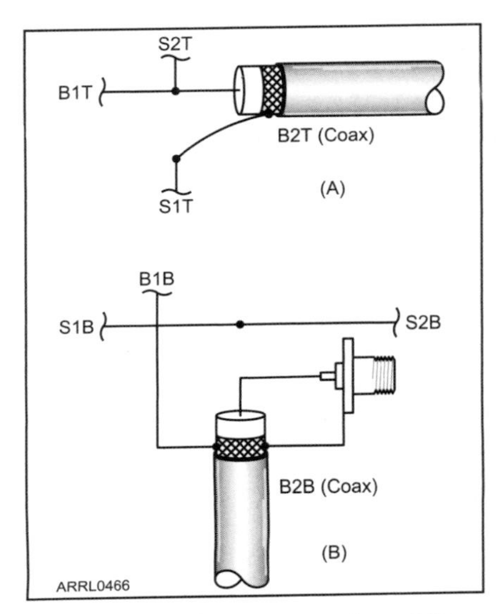

Figure 7—At A, element connections at the top of the antenna. B shows the connections at the bottom of the antenna. The identifiers are those shown in Figure 6 and explained in the text.

of wire and coax used for the elements. Then the cross arms are centered and cemented in place with PVC cement. For the weather-satellite antennas, I use #10 copperclad antenna wire for three of the helices and a length of RG-8 for the balun, which is also the fourth helix. (I do not consider the velocity factor of the coax leg for length calculation.) For the UHF antennas, I use #10 soft-drawn copper wire and RG-58 coax. Copperclad wire is difficult to work with, but holds its shape well. Smaller antennas can be built without the cross arms because the wire is sufficiently self-supporting.

To minimize confusion regarding the connections and to indicate the individual legs of the helices, I label each loop or cylinder as B (for big) and S (for small); T and B indicate top and bottom. See Figures 6 and 7. I split each loop using leg designators as B1T and B1B, B2T and B2B, S1T and S1B and S2T and S2B, with B2 being the length of coax and the other three legs as wires. For right-hand circular polarization (RHCP) I wind the helices *counterclockwise* as viewed from the top. This is contrary to conventional axial-mode helix construction. (For LHCP, the turns rotate *clockwise* as viewed from the top.) See Figure 7 for the proper connections for the top view. When the antenna is completed, the view shows that there are two connections made to the center conductor of the coax (B2) top. These are B1T and S1T, for a total of three wires on one connection. S2T connects to B2T braid. The bottom of the antenna has S1B and S2B soldered together to complete the smaller loop. B1B and the braid of B2B are soldered together. I attach an SO-239 connector to the bottom by soldering the center conductor of B2B to the center of the connector and the braid of B2B to the connector's shell. The bottom now has two connections to the braid: one to leg B1B, the other to the shell of the connector. There's only one connection to the center conductor of B2B that goes to the SO-239 center pin.

Insulator Quality

A question arose concerning the dielectric quality of the tubing and pipe used for the insulating material. Antennas—being reciprocal devices—exhibit losses on a percentage basis, the percentage ratio being the same for transmit and receive. Although signal loss may not be as apparent on receive with a 2-μV signal as with a transmitted signal of 100 W (ie, it would be apparent if dielectric losses caused the PVC cross arms to melt!), signal loss could be a significant factor depending on the quality of the insulating material used in construction. As a test, I popped the pipe into the microwave and "nuked" it for one minute. The white PVC pipe and the tan CPVC tubing showed no significant heating, so I concluded that they're okay for use as insulating materials at 137.5 MHz or thereabouts.

The antennas cost me nothing because the scrap pieces of PVC pipe, tubing and connectors were on hand. Total price for all new materials—including the price of a suitable connector—should be in the neighborhood of $8 or less.

Results

I use a 70-foot section of RG-9 between the receiver and antenna, which is mounted about 12 feet above ground. As with the earlier antennas, I use a preamp in the shack. With AOS (acquisition of signal) on the first scheduled pass of NOAA-14, I was pleasantly surprised to receive the first of many fade-free passes from the weather satellites, including some spectacular pictures from the Russian Meteors! Although the design indicates a 3-dB beamwidth of 140°, an overhead pass provides useful data down to 10° above the horizon. (My location has a poor horizon, being located in a valley with hills in all directions but south.) I've also received almost-full-frame pictures of the West Coast and northern Mexico at a maximum elevation angle of only 12° at my location. (The 70-cm antenna works fine for PACSATs, although Doppler effect makes manual tracking difficult.) The weather-satellite antenna prototype worked better than expected and a number of copies built by others required no significant changes. The quadrifilar helix antenna is *definitely* a winner! And believe me, *it's easy to build!*

Acknowledgments

Thanks to Chris Van Lint, and Tom Loebl, WA1VTA, for supplying me with the necessary technical data to complete this project. A special thanks to Walt Maxwell, W2DU, for his review and technical evaluation and for sharing his technical expertise with the amateur satellite community.

Notes

[1]C. C. Kilgus, "Resonant Quadrafilar Helix," *IEEE Transactions on Antennas and Propagation,* Vol AP-17, May 1969, pp 349 to 351.

[2]M. Walter Maxwell, W2DU, "Reflections, Transmission Lines and Antennas," (Newington: ARRL, 1990). [This book is now out of print.—*Ed.*]

[3]Randolph W. Brickner Jr and Herbert H. Rickert, "An S-Band Resonant Quadrifilar Antenna for Satellite Communication," RCA Corp, Astro-Electronics Division, Princeton, NJ 08540.

Eugene "Buck" Ruperto, W3KH, was first licensed as W3QYG in 1950, then upgraded to Extra Class in 1957. Buck worked for American Airlines as a radio operator and retired from the Federal Aviation Administration, where he worked for 15 years as a data systems specialist, systems analyst and technical writer. Buck was the manager of automation for the ARTS III system (the automated radar terminal system for tracking aircraft) several years before retirement. Still flying, Buck holds a commercial pilot's license and is a certified flight instructor, with ratings for aircraft, instruments and gliders. Flying, he says, accounts for his interest in the weather. Buck's Amateur Radio interests include working the OSCAR satellites, especially those with Molniya orbits like AO-10 and 13. He's an avid island and DX chaser, at the Top of the Honor Roll and holds a W6OWP 40-wpm CW certificate. Buck's a long-time member of MENSA. You can reach Buck at 1035 McGuffey Rd, West Alexander, PA 15376, e-mail ***w3kh@pulsenet.com****.*

Photos by the author

Dual Band Handy Yagi

Thomas M. Hart, AD1B

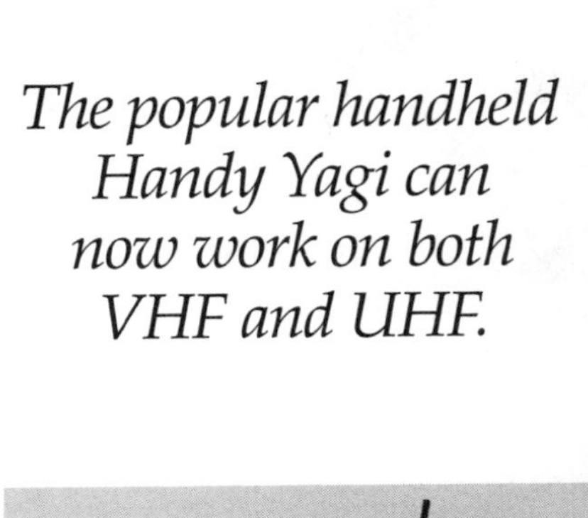

Recently, I started operating through the SO-50 and AO-51 satellites with my dual band Yaesu FT-60 handheld transceiver. The standard flexible antenna worked, but I decided to investigate handheld Yagi antennas to improve conditions. My goal was to build a simple dual band 2 meter and 70 cm Yagi without driven elements, a matching network or a feed line. In short, my plan was to mount the FT-60 on the antenna as I had with the original 70 cm Handy Yagi.[1]

After testing several configurations on Roy Lewallen's *EZNEC* program (see **www.eznec.com**), I settled on a design with seven directors on 70 cm and three on 2 meters. My Yaesu FT-60 dual band handheld serves as the driven element. Figures 1 and 2 show the configuration.

Construction

The final design balances performance and size. The elements for both bands are interlaced and mounted in parallel. There is no driven element or reflector. Instead, the FT-60 and a bicycle handlebar grip occupy the usual reflector and driven element end of the boom. The 2 meter elements can be rotated parallel to the boom to simplify storage. A screw eye at one end can be used to hang the antenna when not in use.

Computer modeling indicated that $\lambda/4$ element spacing works reasonably well on both bands. This allows the use of a 55 inch boom. The handheld is attached firmly in place by its belt clip. A speaker microphone makes transmitting and receiving very simple. See the illustrations for additional details.

All elements were cut from ⅛ inch diameter steel rod. An article by Ron Hege, K3PF, provided the dimensions for the 2 meter elements shown in Table 1. The 70 cm element dimensions are found in *The ARRL Antenna*

[1]Notes appear on page 6-8.

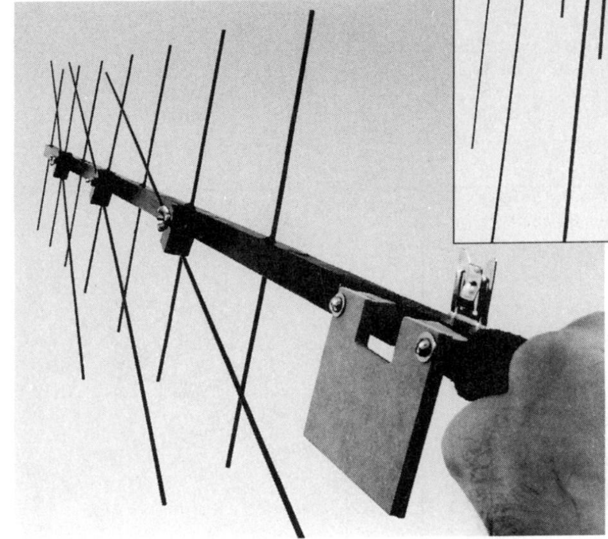

Figure 1 — Dual band Handy Yagi with 2 meter elements folded for storage.

Table 1

Length and Spacing of 2 Meter Elements

All dimensions are in inches.

Director:	*D1*	*D2*	*D3*
Length	37.5	36.375	36.0
Element Spacing	DE to D1	D1 to D2	D2 to D3
Spacing	12	12	12
Cumulative	12	24	36

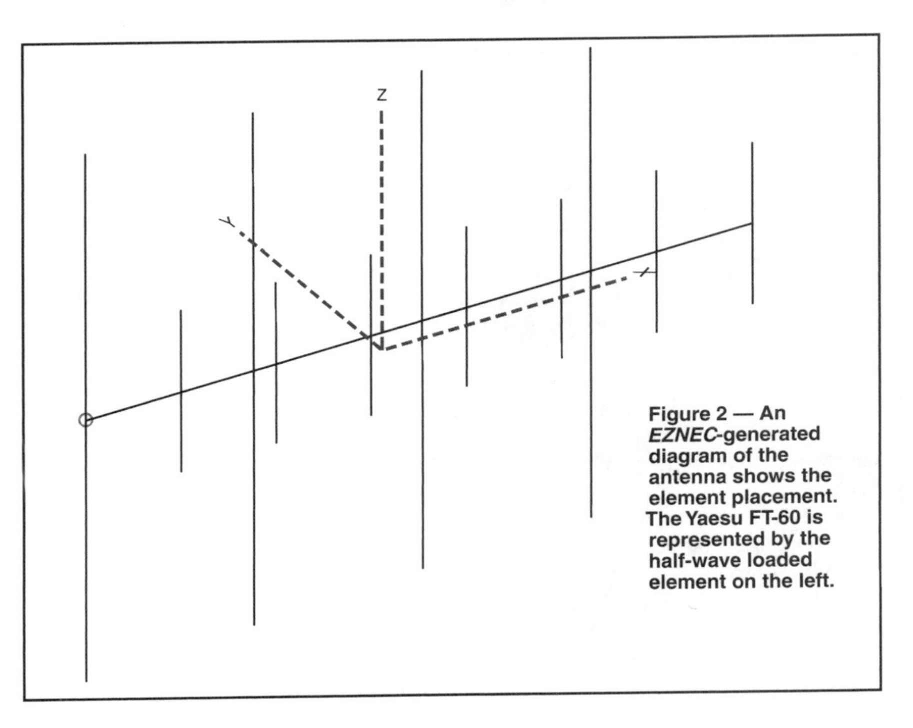

Figure 2 — An *EZNEC*-generated diagram of the antenna shows the element placement. The Yaesu FT-60 is represented by the half-wave loaded element on the left.

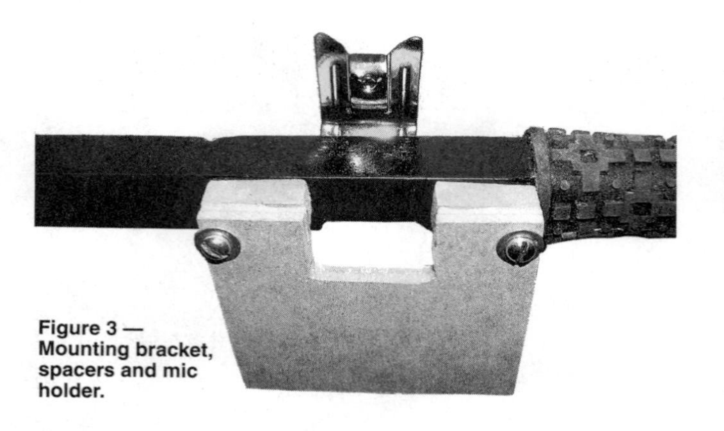

Figure 3 — Mounting bracket, spacers and mic holder.

Figure 4 — The wing nut and bolt allow the 2 meter element to rotate.

Table 2

Length and Spacing of 70 cm Elements

All dimensions are in inches.

Director:	*D1*	*D2*	*D3*	*D4*	*D5*	*D6*	*D7*
Length	11.750	11.688	11.625	11.563	11.500	11.438	11.375
Elements	DE to D1	D1 to D2	D2 to D3	D3 to D4	D4 to D5	D5 to D6	D6 to D7
Spacing	6.78	6.78	6.78	6.78	6.78	6.78	6.78
Cumulative	6.78	13.56	20.34	27.12	33.9	40.67	47.45

Table 3

Dimensions of Other Antenna Assemblies

All dimensions are in inches.

Boom	0.75 × 0.75 × 55
Handle grip	5
Handheld bracket	3.5 × 3.0 × 0.25 (WHD)
Notch for handheld	1.125 W × 0.75 H
Bracket spacers	0.375
2M element mounts	0.75 × 0.75 × 1.5

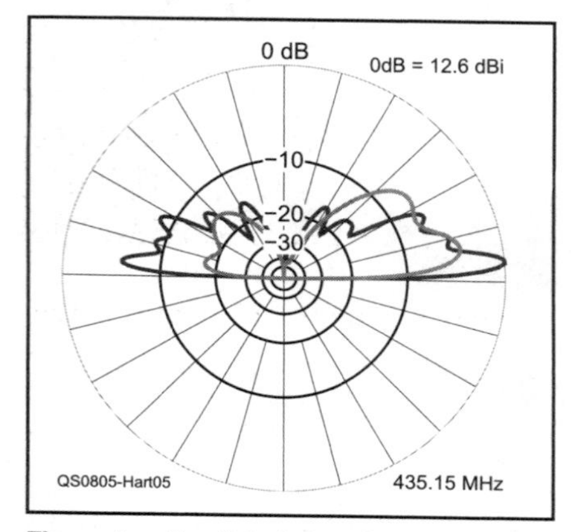

Figure 5 — Predicted elevation patterns of the 2 meter (red) and 70 cm (blue) Yagis.

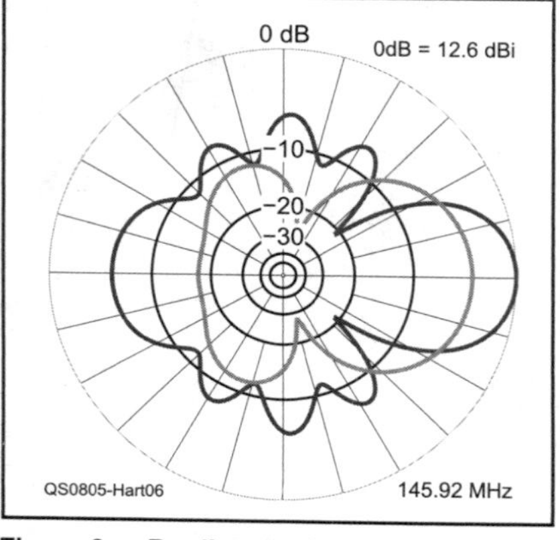

Figure 6 — Predicted azimuth patterns at the peak of the elevation response for the 2 meter (red) and 70 cm (blue) Yagis.

Book and reproduced as Table 2. Other dimensions are shown in Table 3.

Two glued together pieces of ⅛ inch pressed fiberboard form the handheld bracket shown in Figure 3. The elements are held in place with epoxy. After cutting all 10 to length, drill seven holes through the boom and three more through the 2 meter element mounts. Slide the elements into place and apply a bead of epoxy on both sides. Several light applications of epoxy will hold the rods in place.

All 2 meter element mounts require two holes. One secures the element while the second is used for the bolt and wing nut that allow the element to rotate during storage or transport as shown in Figure 4. Finally, two coats of black satin paint give the antenna a more professional appearance.

Testing

Faced with an absence of analytical instruments, the testing process involved *EZNEC* computations and operational observations. *EZNEC* computed the front-to-back ratios as 14 dB for 2 meters and 5.5 dB for 70 cm. *EZNEC* azimuth and elevations patterns are shown in Figures 5 and 6.

Field testing on 2 meters involved contacting the 146.97 MHz repeater in Paxton, Massachusetts. The repeater, according to my *Magellan Topo* 3D GPS software, is 38.0 miles away at a bearing of 251°. The FT-60 helical antenna cannot reach the repeater from my house. With the dual band Handy Yagi, I had no trouble reaching the repeater. Moving the antenna in an arc toward and away from the repeater produced corresponding signal strength changes on the meter corresponding to the predicted directivity.

On 70 centimeters, field tests involved reception of signals from the AO-51 and SO-50 satellites. Signal strength increased dramatically when the antenna approached and centered on the target. My conclusions are that operations on both bands are in agreement with *EZNEC* predictions.

The antenna that I built is designed for satellite contacts. However, the basic concept can be used for portable operations, fox hunting and emergency use as well.

Notes

[1]T. Hart, AD1B, "The Handy Yagi," *QST*, Nov 2007, pp 37-38.

[2]R. Hege, K3PF, "A Five-Element, 2-Meter Yagi for $20," *QST*, Jul 1990, pp 34-36.

[3]R. D. Straw, Editor, *The ARRL Antenna Book*, 21st Edition, p 18-45. Available from your ARRL dealer or the ARRL Bookstore, ARRL order no. 9876. Telephone 860-594-0355, or toll-free in the US 888-277-5289; **www.arrl.org/shop/**; **pubsales@arrl.org**.

All photos by the author.

Tom Hart, AD1B, began listening to short wave broadcasts in 1961. He received his Novice class license, WN1JGG, in 1968 and has been an active on CW, SSB, RTTY, FM and packet ever since. Tom has a BS from Tufts and an MS from Northeastern. He is an accountant who would rather be chasing or giving out counties on 20 meters. You can reach Tom at 54 Hermaine Ave, Dedham, MA 02026 or via **tom.hart@verizon.net**.

An EZ-Lindenblad Antenna for 2 Meters

This easy to build antenna works well for satellite or terrestrial communication, horizontal or vertically polarized.

Anthony Monteiro, AA2TX

Lindenblad is the name of a type of antenna that is circularly polarized yet has an omnidirectional radiation pattern. With most of its gain at low elevation angles, it is ideal for accessing low earth orbit (LEO) Amateur Radio satellites. Because it is omnidirectional, it does not need to be pointed at a satellite so it eliminates the complexity of an azimuth/elevation rotator system. This makes the Lindenblad especially useful for portable or temporary satellite operations. It is also a good general purpose antenna for a home station because its circular polarization is compatible with the linearly polarized antennas used for FM/repeater and SSB or CW operation.

This type of antenna was devised by Nils Lindenblad of the Radio Corporation of America (RCA) around 1940.[1] At that time, he was working on antennas for the then nascent television broadcasting (TV) industry. His idea was to employ four dipoles spaced equally around a $\lambda/3$ diameter circle with each dipole canted 30° from the horizontal. The dipoles are all fed in phase and are fed equal power. The spacing and tilt angles of the dipoles create the desired antenna pattern when the signals are all combined. Unfortunately, the start of World War II halted Lindenblad's TV antenna work.

After the war, George Brown and Oakley Woodward, also of RCA, were tasked with finding ways to reduce fading on ground-to-air radio links at airports.[2] These links used linearly polarized antennas. The maneuverings of the airplanes often caused large signal dropouts if the antennas became cross-polarized. Brown and Woodward realized that using a circularly polarized antenna at the airport could reduce or eliminate this fading so they decided to try Lindenblad's TV antenna concept.

Brown and Woodward designed their antenna using metal tubing for the dipole elements. Each dipole element is attached to a section of shorted open-wire-line, also made from tubing, which serves as a balun transformer. A coaxial cable runs through one side of each open-wire-line to feed each dipole. The four coaxial feed cables meet at a center hub section where they are connected in parallel to provide a four-way, in-phase power-splitting function. This cable junction is connected to another section of coaxial cable that serves as an impedance matching section to get a good match to 50 Ω. While the Brown and Woodward design is clever and worked well, it would be quite difficult for the average ham (including this author!) to duplicate.

The major cause of the difficulty in designing and constructing Lindenblad antennas is the need for the four-way, in-phase, power splitting function. Since we generally want to use 50 Ω coaxial cable to feed the antenna, we have to somehow provide an impedance match from the 50 Ω unbalanced coax to the four 75 Ω balanced dipole loads.

Previous designs have used combinations of folded dipoles, open-wire lines, twin-lead feeds, balun transformers and special impedance matching cables in order to try to get a good match to 50 Ω. These in turn increase the complexity and difficulty of the construction.

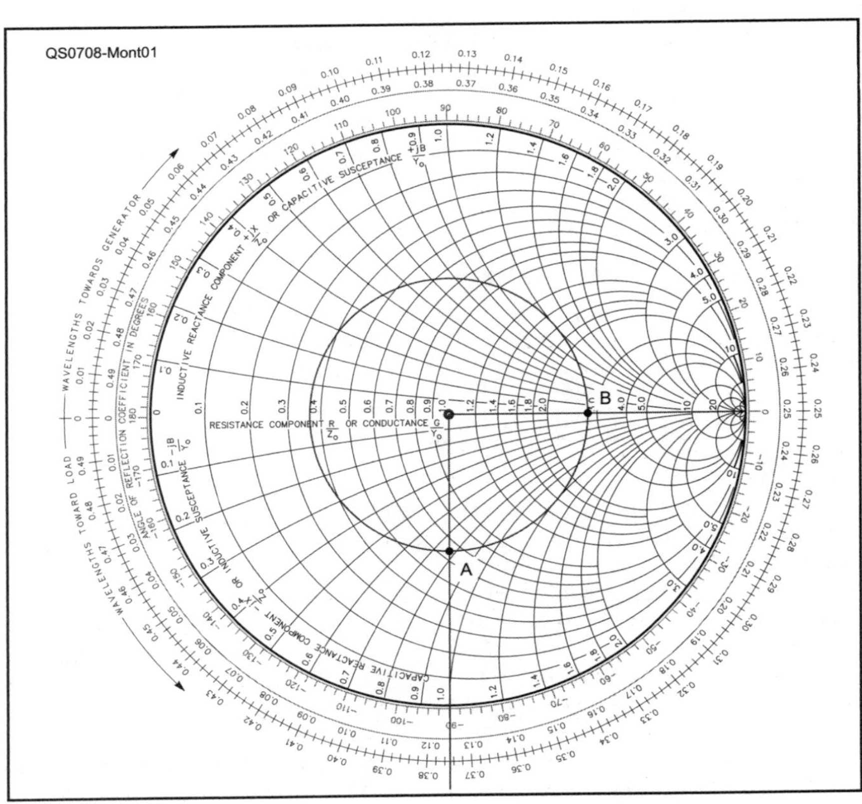

Figure 1 — Smith Chart showing transformation along a 75 Ω transmission line to yield 200 Ω.

[1]Notes appear on page 6-12.

Table 1
Required Materials

Quantity one, unless noted.

Aluminum tubing, 17 gauge, 6 foot length of ¾" outer diameter, quantity 3. Available from Texas Towers, **www.texastowers.com**.
Aluminum angle stock, 8" length of 2" × 2" × 1⁄16".
Aluminum angle stock, 2" length of 2" × 2" × 1⁄16" for mounting connector.
Screws, #8 × ½" aluminum sheet metal, quantity 12.
Screws, #8 × ½" aluminum sheet metal or 3⁄16" aluminum rivets, quantity 12.
PVC insert T-connector, ½" × ½" × ½" grey for irrigation polyethylene tubing. LASCO Fittings, Inc. Part# 1401-005 or equivalent. Available from most plumbing supply and major hardware stores, quantity 4.
Plastic end caps (optional), black ¾", quantity 8.
N-connector for RG-8 cable, single-hole, chassis-mount, female.
Cable ferrite, Fair-Rite part # 2643540002, quantity 4 (Mouser Electronics #623-2643540002).
RG-59 polyethylene foam coax with stranded center conductor, 10' length.
Copper braid, 4" long piece.
Ring terminal, uninsulated 22-18 gauge for 8-10 stud, quantity 4.
Ring terminal, uninsulated 12-10 gauge for 8-10 stud, quantity 4.
Heat shrink tubing for ¼" cable, wire ties, electrical tape, as needed.
Ox-Gard OX-100 grease for aluminum electrical connections.

The EZ-Lindenblad

An antidote to these difficulties is the EZ-Lindenblad. The key concept of the EZ approach is to eliminate anything that is electrically or mechanically difficult, leaving only things that are *easy*. This leads to the idea of just feeding the four dipoles with coax cable and soldering the cables to a connector with no impedance matching devices at all. This would certainly be *easy* but we also want the antenna to work! Without the extra impedance matching devices, how is it possible to get a good match to 50 Ω?

If we could get each of the four coax feed cables to look like 200 Ω at the connector, then the four in parallel would provide a perfect match to 50 Ω. We could do this if we used quarter-wavelength sections of 122 Ω coax to convert each 75 Ω dipole load to 200 Ω. Unfortunately, there is no such coax that is readily available.

But we can accomplish the same thing with ordinary 75 Ω, RG-59 TV type coax if we run the cable with an intentional impedance mismatch. By forcing the standing wave ratio (SWR) on the cable to be equal to 200/75, or about 2.7:1, we can make each cable look exactly like 200 Ω at the connector as long as we make them the right length. It is easy to make the SWR equal 2.7:1 by just making the dipoles a little too short for resonance. An *EZNEC* antenna model can be used to determine the exact dipole dimensions.[3]

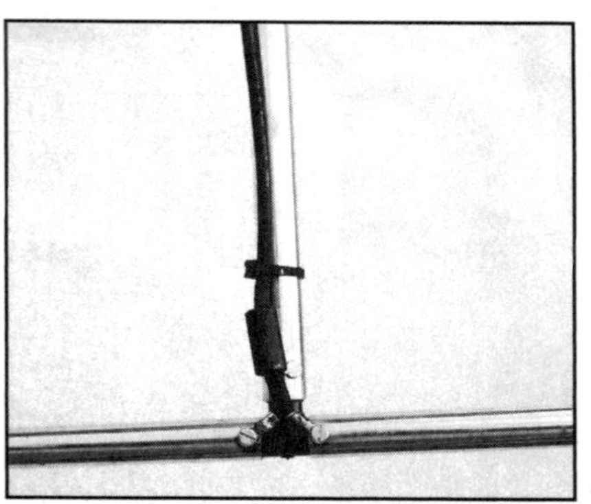

Figure 2 — Close-up of dipole electrical connections.

Figure 3 — View of cross booms mounted to mast.

The conversion from the balanced dipole load to the unbalanced coax cable can be painlessly accomplished by threading each cable through an inexpensive and readily available ferrite sleeve making essentially a choke balun. The only remaining issue is the required length of the feed cables, and this can be easily determined using a *Smith Chart.*

Smith Chart

The Smith Chart was invented by Phillip Smith of The Bell Telephone Laboratories in 1939.[4] As a high school student, Phillip had been a ham radio operator and used the call sign 1ANB. After graduating from Tufts College, he went to work for Bell Labs in the radio research department. As part of his job, he needed to make many impedance calculations that in those days required doing many complex computations by hand. Phillip realized he could create a chart that would allow the solution to be plotted on a graph making his job a lot easier. Phillip Smith's chart was soon adopted by other radio engineers and quickly became a standard engineering tool that is still in use today. Technical information and free downloadable Smith Charts are available via links on the ARRL Web site.[5] Please see the Smith Chart in Figure 1.

With a Smith Chart, we can easily determine the required length of 75 Ω coax to provide a 200 Ω load. The Smith Chart shown is normalized to 1 Ω in the center so we must multiply all impedance values by our coax impedance of 75 Ω. An *EZNEC* antenna model was used to simulate cutting the dipole lengths until the SWR on the line reached 2.7:1. The model showed that the dipole load impedance would then be $49 - j55$ Ω and this is plotted at point A on the chart. The desired 200 Ω impedance at the connector is plotted at point B on the chart. A constant 2.7:1 SWR curve is drawn between the two impedance points. The length of cable needed is read clockwise along the scale labeled, WAVELENGTHS TOWARD GENERATOR from the lines drawn through points A to B. From the chart, the length of the line needed is 0.374 λ.

The EZ-Lindenblad was designed for a center frequency of 145.9 MHz to optimize its performance in the satellite sub band. At 145.9 MHz, a wavelength is about 81 inches and since the coax used has a velocity factor of 0.78, we need to make the feed cables $81 \times 0.374 \times 0.78 = 23.6$ inches long.

There are several computer programs available today that can also be used to do the Smith Chart calculations. These include *TLW*, provided with *The ARRL Antenna Book* and *MicroSmith* formerly offered by the ARRL, which was used as a cross check.[6]

Construction

This antenna was designed to be rugged and reliable yet easy to build using only hand

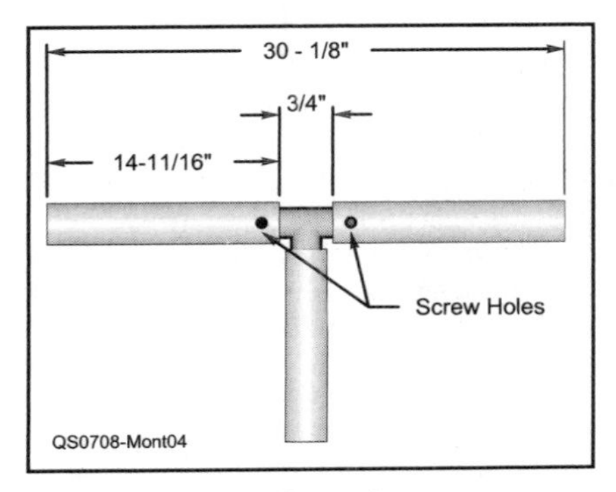

Figure 4 — Dipole dimensions.

Table 2
Aluminum Tubing Lengths

Description	*Quantity*	*Length*
Dipole rods	8	14¹¹⁄₁₆"
Cross booms	2	23"

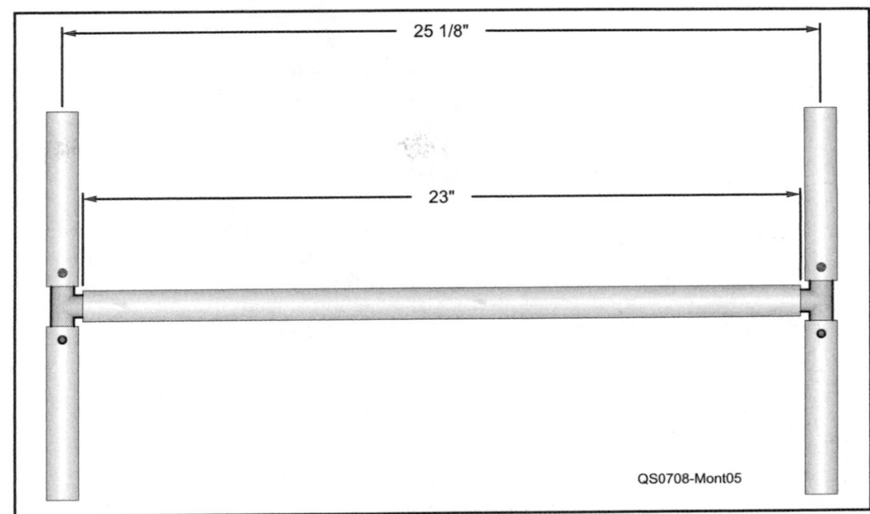

Figure 5 — Dimension of dipole spacing on cross-boom.

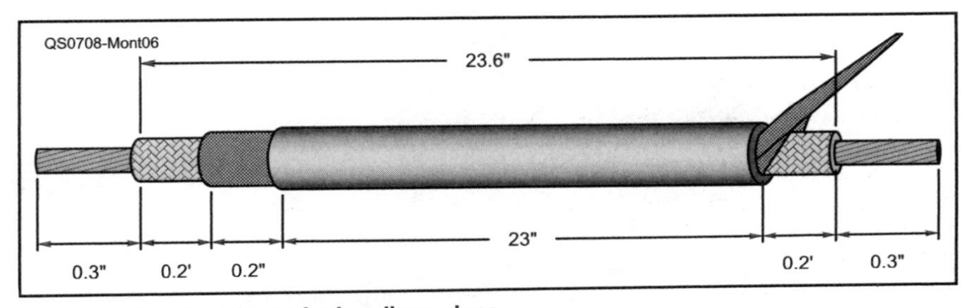

Figure 6 — Feed cables stripping dimensions.

tools with all of the parts readily available as well. Please see the parts list in Table 1. Although not critical, the construction will be easier if the specified 17 gauge aluminum tubing is used since the inner wall of the tubes will be just slightly smaller than the outer wall of the PVC insert Ts used to connect them. If heavier gauge tubing is used, it will be necessary to file down the PVC insert Ts to make them fit inside the aluminum tubes.

Start by making a mounting bracket to mount the N-connector and the cross booms. Cut a ⅝ inch hole in one side of the short piece of angle stock and rivet or screw it to the bottom of the long piece of angle stock. The completed bracket with the connector and cables attached can be seen in Figure 3.

Next, cut the aluminum tubing to make the cross booms and dipole rods as shown in Table 2. Drill holes for the machine screws at each end of the cross booms but do not insert the screws yet. Attach the cross booms to the long section of angle stock with rivets or screws. One cross boom will mount just above the other as can be seen in Figure 3. The cross booms should be perpendicular to the mounting bracket so that they will be horizontal when the antenna is mounted to its mast. Make sure that the centers of the cross booms are aligned with each other so that the ends of the cross booms are all 11½ inches from the center cross.

Make the dipoles by inserting a PVC insert T into two dipole rods. It should be possible to gently tap in the rods with a hammer but it may be necessary to file down the insert T a little if the fit is too tight. Applying a little PVC cement to the insert T will soften the plastic and make it easier to insert into the aluminum tubing if the fit is too tight. The overall dipole length dimension is critical so take care to get this correct as shown in Figure 4.

Drill holes for machine screws in each dipole rod but do not insert the screws yet. The screws will be used to make the electrical connections to the dipoles at the center. The screw holes should be about ⅜ inch from the end of the tubing.

The dipole assemblies are attached by gently tapping the PVC insert-T into the end of each cross-boom with a hammer. The dimensions are shown in Figure 5.

Next, temporarily attach the mounting bracket to a support so that each of the cross booms is perfectly horizontal. Measure this with a protractor. Now, using the protractor, rotate the dipole assemblies to a 30° angle with the right-hand side of the nearest dipole tilting up when you are looking toward the center of the antenna. Drill a small hole through the existing cross-boom holes into the PVC insert-Ts and then use the machine screws to fasten the dipole assemblies into place. For a nice finishing touch, the dipole ends can be fitted with ¾ inch black plastic end caps.

Next, make the four feed cables by cutting and stripping the RG-59 coax as shown in Figure 6. On the dipole connection side, unwrap the braid and form a wire lead. Apply the smaller ring terminal to the center conductor and use the larger ring terminal for the braid. At the other end of the cable, do not unwrap the braid but strip off the outer insulation. Slip a 1 inch piece of shrink wrap over the coax and apply to the dipole side. Next,

Figure 7 — The EZ-Lindenblad as a portable or Field Day antenna. A 70 cm antenna is on the top.

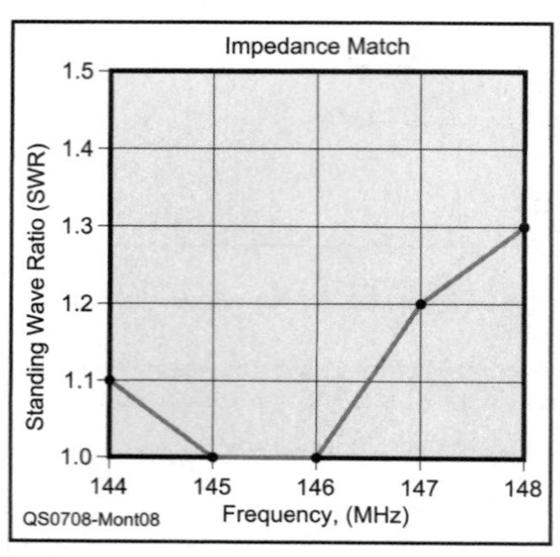

Figure 8 — Lindenblad antenna SWR at 50 Ω.

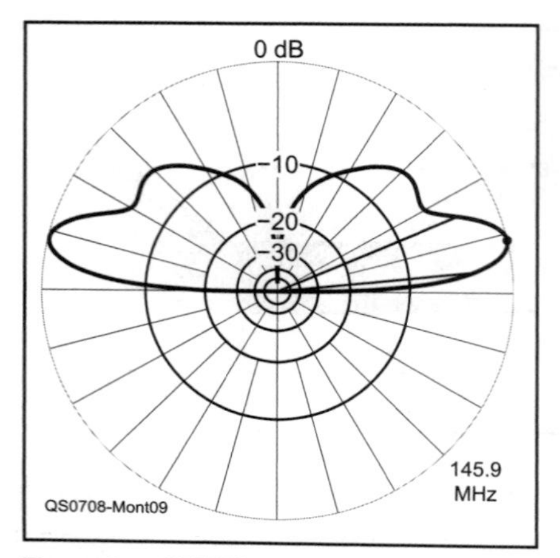

Figure 9 — *EZNEC* radiation pattern of Lindenblad antenna.

slip a cable ferrite over the cable and push all the way to the dipole end as far as it will go (ie, up to the heat-shrink tubing.) The fit will be snug and you may need to put a little grease on the cable jacket to get it started.

Prepare each dipole for its feed cable by first cleaning the area around the screw holes with steel wool and then applying Ox-Gard grease. This is to ensure a good electrical connection. The coax center conductor goes to the up side of the dipole and the braid goes to the down side. To make a connection, put a machine screw through the ring terminal and gently screw into the dipole tubing. Do not overtighten the screws or you will strip the tubing.

Apply Ox-Gard or other corrosion resistant electrical grease around the hole for the N connector. Take the 4 inch piece of braid and put the end of it through the hole for the N connector. This provides the ground connection. Secure the N connector in the mounting hole to clamp the braid. Use a wire tie or tape to hold the four feed cables together at the connector ends. Make sure to align the cables so that all the ground braids are together and the center conductors all extend out the same amount. Do not twist the center conductors together. Carefully push the four cable center conductors into the center terminal of the N connector and solder them in place. Wrap the exposed center conductors of the cables and the connector with electrical tape.

Take the piece of braid that is clamped to the N connector and wrap it around the four exposed ground braids of the coax cables. Solder them all together. This will take a fair amount of heat but be careful not to melt the insulation. After this cools, apply electrical tape over all the exposed braid and fix with wire ties. The cables should be secured to the cross booms with wire ties.

The mounting bracket provides a way to attach the antenna to a mast using whatever clamping mechanism is convenient. For a permanent mount, drill holes in the bracket to accept a pair of U bolts. The author's antenna was intended for portable operation and the bracket was drilled to accept two #8 stainless steel screws. These screws pass through a portable mast and the antenna is secured with stainless steel thumbscrews. This allows the antenna to be set up or taken down in less than a minute. The completed portable antenna as used for Field Day 2006 is shown in Figure 7. The little antenna at the top is for 70 cm.

Standing Wave Ratio

The antenna impedance match to 50 Ω was tested using an MFJ-259B SWR meter, which has a digital readout of standing wave ratio (SWR) and frequency. The frequency accuracy was verified using an external frequency counter and the 1.0:1 calibration was checked with a Narda precision 50 Ω load.

The antenna was connected to the SWR meter with a 6 foot coax jumper made of Belden 9913F7, which has very low loss. The SWR was measured at 1 MHz intervals over the 144-148 MHz range. As can be seen in the chart of Figure 8, the antenna provides an excellent match over the entire 2 meter band.

Power Handling Capability

This antenna was designed to safely handle any of the currently available VHF transceivers. The power handling capability was tested by applying a 200 W CW, signal, key down for 9½ minutes. Immediately after the test, the ferrites and cables were checked and there was no noticeable temperature rise.

Radiation Pattern

The antenna radiation pattern predicted by the *EZNEC* model is shown in Figure 9. This is the elevation plot with the antenna mounted at six feet above ground although it can be mounted higher if desired for better coverage to the horizon. As shown in the plot, the pattern favors the lower elevation angles. The –3 dB points are at 5° and 25° with the maximum gain of 4.8 dBic (with respect to an isotropic circularly polarized antenna) at around 13°. This is an excellent pattern for accessing LEO satellites. Most of the satellite pass elevations will be in this range and it is also the elevation at which the satellite provides the best chance for DX contacts.

The antenna radiation is right-hand circularly polarized, which will work with virtually any LEO satellite that uses the 2 meter band. The circularity was checked by measuring the difference between the horizontal and vertical radiation components. This was done using a linearly polarized sense antenna mounted 100 feet away feeding into an FT-817 radio with the AGC switched off. The radio audio was connected to an ac voltmeter with a dB scale. The test showed a difference of less than 3 dB, which is very good for an omnidirectional antenna. A reference horizontally polarized antenna measured nearly a 30 dB difference.

On the Air

The EZ-Lindenblad antenna has been used for SSB, FM and packet operation on the AO-07, FO-29, SO-50, AO-51, VO-52, NO-44, NO-60 and NO-61 satellites. The portable setup, as seen in Figure 5, was used for the satellite station at the North Shore Radio Association, NS1RA, 2006 Field Day effort. Field Day is an excellent test of any antenna as it is probably the busiest 48 hours of the year on the satellites and the EZ-Lindenblad performed well.

An earlier version of this antenna was published in the *AMSAT Space Symposium Proceedings* of October 2006.

Notes

[1]G. Brown and O. Woodward Jr. *Circularly Polarized Omnidirectional Antenna*, RCA Review, Vol 8, no. 2, Jun 1947, pp 259-269.
[2]See Note 1.
[3]*EZNEC+ V4* antenna analysis software by Roy W. Lewallen, W7EL, available from **www.eznec.com**.
[4]F. Polkinghorn, "Phillip H. Smith, Electrical Engineer, an Oral History," IEEE History Center, Rutgers University, New Brunswick, NJ. Available at **www.ieee.org**.
[5]Smith Chart information is available at **www.arrl.org/tis/info/chart.html**.
[6]R. D. Straw, Editor, *The ARRL Antenna Book*, 21st Edition. Available from your ARRL dealer or the ARRL Bookstore, ARRL order no. 9876. Telephone 860-594-0355, or toll-free in the US 888-277-5289; **www.arrl.org/shop/**; **pubsales@arrl.org**.

Tony Monteiro, AA2TX, was first licensed in 1973 as WN2RBM. He has worked in the electronics industry for over 25 years starting as a member of the technical staff at Bell Laboratories and later as an engineering director at several telecommunications companies. You can reach Tony at 25 Carriage Chase Rd, North Andover, MA 01845 or at **aa2tx@amsat.org**.

Low-Profile Helix Feed for Phase 3E Satellites: System Simulation and Measurements

The authors modeled, built and measured the performance of a helix feed system for a parabolic dish suitable for 2.4 GHz satellite operation.

Paolo Antoniazzi, IW2ACD and Marco Arecco, IK2WAQ

Introduction

An important push toward the intensive use of helix antennas or helix-fed disks was born with a revolutionary new generation of satellites: the Phase 3 class. At the beginning of 2004, unfortunately, the Phase 3D primary battery failed, and AO40 is nonoperational. The new Phase3E (see Figure 1) is being built in Germany by AMSAT DL, and is expected to launch within the next couple of years.[1, 2, 3] The declared goal of the P3E is to offer newcomers access to high-orbit satellite communications as well as a working platform for those already operating through AO-40 in S and K modes.

On top of that, future communication links to a P5A Mars station at 10.5 GHz will be simulated and tested, which means more antennas and modules for microwave bands.[4] Starting from these considerations and specifically for people interested in communicating with the new satellites, for the first time a complete antenna system (disk plus helix feeder) is mechanically designed by *AutoCAD* (Figure 2), simulated by NEC2 and NEC Synth and finally measured in the field.

[1]Notes appear on page 6-18.

For circular polarization applications, the axial-mode helix antenna is a natural candidate because its good polarization performance is an inherent attribute of the antenna shape, without the need for a special feeding arrangement. Polarization properties of the helix have been the subject of several publications since the early work of Kraus.[5, 6] Traditionally, the pitch angle — an important parameter of the helix — may range from about 12° to 16°. The pitch angle, α, is the angle that a line tangent to the helix wire makes with the plane perpendicular to the axis of the helix.

QX0605-Anton01

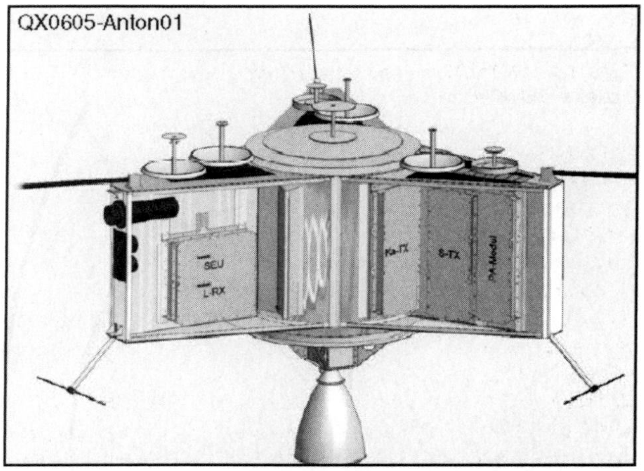

Figure 1 — A new satellite scheduled from Amsat : Phase 3E.

In the past, low-pitch helices have been recognized as ineffective radiating elements for a circularly polarized wave. Numerical results using *NEC-Win Pro* and intensive field measurements, however, lead to some low-pitch helices with gains comparable to that of a conventionally long helix.[7] The proposed complete system for the 2.4 GHz band includes a very short feed helix and a 60-cm diameter disk. Moreover, the originality of this project includes the use of the uniform segmentation technique to simulate both the helical feeder ground plane and the surface of the disk

Paolo Antoniazzi, IW2ACD
Via Roma 18, 20050 Sulbiate MI, Italy
paolo.antoniazzi@tin.it

Marco Arecco, IK2WAQ
Via Luigi Einaudi 6, 10093 Cologno Monzese MI, Italy
ik2waq@aliceposta.it

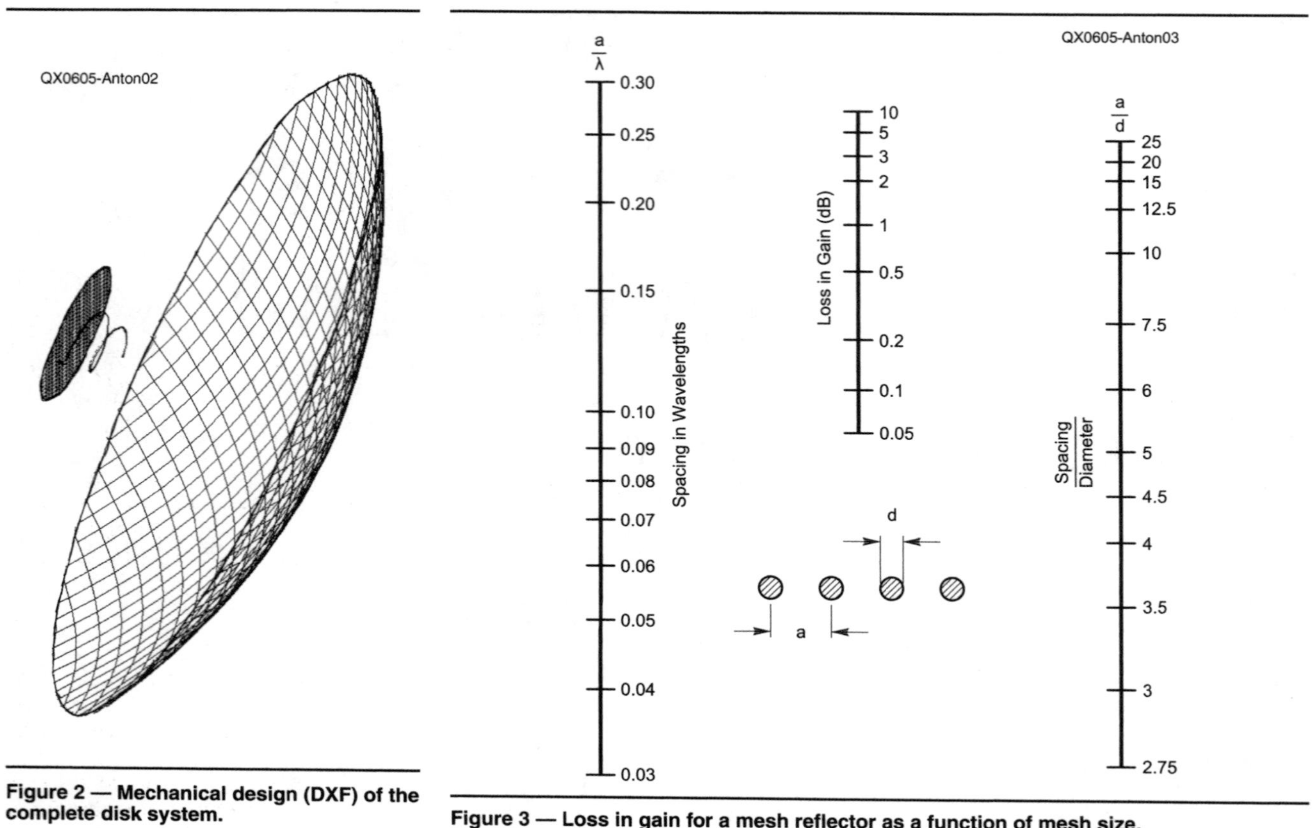

Figure 2 — Mechanical design (DXF) of the complete disk system.

Figure 3 — Loss in gain for a mesh reflector as a function of mesh size.

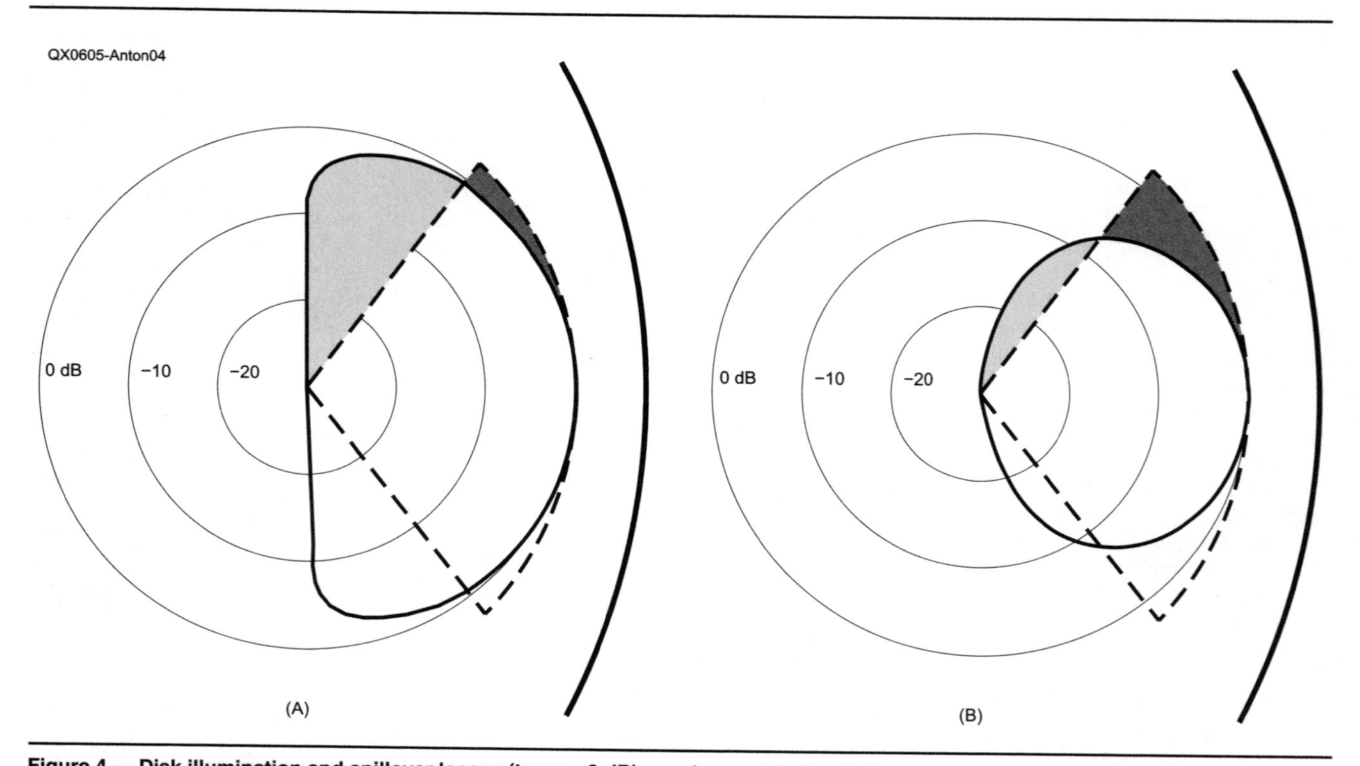

Figure 4 — Disk illumination and spillover losses (taper = 3 dB) are shown at A. Part B shows the disk illumination and spillover losses (taper = 10 dB).

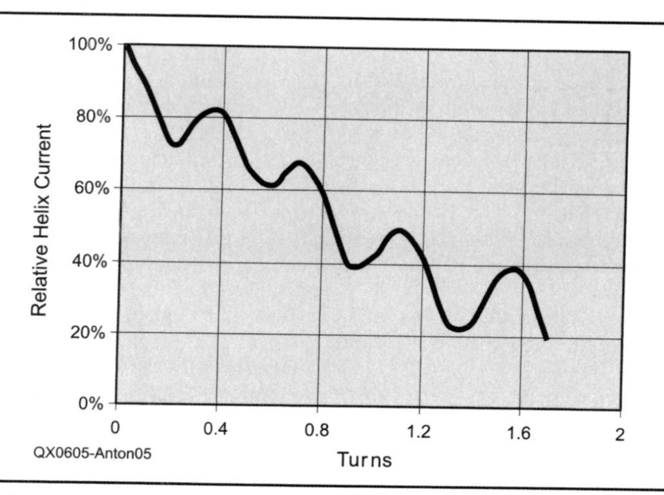

Figure 5 — Current simulation (low profile helix) using NEC-Win Pro.

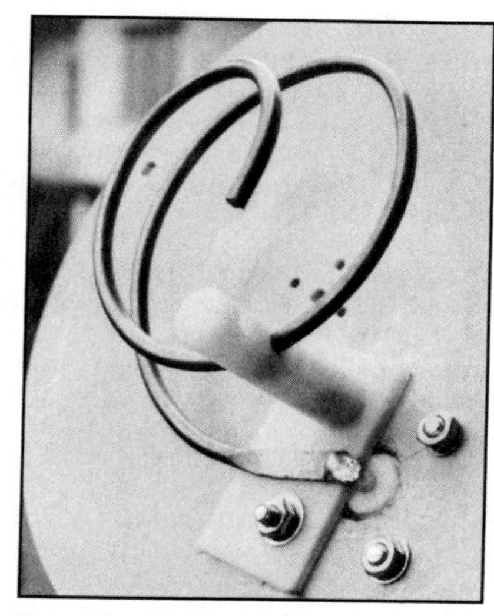

Figure 6 — A zoom on the helix feeder.

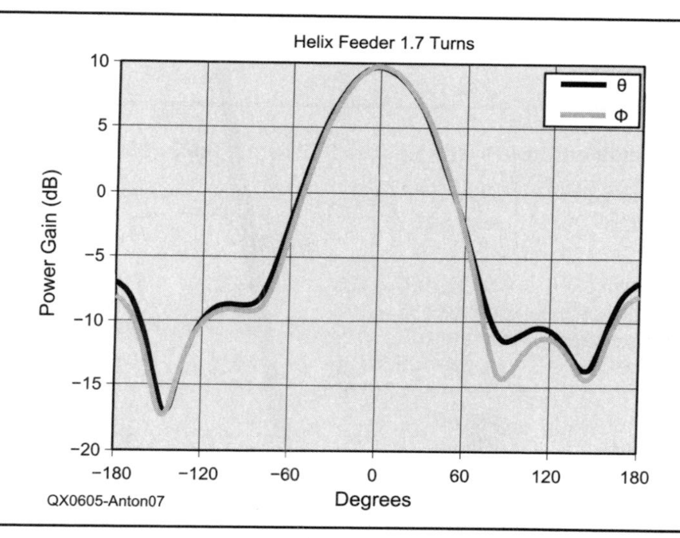

Figure 7 — Simulated radiation diagram of the low-profile feeder.

reflector to avoid the program limitation due to the nearness of the wires in the center of round geometric figures.

The Disk Reflector

The parabolic dish antenna can provide very high power gains at microwave frequencies. At a well-defined frequency, the gain is only limited by the size and accuracy of the parabolic reflector, if the antennas are properly implemented. A dish is a parabola of rotation: a parabolic line rotated around the axis that passes through both the focus and the center of the curve. The choice of a parabola with a central focus instead of one with offset has been made considering the easiness of finding it and of exactly centering the feeder without problems.

The physical dimensions of the reflector we considered are the diameter of the disk, $D = 600$ mm, and its depth in the center, $c = 94$ mm. Using these two dimensions, we can calculate the focal length f and the f / D ratio (that is an easy way to describe our parabola), and the power gain G in dBi (referred to the isotropic):

$f = D / (16\ c) = 239.4$ mm and $f / D \approx 0.4$
$G = 10 \log_{10} [(\pi^2 D^2 \eta) / \lambda^2]$
where: λ = wavelength in millimeters
η = aerial efficiency (< 1)

This last equation comes from the capture area of an antenna and is a re-elaboration of it.

It is a simple geometric computation to consider a dish antenna pointed at an isotropic antenna that is transmitting from some distance away. The assumed isotropic antenna is a point source that radiates equally in every direction.

The efficiency is the ratio between the power received at the electrical connection to the feed, and the power that actually arrives. By hypothesis, considering the unrealistic case of a perfect lossless antenna ($\eta = 1$), the power gain can grow until $G_{MAX} = 23.6$ dBi. A more practical value of the gain should be $G = 20.6$ dBi, assuming that its efficiency drops to 50% ($\eta = 0.5$). This point will be analyzed better again, after the evaluation of the simulation results of the complete system.

As already stated in the introduction, to avoid the error problems related to the model included in *NEC-Win Synth* software, we decided to use *AutoCAD* to prepare the file for the simulation program.[7] The equation used was the classic one of the parabola of rotation — easy to find in every manual of analytic geometry:

$y = x^2 / (4 f) + z^2 / (4 f)$

where: x, y and z are the three coordinates in the space of the reflector expressed in millimeters.

At this point we used the *NEC-Win Synth* software for a double purpose: to translate the .DXF file to a .NEC file and to remove the possible segmentation errors caused by *AutoCAD*. It is mandatory to perform this kind of translation because .NEC is the only file format acknowledged by *NEC-Win Pro*.

To have a good trade-off between the simulation accuracy and the computation time (the number of segments of the complete system is 2102) it was decided to use a mesh reflector pitch $a = 20$ mm combined with a wire diameter d = 6 mm. This matching, with the aid of the chart in Figure 3, allows us to calculate the gain loss caused by the use of a non-solid reflector.[8] The values to be entered in the graph are $a / \lambda = 0.16$ and $a / d = 3.33$, which is equivalent a very low loss of gain: less than 0.05 dB (off the scale of Figure 3).

For your information, *Microsoft Excel* can also be used, but the procedure to obtain the .NEC file is more complicated because some operations must be done manually. The first step consists in the preparation of a table, using a format compatible to *NEC-Win Pro*, but limited to one half of the parabola. The resulting file is to be converted by *Excel* to text and then to .NEC. To generate the entire disk, *NEC-Win Pro* allows us to perform a rotation by 90° and a three times duplication. Another inconven-ience of this procedure is the impossibility to use *NEC-Win Synth* to remove possible segmentation errors because this software does not acknowledge the "GM" instruction.

Low-Profile Helix Feeder

As a feeder for the 60-cm diameter disk we need an RF source, placed at the parabola focus, having a beam width given by the following:

$\theta = 2 \arctan [D / 2 (f - c)] \approx 128°$

Almost all feeders will provide less energy at the edge than in the center of the disk.[9] The difference in power at the edge of the disk is defined as the "edge taper." Changing the shape of the feeder radiation pattern, we can change the edge taper illuminating the disk. Different edge tapers will produce different amounts of illumination and spillover losses: a small edge taper results in a larger spillover loss, while a large edge taper reduces the spillover loss at the expense of an increase of the illumination loss. A good tradeoff, to have the maximum efficiency of the antenna, is obtained with an edge or illumination taper of about 10 dB.

Looking closely at the parabolic surface, we find that the focus distance (f = 239.4 mm) at the center is smaller than the distance d between the focus and the disk edge:

$d = \text{sqrt}\ [(D / 2)^2 + (f - c)^2] = 333$ mm

So, the wave that follows the edge way has been submitted to an additional "ΔA" attenuation:

$\Delta A = 20 \log_{10} (d / f) = 2.9$ dB

This equation comes from the free space attenuation formula where the variables, frequency and gain, are constant (also defined as inverse square law) and so do not appear in the formula.

Considering the additional attenuation just analyzed, the optimum illumination taper becomes: 7.1 dB (10 dB – 2.9 dB). The feeder comprises every microwave device able to concentrate the beam, on the reflector, within the already calculated beamwidth θ: dipole, horn, waveguide, helix and so on. We decided to use the helix, considering our previous experience in simulating and manufacturing this kind of antenna.[10, 11, 12] In Figure 4 we can see the spillover and illumination losses in two different situations (3 dB taper and 10 dB taper).

The behavior of the current versus length of a typical helix shows three different regions:

1) Near the feed point, where the current decay is exponential.

2) Near the open end, with visible standing waves.

3) Between the two helix ends, where there is a relatively uniform current and small SWR (transmission line).

There are two ways to obtain a good circular polarization helix:

1) Tapering the helical turns near the open end, to reduce the reflected current from the arm end, and

2) Using only the first helical turns where the decaying current travels from the feed point to the first minimum point (see Figure 5).

Starting from these considerations our final low-profile helix uses a pitch $a = 0.16\ \lambda$ (20 mm at 2.4 GHz) and is both conically wound with conic 62 / 41 mm diameters and very short (only 1.7 turns — as shown in the photo of Figure 6).

Let us summarize the mechanical dimensions of the feeder chosen to supply our 60 cm disk:

1) Number of turns: 1.7 (10 segments per turn).

2) Pitch between contiguous turns: 20 mm.

3) Starting diameter: 62 mm.

4) Ending diameter: 41 mm.

5) Helix wire diameter 3 mm.

6) Stub: 6 mm long, 2 mm wire diameter,

Table 1
Different Simulated Low-Profile Helices (1.7 Turns, Pitch=20 mm)

Helix Diameter and Shape (mm)	*Reflector Type and Size (mm)*	*Gain (dB)* f = ***2.2*** *GHz*	*Gain (dB)* f = ***2.4*** *GHz*	*Gain (dB)* f = ***2.6*** *GHz*
67.5 / 45.0 Conical	Square 125 × 125	9.49	9.64	8.80
62.0 / 41.0 Conical	Square 125 × 125	9.35	9.73	9.84
62.0 / 41.0 Conical	**Circular** (dia = 124)	**9.31**	**9.59**	**9.55**
56.0 Cylindrical	Square 125 × 125	9.38	9.68	8.52
67.5 / 45.0 Conical	Square 75 × 75	9.15	9.28	8.36
62.0 / 41.0 Conical	Square 75 × 75	8.84	9.29	9.39
56.0 Cylindrical	Square 75 × 75	8.90	9.19	8.38

Table 2
Simulation of Gain and F/B @ 2.4 GHz of the Complete System vs Feeder Positioning

Distance between Feeder Reflector and Parabola Surface (mm)	*Power* Gain (dB)	*–3db Radiation* Angle (degrees)	*Front to Back* Ratio (dB)	*Notes*
220	20.33	16	32.24	High Illumination Loss and Very Low Spillover Loss
230	20.23	16	30.98	
240	20.19	15	30.34	
250	20.57	15	30.08	
260	21.25	16	28.6	
265	**21.38**	**16**	**27.87**	Best power Gain
270	21.35	16	27.45	
280	21.05	15	26.85	
290	20.90	16	26.30	

Feeder Reflector Diameter = 125 mm

only for simulation purposes to contain the current in the screen.

7) Screen diameter: 125 mm, wire diameter 1 mm.

The just-defined feeder has been accomplished using 674 segments: 18 for the helix and the stub and the remaining 654 to define the screen with a 6.25 mm segmentation pitch.

The simulation of the feeder, using *NEC Win Pro* software (see Note 10 and Table 1), gave the following results at the frequency of 2.4 GHz:

Power gain: $G = 9.6$ dB

Radiation angle: 59° at –3 dB, 91° at –7.1 dB, 107° at –10 dB, 128° at –14 dB, 146° at –17 dB.

Front to back ratio: 16.6 dB.

The power gain versus the angles θ and ϕ that are in two planes perpendicular between them and parallel to the wave propagation direction is plotted in Figure 7.

As you can see from the simulation results, our 21-dB-gain disk has a high illumination loss, but this is an advantage from the F/B, secondary lobes and spillover point of view.

The Complete System

During the simulation phase, we made some changes in the original file to optimize the performance of our antenna. The first one involved the diameter of the feeder reflector. We reduced it from 125 to 100 mm in order to reduce its masking effect on the beam reflected by the disk surface by 36%. The advantage of this operation was insignificant both for power gain (–0.05 dB) and for front-to-back ratio (–1 dB) so we decided to come back to the original screen: 125 mm diameter.

The second action involved increasing the distance between the feeder ground plane and the parabola surface from 220 to 290 mm. The best distance, from the power gain point of view, which is the most important parameter for the amateur, was 265 mm (see Table 2).

Theoretically, the disk receiving antenna works as follows: the electromagnetic waves coming on a parallel path from a distant source are reflected by the parabola to a common point: the parabola focus point. A transmitting antenna reverses the path: the radio wave generated by a point source placed at the focus point is reflected into a beam of rays parallel to the axis of the parabola.

The differences between the theory and the practice are related to the following:

1) The focus is only a geometric point and the feeder is bigger than that.

2) It is not so easy to find the electrical center of a helix having a truncated cone shape. Considering the simulation results, and the calculated focal length, it can be put at about ¾ of the total feeder length (34 mm).

After the implementation of the whole optimization described, the simulation results at the frequency of 2.4GHz are summarized here:

Power Gain: 21.4 dBi

Beamwidth: 16° at –3 dB

Front to back ratio: 29.7 dB

Figure 8 shows the power gain graphs, as a function both of the azimuthal and elevation angles ϕ and θ.

Now we are able to calculate the antenna efficiency η that is the ratio between the gain, obtained from the simulation, and the maximum possible gain G_{MAX}, already computed using the equation of aerial capture area:

$$\eta = (10^{G/10}) / [(\pi D) / \lambda]^2 \approx 60\%$$

The efficiency loss of our complete system (~2.2 dB) can be referred to the disk illumination (see Table 2). We consider a uniform illumination of the disk and so the increased attenuation caused by the different path that the electromagnetic wave covers at the periphery of the reflector with respect to the center. The spillover and side-lobe losses can be neglected.

Measuring the Disk

The complete system is shown in Figure 9. In our experience, it's not very difficult to design and make helices for different working frequencies and gains; but more difficult, for the serious experimenter, is a way of precise measurements.

Before starting with measurements, it is important to define the polarization sense of the helices used in the test setup. The IEEE definition that has become the standard is that in viewing the antenna from the feed-point end, a clockwise wind results in right-hand circular polarization (RHC), and a counterclockwise wind results in left-hand circular polarization (LHC). This is important, because when two stations use helical antennas *over a nonreflective path*, both must use antennas with the same polarization sense. If antennas of opposite sense are used, a signal loss of at least 20 to 30 dB results from cross polarization alone. Remember that the complete disk produces circular polarization with a sense *opposite* that of the feed helix

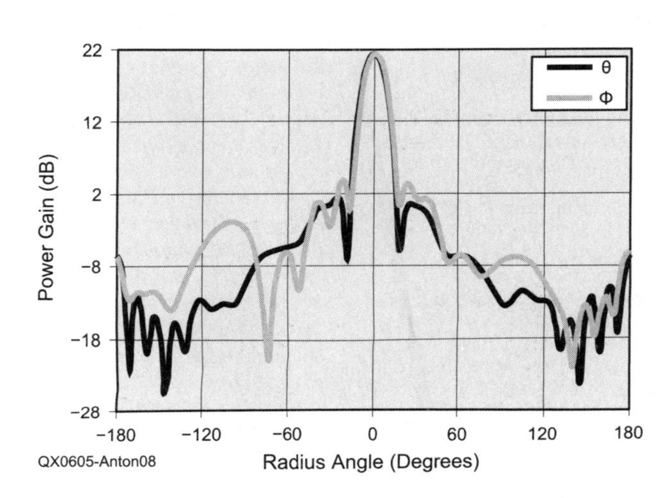

Figure 8 — Simulated radiation diagram of the complete system.

Figure 9 — The complete 60 cm disk antenna.

(because of the reflection of the wave from the parabolic surface).

For almost all tests, we used a generator from 2.2 to 2.7 GHz composed of a VCO, JTOS-3000 by Mini-Circuits, followed by the MNA-6 broadband amplifier (3 × 3 mm), or the Keps 0535AN2. The output level from the oscillator is very high (+10 dBm), but some attenuation must be included for stability (the wide-bandwidth amplifiers oscillate easily with loads not exactly 50 Ω). Using a Boonton RF Millivoltmeter (mod 92B) as a detector, we have a sensitivity loss of about 10 dB at 2.4 GHz (referred to the maximum suggested operating frequency). More suited to the measurements is the classic HP 431B power meter.

To limit the measurement errors, the distance between transmitting and receiving antennas has to be considered. To determine this distance, you need to be able to measure the signal level easily with a filtered RF voltmeter having a 20 to 30 dB dynamic range. Also, the wave reaching the receiving antenna should be as planar as possible.

The first condition can be easily established, starting with the received power and calculating the attenuation experienced by the wave in open space:

$A = 32.4 + 20 \log (f) + 20 \log (d) \quad G_t - G_r$

where A is the attenuation in decibels, f is the frequency in megahertz, d is the distance in km, G_t is the gain of the transmitting antenna in dBi and G_r is the gain of the receiving antenna, also in dBi, obtained by simulation.

There is also a simple, easy to remember method of calculating the free-space attenuation by considering the distance between the two antennas in terms of wavelengths. When the distance between two isotropic antennas is $d = \lambda$, the free-space attenuation is always 22 dB. This distance is 12.5 cm at 2400 MHz. The attenuation increases by 6 dB for each doubling of the path distance. This means that the free-space attenuation is 22 dB at 0.125 m, 28 dB at 0.25 m, 34 dB at 0.5 m, and so on. To make the wave reaching the receiving antenna as planar as possible, the capture area in square meters of the receiving antenna is:

$$A_c = G_r \lambda^2 / 4 \pi$$

This expression is valid for an antenna with no thermal losses and was certainly useful for our experiments. With a circular capture area, the minimum distance in meters between the antennas will be:

$$d > n D^2 / \lambda$$

where D: disk diameter in meters.

A maximum acceptable phase error will also be considered. For a phase error of 22.5°, which is usually enough, $n = 2$. If a phase error of only 5° is required, $n = 9$.[13, 14] See Figure 10 for free space attenuation at 2.4 GHz. The site (in the garden) we selected is particularly useful for all our helix measurements (d = 4 m = 32 λ at 2400 MHz).

Table 3 compares the simulation results with our measurements.

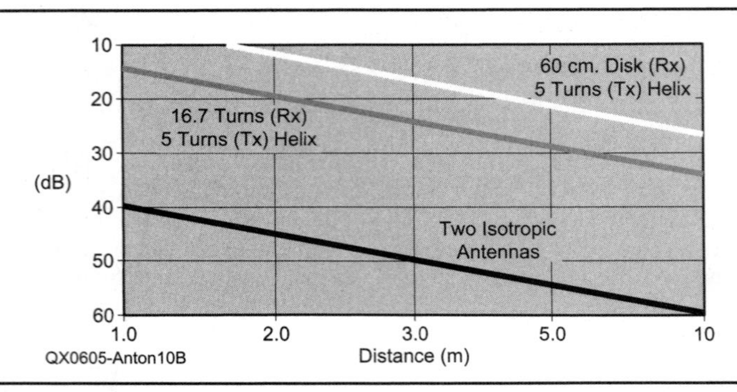

Figure10 — Free space attenuation @ 2.4 GHz.

Table 3
Comparison Between 60 cm Diameter Disk and Standard 16.7 Turns Helix

	Simulation Results			*Measurements*	
Type of Antenna	*Power gain (dBi)*	*Radiation Angle –3 dB*	*Front to back Ratio (dB)*	*Radiation Angle –3 dB*	*Relative Gain (dB)*
60 *cm disk*	21.4	16	27.9	14	+7
16.7 turn helix	14.5	26	18.5	30	0

Notes

[1]"P3-Express — An Intermediate Report," *AMSAT-DL Journal*, Dec/Feb 2003/2004.

[2]"AMSAT-Phase 3 EXPRESS," *AMSAT-DL Journal*, Sep/Nov 2003.

[3]"P3E News," *AMSAT-DL Journal*, Mar/May 2004.

[4]P. Antoniazzi and M. Arecco, "Chaparral Style 10 GHz Feed Simulation," *DUBUS*, N.2/2005, pp 38 to 50.

[5]John D.Kraus, The Helical Antenna, pp 265 to 339, *Antennas*, McGraw-Hill, 2nd Edition, 1988.

[6]H. Nakano,Y. Samada and J. Yamauchi, "Axial Mode Helical Antennas," *IEEE Transactions on Antennas and Propagation*, 1986, pp 1143 to 1148.

[7]**www.nittany-scientific.com** (NEC-Win Pro Rel 1.4 and NEC-Win Synth Rel 1.0).

[8]*Microwave Handbook*, Volume 1, RSGB, 1989, p 4.21

[9]W1GHz, *Microwave Antenna Handbook Online*, **www.w1ghz.org**.

[10]P. Antoniazzi and M. Arecco, "Simulating and Making Low Profile WiFi Antennae," *Electronics World*, Dec 2004, pp 22 to 25.

[11]P. Antoniazzi and M. Arecco, "Measuring 2.4 GHz Helix Antennas using Slotted Lines," *Electronics World*, Jul 2003, pp 47 to 52.

[12]P. Antoniazzi and M. Arecco, "Measuring 2.4 GHz Helix Antennas," *QEX*, May/Jun 2004, pp 14 to 22.

[13]P. Antoniazzi and M. Arecco, "Measuring Yagis," *Electronics World*, Dec 1998, pp 1002 to 1006.

[14]D. R. J. White, *Electromagnetic Interference and Compatibility*, Vol 2, D.White Consult, Inc, 1980, pp 2.11 to 2.13.

Paolo Antoniazzi was licensed as IW2ACD and became a member of the Associazione Radiamatori Italiani (ARI) in Italy in 1961. He worked for Siemens, GTE, Sprague and ST Microelectronics in the fields of RF and telecommunications marketing and applications until his retirement in November 2003. Paolo is particularly interested in simulation and measurements (and writing papers) about narrow-band LF, 2.4 GHz, and more recently, the 10 GHz band. Space communications with the future P3E and P5A satellites is a priority interest for the future.

Marco Arecco, IK2WAQ, matured his professional experience in semiconductor manufacturing, working for STMicroelectronics for more than 35 years. His last position, before retirement in early 2004, was as the Nonvolatile Memory Product Engineering Team leader. Marco is a member of ARI. He has been licensed 12 years and his Amateur Radio interests include LF operation and VHF, UHF and microwave antenna simulations and measurements.

Mark Spencer, WA8SME

A Satellite Tracker Interface

Build your own computer rotator control interface for ARISS and OSCAR-style operation and save a bundle.

Bringing space into the classroom is an incredibly powerful learning experience for students. The marriage between wireless technology literacy and space literacy is a strong one that produces a portfolio of activities that can engage virtually all students at some ability and interest level. This portfolio of activities can range from simply having a satellite predication software package running in the back of the classroom on an old, salvaged computer to a full-up ground station to communicate with the astronauts in the International Space Station through the ARISS program.[1] Planning and preparing to bring space into the classroom is hard work, and it can be expensive depending on the level of material and experiences that the teacher wants their students to share.

One of the station requirements to be scheduled for an ARISS contact is to have an antenna system consisting of an OSCAR style circular polarized beam antenna and an azimuth/elevation rotator control. Because of the orbital characteristics of the ISS, manually controlling the azimuth/elevation rotator system would be difficult, at best, and could jeopardize the success of the contact. Consequently, a computer controlled rotator system would be preferable (not only for an ISS contact but for all low orbiting satellite work). The only azimuth/elevation rotator system for the ham radio market is the Yaesu G-5500. With a price tag of about $600, this is an expensive package. The rotator to computer interface accessory is an additional $500.

In this article I describe a simple rotator-to-computer interface that can be built for a fraction of the cost of the commercial accessory. Not only will the builder save the cost difference that could be applied to other station enhancements, but when you build your own, you will learn volumes about orbital mechanics and antenna orientation strategies, and you will become familiar with the inherent software of the interface to make adjustments as needed to keep the antenna pointing system in peak condition.

This particular project is designed to interface the Yaesu G-5500 rotator system to a PC running the *NOVA* satellite tracking software. The project is by no means unique, numerous interface designs can be found on the Web. (This circuit design and software is an adaptation of a similar system developed by Gene Brigman, KC4SA, and Mark Hammond, N8MH.)[2] This project, however, is a no frills design that does not duplicate the antenna pointing indications that are available from the G-5500 console (therefore reducing costs). Also, the programming is straightforward and "airy," allowing for lots of modifications to suit individual applications. The *NOVA* satellite tracking software package was chosen because it contains many powerful features for not only pointing antennas but also display options that are invaluable to the teacher instructing orbital fundamentals (I particularly like the map graphics and the 3D global view options).[3]

NOVA Data Output Architecture

Along with displaying satellite position and antenna pointing information on the computer screen, *NOVA* also sends the antenna pointing information to the serial port using the EASYCOM1 protocol.[4] Rotator parameters and limitations are set into drop-down user interface screens so that the rotator positioning is adjusted to the particular rotator capabilities. The antenna rotator setup for this interface should be as shown in Figure 3 (set the appropriate COM port).

G-5500 Architecture

Four momentary switches control the G-5500 azimuth/elevation rotator system: up, down, clockwise and counter-clockwise. The G-5500 rotator system has variable resistors connected to the azimuth and elevation rotators that are turned along with the antenna. These resistors produce a variable voltage within the rotator

[1]Notes appear on page 6-22.

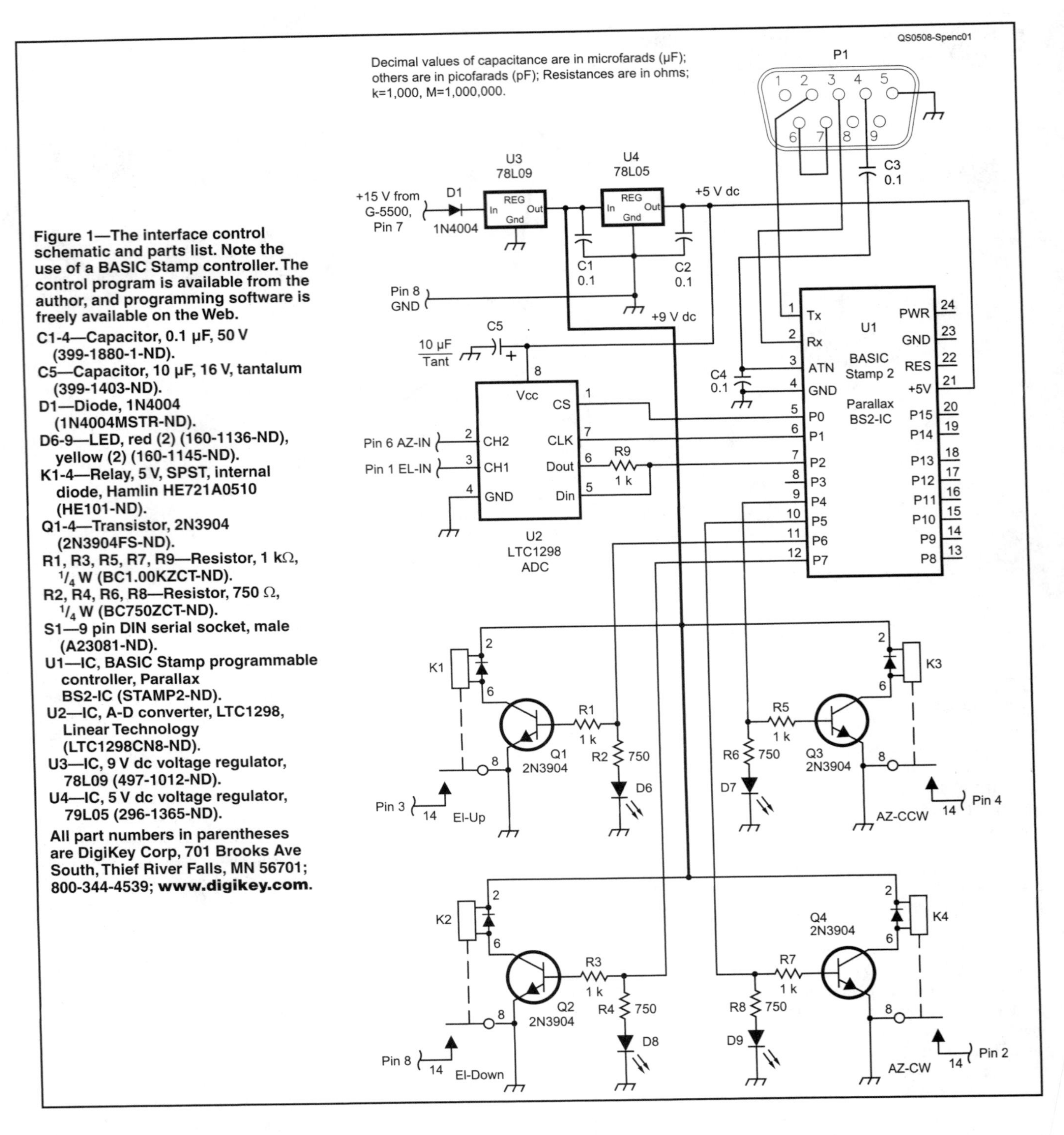

Figure 1—The interface control schematic and parts list. Note the use of a BASIC Stamp controller. The control program is available from the author, and programming software is freely available on the Web.

C1-4—Capacitor, 0.1 µF, 50 V (399-1880-1-ND).
C5—Capacitor, 10 µF, 16 V, tantalum (399-1403-ND).
D1—Diode, 1N4004 (1N4004MSTR-ND).
D6-9—LED, red (2) (160-1136-ND), yellow (2) (160-1145-ND).
K1-4—Relay, 5 V, SPST, internal diode, Hamlin HE721A0510 (HE101-ND).
Q1-4—Transistor, 2N3904 (2N3904FS-ND).
R1, R3, R5, R7, R9—Resistor, 1 kΩ, 1/4 W (BC1.00KZCT-ND).
R2, R4, R6, R8—Resistor, 750 Ω, 1/4 W (BC750ZCT-ND).
S1—9 pin DIN serial socket, male (A23081-ND).
U1—IC, BASIC Stamp programmable controller, Parallax BS2-IC (STAMP2-ND).
U2—IC, A-D converter, LTC1298, Linear Technology (LTC1298CN8-ND).
U3—IC, 9 V dc voltage regulator, 78L09 (497-1012-ND).
U4—IC, 5 V dc voltage regulator, 79L05 (296-1365-ND).

All part numbers in parentheses are DigiKey Corp, 701 Brooks Ave South, Thief River Falls, MN 56701; 800-344-4539; www.digikey.com.

control box from 0 V at 0° to a user set maximum, in this case 4.5 V dc at either 450° for azimuth or 180° for elevation. The control switches and antenna position voltages are available through an 8 pin external control jack on the G-5500 console. Set up your rotator as detailed in the instruction manual so that at maximum azimuth and elevation the voltages are 4.5 V dc.

Interface Functions

As detailed in the schematic of Figure 1, there are two primary components in the interface; the BASIC Stamp microcontroller and a 12 bit analog to digital converter (ADC). The satellite interface uses the ADC to read the position voltages from the G-5500 and converts them to a digital word value between 0 and 4096. Satellite position information is in degrees, antenna position information is in a digital word, one needs to be converted so that the BASIC Stamp microcontroller can compare the two values and send appropriate commands to move the rotators. The BASIC Stamp uses the ADC to read the antenna position from the G-5500, and uses the antenna position information sent from *NOVA*, converts the number of degrees into a scaled digital value, and then compares the digital values (where it is supposed to be compared to where the antenna is). Finally, the BASIC Stamp determines which direction to turn the antenna—up/down, left/right—and closes appropriate relays that are connected in parallel with the direction control switches in the G-5500 console to turn the rotators. The relays remain closed until the antenna is detected to be in the correct position (through reading the ADC). *NOVA* generated satellite position data is then monitored for the next antenna movement command.

When Am I Ever Going To Use This Stuff?

I don't know how many times I have heard this lament during my teaching career. Well, to understand the conversion of degree data to scaled digital data, you have to use basic algebra skills. Here is a use for this stuff!

One complicating factor in making this conversion is that the BASIC Stamp is limited to integer mathematics. This means that even for simple mathematical operations, decimal parts of the computations are truncated and the resulting answers are rounded down to the nearest integer. This can inject unacceptable errors into computations. I will deal with the integer math limitations later.

To make the conversion from degrees to digital, you will have to dust off some basic algebra skills for linear equations. The output of the ADC is linear and follows the algorithm:

$$Y = mx + b$$

where m is the slope of the line $(Y_2 - Y_1) / (X_2 - X_1)$ and b is the Y-axis intercept.

In terms of satellite position in degrees and digital word, Y is the converted value in digital word and X is the position in degrees. By using the little program called *ADC_Checkout.bs2* (available from the author), you can determine the digital word value for the end points (0° and 450° or 180°, as appropriate) of the rotator. Once armed with these two data points, you can determine the slope and Y intercept values for the two equations needed to convert the azimuth and elevation degrees into digital words.

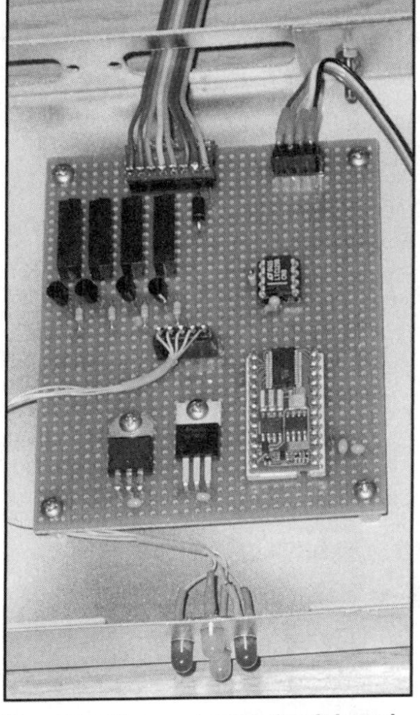

Figure 2—The controller circuit board within its housing. The controller is powered from the rotator. LEDs are arranged to indicate up, down, left and right commands.

The following computation example uses the data gathered with the prototype satellite tracker interface. You should not need to go through these steps to calibrate your interface; this is provided only for those who have the desire to know.

Satellite Tracker Interface ADC Data

AZ Degrees		*AZ Digital Word*	
X1	0	Y1	30
X2	450	Y2	3710
EL Degrees		*EL Digital Word*	
X1	0	Y1	12
X2	180	Y2	3712

$$\text{Azimuth: slope } m = \frac{3710-30}{450} = 8.18 \quad \text{Y- intercept b=30}$$

$$\text{Elevation: slope } m = \frac{3712-12}{180} = 20.56 \quad \text{Y-intercept b=12}$$

Because the BASIC Stamp uses integer math, if we were to multiply the number of degrees by these slope values, 8.18 or 20.56, the fractional part would be truncated and result in error. The BASIC Stamp has a math operator that takes care of this situation, the *multiply middle* operator (*/). The *multiply middle* operator multiplies variables and/or constants, returning the middle 16 bits

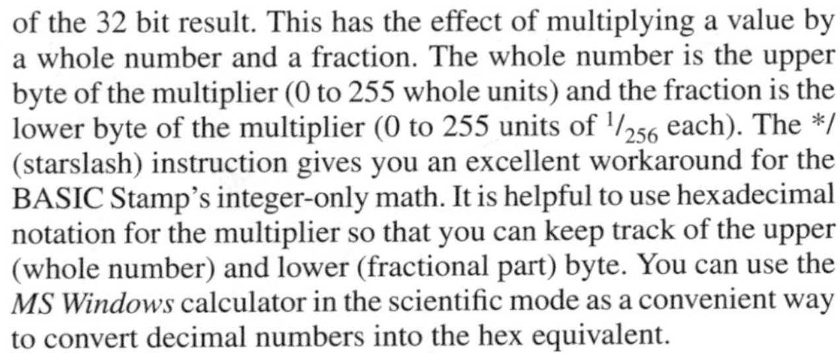

of the 32 bit result. This has the effect of multiplying a value by a whole number and a fraction. The whole number is the upper byte of the multiplier (0 to 255 whole units) and the fraction is the lower byte of the multiplier (0 to 255 units of $^1/_{256}$ each). The */ (starslash) instruction gives you an excellent workaround for the BASIC Stamp's integer-only math. It is helpful to use hexadecimal notation for the multiplier so that you can keep track of the upper (whole number) and lower (fractional part) byte. You can use the *MS Windows* calculator in the scientific mode as a convenient way to convert decimal numbers into the hex equivalent.

For example, to convert the azimuth value of degrees into the ADC digital word equivalent, we need to multiply the number of degrees by the slope of the line and add the Y intercept:

$$Y = 8.18X + 30$$

8 = 08 in hex.
0.18*256 =46.08 rounded to 46 = 2e in hex. The multiplier in hex is $082e; therefore, Y = $082e * X + 30. The line of code to reflect this algorithm is: **new_az= new_az*/$082e+30**. In a like manner, the multiplier for the elevation conversion is $148f (20 decimal in hex is 14). The line of code to reflect this conversion is: **new_el=new_el*/$148f+12**.

Construction

The layout and construction of the satellite interface is not critical. As shown in Figure 2, the interface is constructed on a standard prototyping board using point to point wiring. Wiring the serial port as indicated in the schematic (Figure 1) will allow the same computer to be used to program the BASIC Stamp and to access the serial data from *NOVA*. I arranged the LED display in a pattern to give a visual indication of rotator direction. The power for the interface comes from the G-5500. This voltage source is around 14 V dc with no load, but when the rotators are engaged, there is a significant voltage drop. To provide some level of stability, the voltage from the G-5500 is run through a 9 V dc regulator (the power source for the relays) before the voltage is further reduced and regulated by the 5 V dc regulator to power the BASIC Stamp and the ADC.

I suggest the interface be constructed and tested in stages. Begin first by wiring and testing the voltage regulators. Next add the BASIC Stamp and serial port. The BASIC Stamp is programmed using the BASIC Stamp editor program available from Parallax.[5] You can test this portion of the circuit by using the features of the editor.

Next, wire-up the ADC and the position voltage connections to the G-5500. Before you test this portion of the circuit, ensure that you set the full-scale voltage for the 180° and 450° rotator positions to 4.5 V dc, as outlined in the |G-5500 manual. Run the program *ADC_ Checkout.bs2* to test the ADC portion of the circuit. Note the ADC values as you rotate the rotators through their full travel, the values for 0° and 180° and 450° should be close to the values listed in the table. If they are, you should be able to use the main software without adjustment; if not, and the wiring is correct, you will have to do a little math and program modifications as detailed above. Finally, wire up the relays and do the final smoke check with the main program (*Sat_Tracker_Final.bs2*) uploaded to the BASIC Stamp.

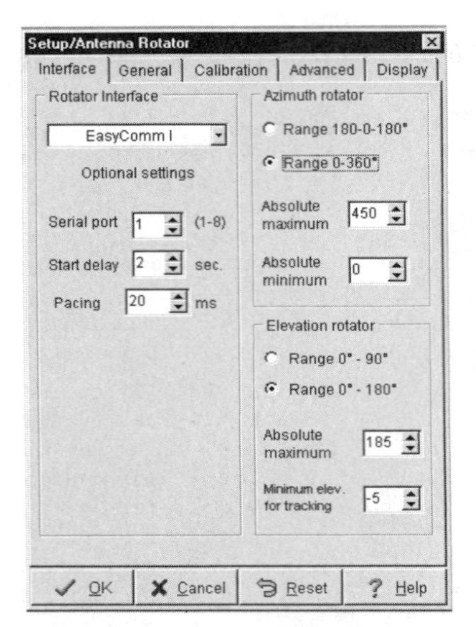

Figure 3—The rotator software setup screen.

The Software

The real advantage to making your own interface, aside from the obvious fact of reduced costs, is that you have control of the interface through the software. The software is well documented and should be easy to follow. If the ADC values for full scale are significantly different from the values listed in the table above, you will probably want to make adjustments to the constants in the conversion algorithms. Also, if the rotator system chatters in operation due to antenna torque movement, adjust the tolerance value to alleviate the problem.

Operation

Connect the interface to the G-5500 and to the computer serial port. Run the *NOVA* software; select the satellite to be tracked, select auto-tracking and the rotators will be commanded to the position for the start of the next pass. When the satellite is within range, the tracker will respond by updating the rotator position until the pass is completed. Aside from manipulating the *NOVA* software, the satellite tracker interface should take care of the rest. A typical software setup screen is shown in Figure 3. The interface can be conveniently placed beneath the rotator control box, as shown in Figure 4. Figure 5 shows young Sean Harlow, KB1LUZ, working with the weather satellite system.

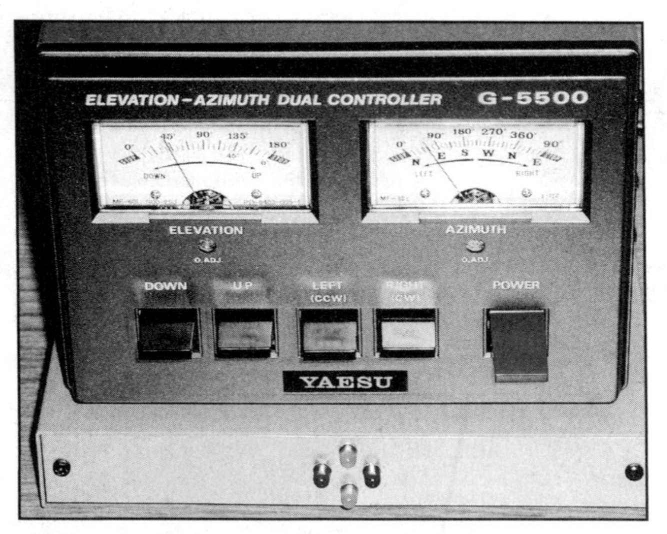

Figure 4—The rotator console sits conveniently on the interface controller.

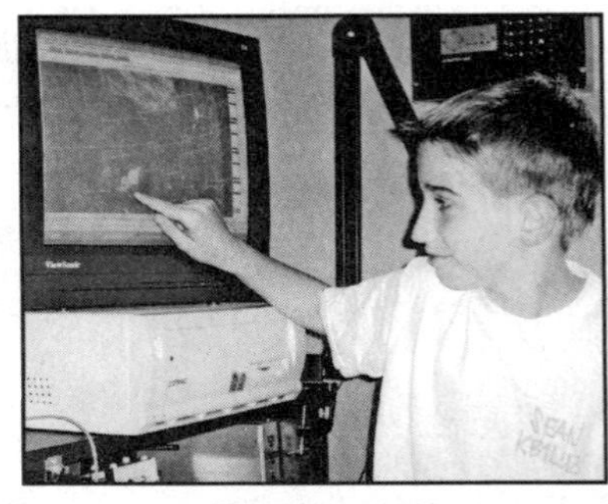

Figure 5—Sean Harlow, KB1LUZ, notes that Texas is getting hit with severe thunderstorms.

Final Thoughts

For the system to perform successfully, you will have to become familiar with the G-5500 rotator system, the operation of the *NOVA* satellite tracking software, and the BASIC Stamp editor for programming the Stamp. They all come together in the interface. So this project goes beyond simply wiring a duplicate circuit—that's the easy part. The fun part is learning how to make all the components of a satellite antenna tracking system work together.

Though this project specifically focused on the G-5500 and *NOVA* components, the interface could easily be adapted to other rotator configurations. It would just take a little homework to adjust the interface software.

Notes

[1]For more about the Amateur Radio on the International Space Station program, see **www.arrl.org/ARISS/**.

[2]**webpages.charter.net/mlhammond/ham/SAEBRTrack/**.

[3]**www.nlsa.com/**.

[4]**www.mustbeart.com/software/easycomm.txt**.

[5]**www.arrl.org/product-notes**. Look for the Satellite Handbook section and click the link to download the software.

All photos by the author.

Mark Spencer, WA8SME, was first licensed in 1965. He's also held the calls G5EPV, DA1OY and HL9AW. Mark has a BSE in Metallurgical Engineering from the University of Michigan and an MA in Communications from the University of Northern Colorado. He retired from the US Air Force after an extensive flying career in T-38, B-52, TR-1 and U-2 aircraft and is a decorated combat veteran of Desert Storm. Mark has taught high school mathematics, computer programming, physical science and physics, and has been an elementary and high school administrator. Currently ARRL Education and Technology Program Coordinator, Mark can be contacted at **mspencer@arrl.org**.

Helix Feed for an Offset Dish Antenna

The surplus PrimeStar offset fed dish antenna with its seven-turn helix feed antenna, Fig 23.50, is described in this section. When the feed antenna is directly coupled with a preamp/down-converter system, this antenna provides superb reception of S-band signals with the satellite transponder noise floor often being the noise-limiting factor in the downlink. This performance is as had been predicted by the W3PM spreadsheet analysis, **Fig 23.54**, and actual operating experience. Operating experience also demonstrates that this antenna can receive the Sun noise 5 dB above the sky noise. Don't try to receive the Sun noise with the antenna looking near the horizon, as terrestrial noise will be greater than 5 dB, at least in a big-city environment. The operator, NØNSV, who provided this dish was rewarded for his effort with a second feed antenna, and he in turn provided new labels for the dish, titling it "FABStar".

The reflector of this dish is a bit out of the ordinary, with a horizontal ellipse shape. It is still a single paraboloid that was illuminated with an unusual feedhorn. At 2401 MHz we must be satisfied with a more conventional feed arrangement. A choice must be made to under-illuminate the sides of the dish while properly feeding the central section, or over-illuminating the center while properly feeding the sides. For the application shown here, the former choice was made. The W1GHZ water-bowl measurements showed this to be a dish with a focal point of f = 500.6 mm and requiring a feed for an f/D = 0.79. The total illumination angle of the feed is 69.8° in the vertical direction and a feedhorn with a 3-dB beamwidth of 40.3°. At 50 percent efficiency this antenna was calculated to provide a gain of 21.9 dBi. A seven-turn helix feed antenna was estimated to provide the needed characteristics for this dish and is shown in **Fig 23.55**.

The helix is basically constructed as described for the G3RUH parabolic dish. A matching section for the first $\lambda/4$ turn of the helix is spaced from the reflector at 2 mm at the start and 8 mm at the end of that fractional turn. Modifications of the design include the use of a cup reflector. For the reflector, a 2-mm thick circular plate is cut for a ϕ94 mm (0.75λ) with a thin aluminum sheet metal cup, formed with a depth of 47 mm. Employment of the cup enhances the performance of the reflector for a dish feed.

The important information for this seven-turn helix antenna is: Boom, 12.7 mm square tube or "C" channel; Element, $\phi^1/_8$-inch copper wire or tubing; Close wind element on a ϕ1.50-inch tube or rod; Finished winding is ϕ40 mm spaced to a helical angle of 12.3°, or 28-mm spacing. These dimensions work out to have the element centerline to be of a cylindrical circumference of 1.0λ.

When WD4FAB tackled this antenna, he felt that the small amount of helical element support that James Miller used was inadequate, in view of the real life bird traffic on the antennas at his QTH. He chose to use PTFE (Teflon") support posts at every ½ turn. This closer spacing of posts also permitted a careful control of the helix winding diameter and spacing making this antenna *very* robust. A fixture was set up on the drill press to uniformly predrill the holes for the element spacers and boom. Attachment of the reflector is through three very small aluminum angle brackets on the element side of the boom.

Mounting of the helix to the dish requires modification of the dish's receiver mounting boom. **Fig 23.56** shows these modifications using a machined mount. NM2A has constructed one of these antennas and shown that a machine shop is not needed for this construction. He has made a "Z" shaped mount from aluminum angle plate and then used a spacer from a block of acrylic sheet. The key here is to get the dish focal point at the 1.5-turn point of the feed antenna, which is also at about the lip of the reflector cup. The

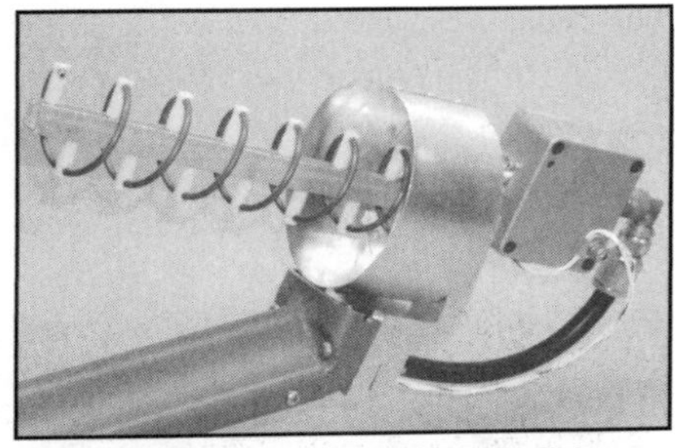

Fig 23.55—Seven-turn LHCP helix dish feed antenna with DEM preamp.

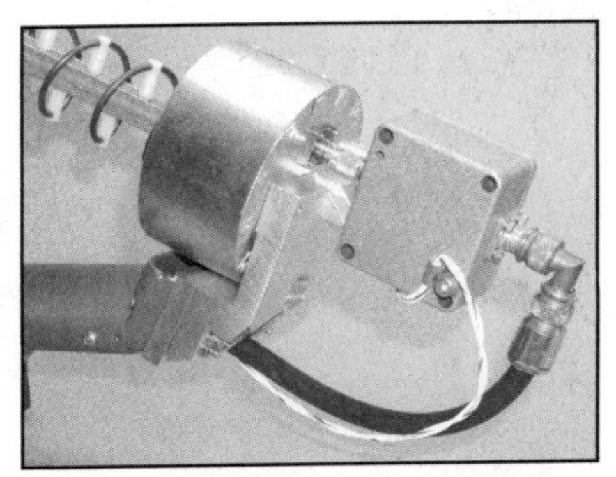

Fig 23.56—Mounting details of seven-turn helix and preamp.

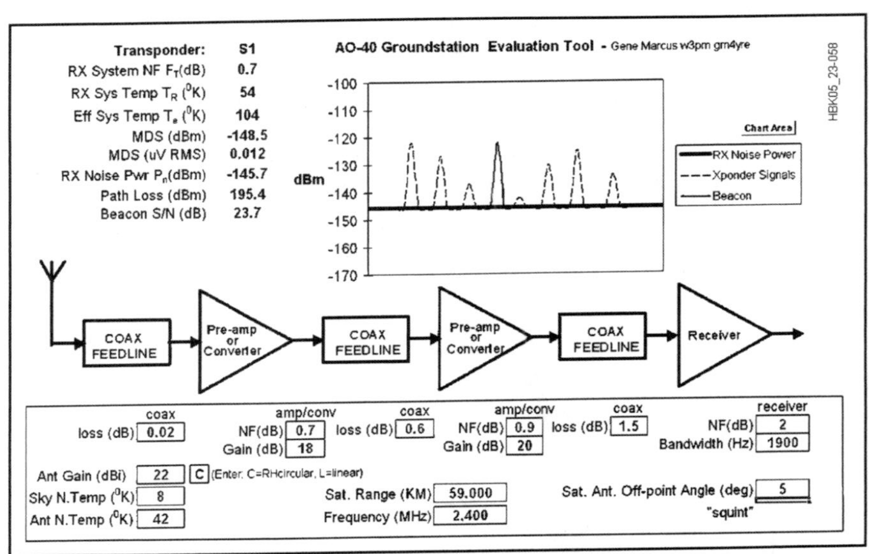

Fig 23.54—Screen display of W3PM's spreadsheet evaluation of ground station operation with AO-40.

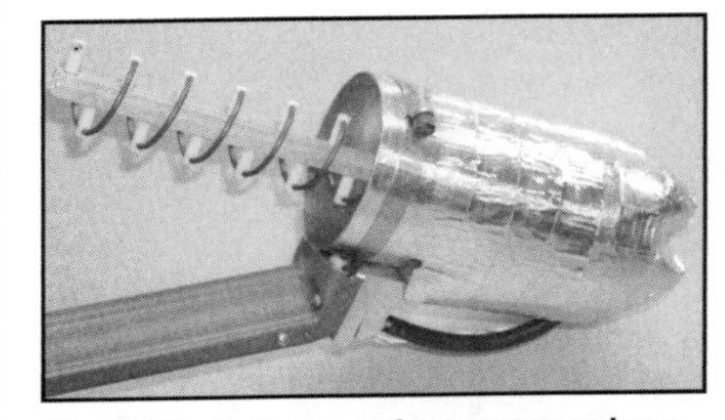

Fig 23.57—Rain cover for preamp using a two-liter soft-drink bottle with aluminum foil tape for protection from sun damage.

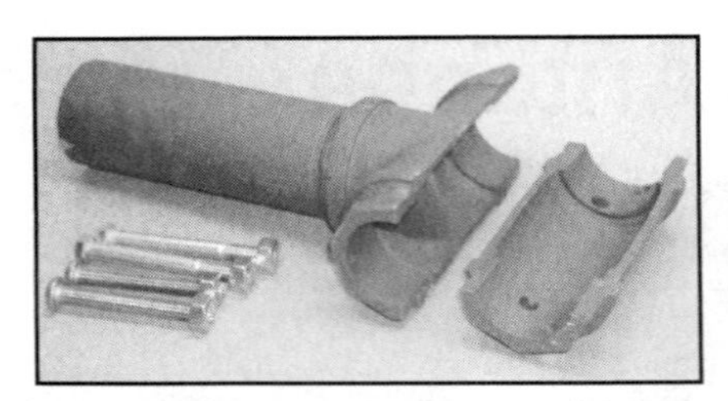

Fig 23.58—Welded pipe fitting mount bracket for FABStar dish antenna.

W1GHZ data for this focal point is 500.6 mm from the bottom edge of the dish and 744.4 mm from the top edge. A two-string measurement of this point can confirm the focal point, all as shown by Wade in his writings. When mounting this feed antenna the constructor must be cautious to aim the feed at the *beam-center* of the dish, and not the geometric center, as the original microwave horn antennas were constructed. Taking the illumination angle information noted above, the helix feed antenna should be aimed 5.5° down from the geometric center of the dish.

As illustrated in Fig 23.56, a DEM preamp was directly mounted to the feed helix, using TNC connector that had been chosen for this case, as an N connector is quite large for the S-band helix. A male chassis mount connector should be mounted on either the preamp or the antenna so that the preamp can be directly connected to the antenna without any adaptors. This photo also illustrates how the reflector cup walls were riveted to the reflector plate. Exposed connectors must be protected from rainwater. Commonly, materials such as messy Vinyl Mastic Pads (3M 2200) or Hand Moldable Plastic (Coax Seal) are used. Since this is a tight location for such mastic applications, a rain cover was made instead from a two-liter soft-drink bottle as shown in **Fig 23.57**. Properly cutting off the top of the bottle allows it to be slid over the helix reflector cup and secured with a large hose clamp. Sun-damage protection of the plastic bottle must be provided and that was done with a wrapping of aluminum foil pressure-sensitive-adhesive tape.

There are many methods for mounting this dish antenna to your elevation boom. Constructors must give consideration to the placement of the dish to reduce the wind loading and off-balance to the rotator system by this mounting. In the illustrated installation, the off-balance issue was not a major factor, but the dish was placed near the center of the elevation boom, between the pillow block bearing supports. As there is already sizeable aluminum plate for these bearings, the dish was located to "cover" part of that plate, so as to not add measurably to the existing wind-loading area of the overall assembly. A mounting bracket provided with the stock dish clamps to the end of a standard 2-inch pipe (actual measure: ϕ2.38 inch) stanchion. This bracket was turned around on the dish and clamped to the leg of a welded pipe Tee assembly, see **Fig 23.58**. Pipe reducing fittings were machined and fitted in the Tee top bar, which was cut in half for clamping over the 1½-inch pipe used for the elevation boom. Bolts were installed through drilled holes and used to clamp this assembly.

By Lilburn Smith, W5KQJ

An Affordable Az-El Positioner for Small Antennas

Build yourself an az-el antenna positioner for one third the cost of a commercial unit and make it easy to point that antenna.

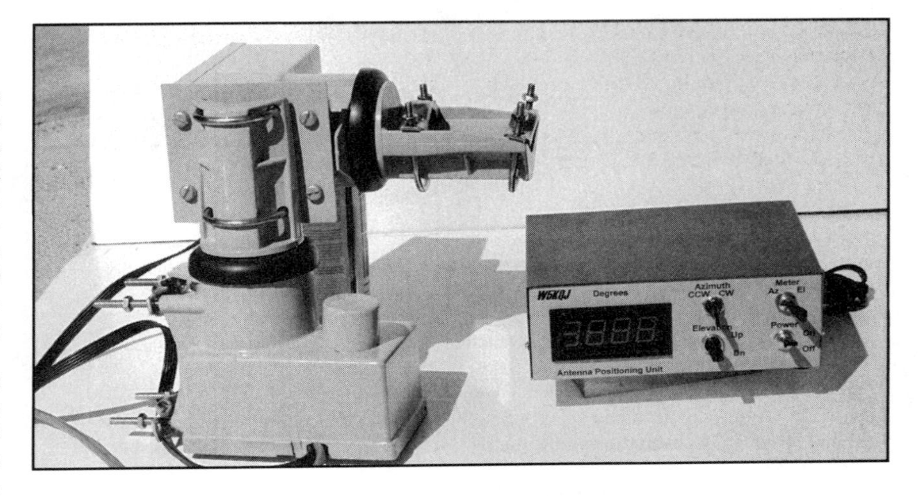

Why an Az-El Positioner?

Interest in the UHF and microwave bands is at an all-time high. This is probably due to AO-40 and other satellites, and to the increased offering of commercial gear that includes the frequencies through 1296 MHz. In addition, several firms now market transverters for amateur use up to 10 GHz. The surplus market has yielded a number of 10 GHz and 24 GHz components. More technically minded amateurs are experimenting with even higher frequencies.

All the bands above 430 MHz benefit from small, directional antennas. Although it is possible to make microwave contacts without positioning a high-gain antenna accurately, it is too much work to be fun. When you get above 10 GHz, it is almost impossible to make contacts reliably without an antenna positioner.

The AO-40 satellite has two working transmitters: S band (2.4 GHz) and K band (24 GHz). It utilizes two receivers: U (435 MHz) and L (1260 MHz). Although the AO-40 orbit is such that it does not move in azimuth and elevation rapidly, accurate aiming of the antenna is preferred. A small az-el antenna rotator that can handle a UHF or an L band transmit antenna and an S band or K band receive antenna adds to the enjoyment of working the satellite. And, for polar LEO satellites, an az-el positioner is a must.

Why Build Your Own Rotator?

An affordable azimuth antenna rotator is not difficult to find. Cornell-Dubilier Electronics (CDE) and its successors built thousands of the *AR* and *Ham* series rotators. Rugged and reliable, they are practically indestructible, as long as water does not get into the case. Parts are readily available, although expensive. Every hamfest has a few rotators and control boxes. Used, the rotator and box sell for about $100, and a new set of bearings and a new position potentiometer bring the cost up to around $150. The elevation axis is a different story. Although many good rotators were manufactured that allowed the mast to extend all the way through the rotator, they are now practically impossible to find, even used. New az-el antenna rotators are available, but expensive. For example, the Yaesu 5500 is convenient and works well, but will set you back $700.

One by one, the former rotator manufacturers ceased to manufacture, until all the inexpensive rotators now being sold new appear to be coming from the same factory. The approach discussed in this article is not new, but the rotators used in past years are no longer available. The RadioShack *Archerotor* and the Channel Master *Colorotor* are actually the same unit. Both models have eliminated the position potentiometer. Also, the rotator does not allow a mast to be run through the unit for the elevation axis. Although there are several ways around the mast problem, the position potentiometer is an absolute must. The rotators sold use a control box with a dial rotated by a motor that runs at the same speed as the rotator. The dial just rotates for the same time interval as the rotator. Synchronization is maintained by occasionally rotating the antenna and dial to their mechanical stops: first clockwise, then counterclockwise. The error builds up until resynchronization is necessary. Although tolerable as a TV antenna rotator, the position-indicating system is unsatisfactory for amateur use. Switching rotation directions rapidly while trying to peak on a satellite or distant station just confuses the indicator.

The latest RadioShack and Channel Master catalogs list only a remotely controlled rotator. Azimuth positions that correspond to TV channel numbers are stored in a microprocessor and are transmitted to the control box by an infrared link. Obviously, this scheme has no application to satellite tracking.

This article will address the mechanical mounting of the antennas in an az-el configuration and the addition of a position feedback potentiometer and a simple control box with a digital readout. Although limited in performance, the whole system can be built for $250 if all new parts are purchased. If flea market parts are used, the whole thing can be built for under $100. The rotators are readily available at hamfests and flea markets because they were so poorly suited to amateur use without the position potentiometer. I bought two brand new ones for $20 each. This az-el positioner will handle any small satellite array and the cost is reasonable.

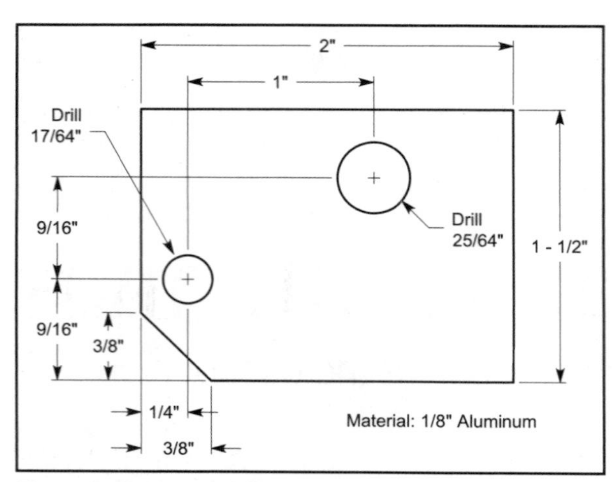

Figure 1—Potentiometer mounting plate. The position of the potentiometer mounting hole is critical for proper gear-meshing. Drill the plate accurately and mark the blank plate before drilling.

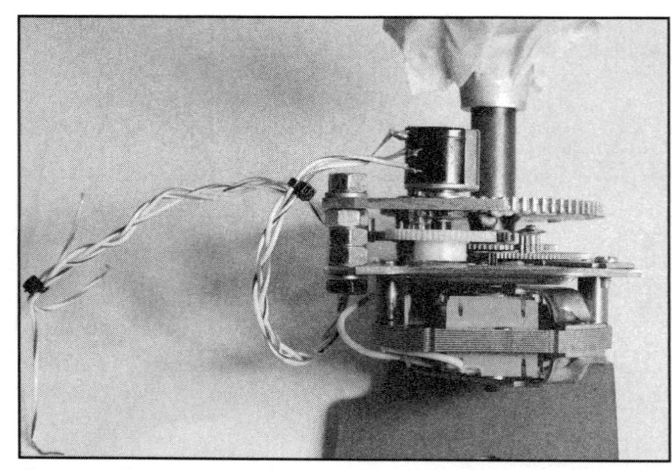

Figure 2—Potentiometer mounting details. Make sure the gears mesh properly, the potentiometer shaft turns freely and the shaft doesn't hit the potentiometer mechanical stops before the rotator shaft stops.

Modifying the Rotator for Positional Feedback

The position feedback potentiometer will be added to make the rotator useful for ham radio. Without it, it has little or no application. Doug Braun, NA1DB, worked out a simple but elegant method to add a 10-turn position pot to the rotator. The potentiometer is easy to add, but it does add about $16 to the cost. The two items to be added are a gear and a 10-turn potentiometer. The gear is available for about $7 and the potentiometer for about $9.[1]

Make a potentiometer mounting plate from scrap ⅛ inch aluminum sheet to the dimensions shown in Figure 1. Take care to locate the two holes accurately. A good source of ⅛ inch aluminum for your projects is salvaged front panels from commercial gear. These are available at hamfests for practically nothing. Cut the plate to size with a hacksaw or jigsaw and drill the mounting holes. File the edges smooth and square to dress up your work and sand the plate to deburr it and remove any old paint. See the sidebar (p 6-30) for metal-working tips.

Using the control box, rotate the rotator CW to the stop. Remove all the external mast mounting hardware from the rotator and save it for future use, including the ¼-20 mast-mounting studs. Gently remove the weather seal. The rotator is held together by three black finished hex-head bolts. One of the bolts is under a fiber insulator. Do not remove the insulator. Just bend it up slightly to allow access to the third bolt. The bolts are torqued down heavily. Use a ⅜ inch socket to remove them. Remove the gear mechanism from the housing.

Rotate the 10-turn potentiometer CW to the stop and mount the potentiometer to the plate. Place the gear on the potentiometer shaft, with the hub away from the potentiometer. Then mount the plate through the existing hole, as shown in Figure 2, using a 1½ inch ¼-20 machine bolt, with extra nuts as spacers. Align the gear to engage with the existing drive gear. Once the gear is properly engaged, torque the mounting nut down until the split lock washer is fully compressed to prevent any movement of the plate. Check the gear again.

Wire the potentiometer, and route the wires down to the terminal area so that moving parts do not trap them. A dab of clear RTV will hold them in place.

Connect the rotator to the control box and rotate the rotator counterclockwise. Observe that the potentiometer gear is rotating freely. Run the rotator into the counterclockwise stop. It will be necessary to support the rotator main shaft when the rotator is turned outside the housing. The potentiometer gear should slip on the pot shaft when the pot stop is reached. The potentiometer should indicate zero just as the clockwise actuator stop is reached. Rotate the actuator from stop to stop and ensure the 10-turn potentiometer is not hitting its stops. When the potentiometer is positioned correctly, put one drop of superglue on the pot shaft and gear hub. Be careful not to get it on the potentiometer bushing, which will lock the shaft.

Drill 1 - 1/8"

1 - 1/4"

1 - 3/8"

Figure 3—Cut a hole in the rotator housing to clear the potentiometer. This hole is difficult to measure. Drill a small hole after the location is marked to check the placement. Then punch a 1⅛ inch hole and check for clearance. Finally, epoxy a 1 inch PVC pipe cap over the potentiometer and paint.

Cut a hole in the rotator housing to clear the potentiometer, as shown in Figure 3. Note that the walls of the casting slope slightly. Measure the location of the hole from the outside of the casting with it resting on a flat surface. Use a square to project the corner into the plane of the top of the housing and then measure from the blade of the square. Otherwise the slope will throw the hole location off a little. A couple of triangles will help. If the hole is just slightly off, it is big enough to clear the potentiometer. Mark the hole and drill it with a small drill. Then make a trial assembly of the gear mechanism into the housing and check to see if your small hole is in the center of the potentiometer. If not, move it. When you get it right, drill or punch out the small marker hole to 1⅛ inch. If you do not have a way to drill or punch the hole, drill the biggest one you can and enlarge the hole to 1⅛ inch with a round file. Check, from time to time, to see that the potentiometer will be centered in the hole when it is mounted. Notice that the solder lugs will also have to clear.

Assemble the gear mechanism into the housing. Check the pot wiring to ensure it is clear of rotating parts and not crimped under something. Reinstall the black bolts and torque them down securely. Make certain the shaft is not binding on the housing and adjust the position slightly before tightening the black bolts.

Cement a 1 inch PVC pipe cap over the potentiometer, using clear epoxy cement.

[1]Notes appear on page 31.

Check the rotator operation and the potentiometer wiper resistance. It should vary from nearly zero to about 485 Ω as the rotator goes from stop to stop. If the potentiometer resistance stops changing before the rotator hits the stop, something has gone awry and should be corrected before proceeding.

When everything has been checked, paint the rotator and the PVC pipe cap. Both the PVC and the epoxy cement will deteriorate in the sun unless they are painted. Besides, a painted rotator will make the job look more professional.

Building the Control Box

The control box has two simple jobs to do:

- Power the rotator motor in the correct direction.
- Convert the potentiometer resistance to a display in degrees.

Although a few rotators were manufactured with dc motors, all the units currently in production use split-phase ac motors. A reversible split-phase motor has two identical windings. One winding is fed with 30 V ac. The other winding is fed 30 V ac shifted in phase by 90°. The result is a rotating field with both starting and running torque. To reverse direction, simply change the windings that have the respective voltages. The 90° phase shift is obtained by putting a large value capacitor in series with the winding. A DPDT center-off switch will control the motor direction.

The voltage on the wiper of the pot represents the position of the rotator. If the voltage across the pot is adjusted so that a full rotation of the rotator represents 3.60 V, a voltmeter will read directly in degrees. A small error at zero will result because the resistance of the ground wire from the rotator to the control box is in series with the pot. The error resulting from the simplified version of the circuit is only a degree or so, at worse, with a reasonable length of rotator cable. If the zero error is unacceptable, several ways exist to null it out. These will be discussed later.

The digital panel meter used in the prototype is a C+C Model PM-1029B.[2] The C+C series of meters are based on the Intersil 7107 IC chip. Built in China, the meters are simply copies of the Intersil application note circuit. They are low cost and accurate and have numerous features including range selection, automatic polarity display and independent selection of the decimal point. The unit chosen requires +5 V dc at approximately 80 mA. The LED

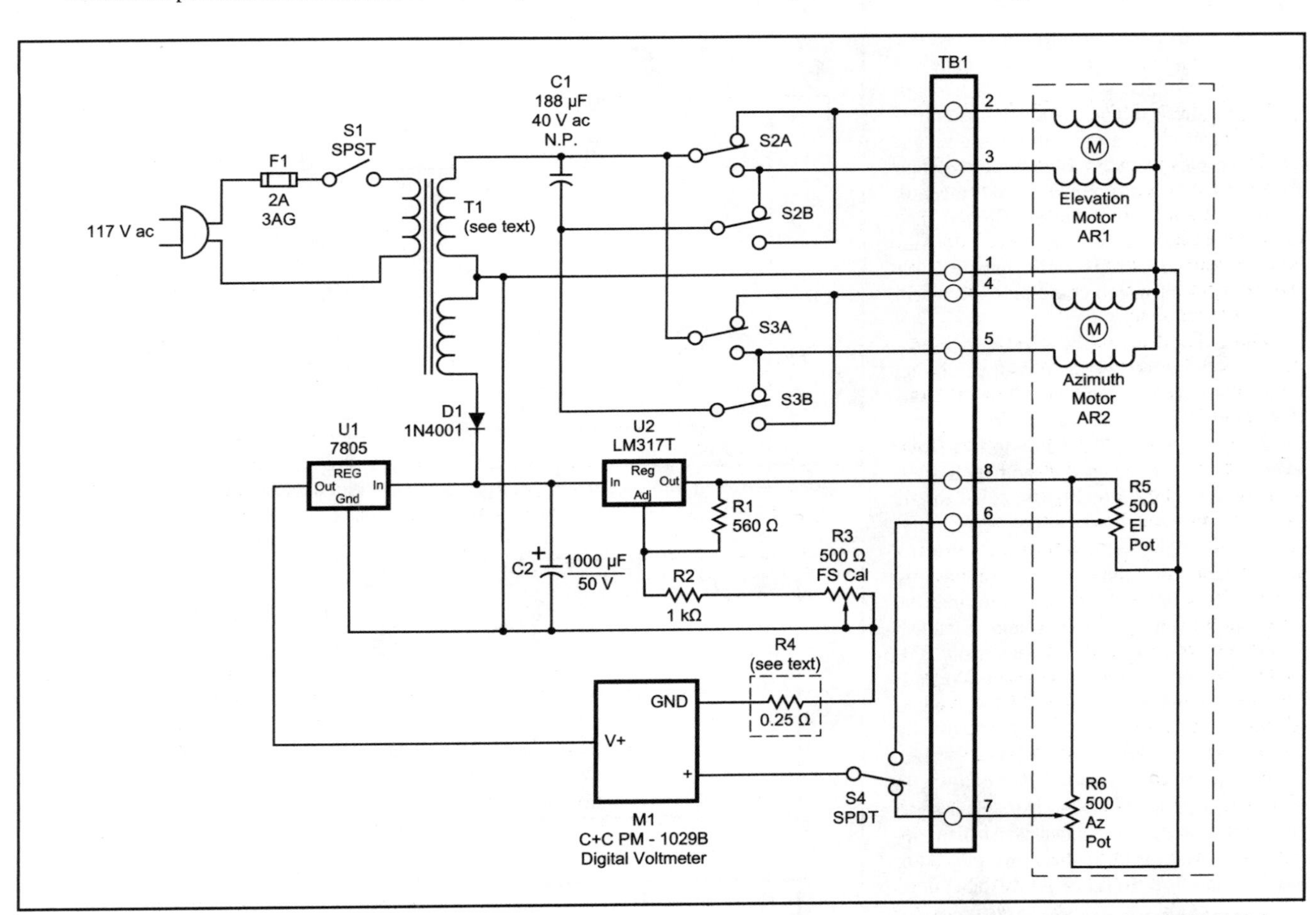

Figure 4—The control box schematic and rotator parts list. RS denotes RadioShack part numbers. Mouser Electronics, 1000 N Main St, Mansfield, TX 76063; tel 800 346-6873; **www.mouser.com/**.

AR1, AR2—Antenna rotators, RS 15-1245.
C1—188 µF, 40 V ac, non-polarized (salvaged from old rotator control box).
C2—1000 µF, 50 V, RS 272-1047.
D1—1N4001, 1 A, 50 V, RS 276-1101.
F1—Fuse holder, RS 270-364.
M1—Voltmeter, digital, C+C, PM-1029B (see text, Note 2).
R1—560 Ω, ¼ W, RS 271-1116.
R2—1000 Ω, ¼ W, RS 271-1321.
R3—500 Ω potentiometer, Mouser 31VA205.
R4—0.25 Ω (see text) #30 AWG wire on ½ W resistor.
R5, R6—500 Ω, 10 turn potentiometer, Mouser 594-53611501.
S1—Switch, SPST, toggle, RS 275-651.
S2, S3—Switch, DPDT, toggle, RS 275-709.
S4—Switch, SPDT, toggle, RS 275-652.
T1—Transformer, 117 V ac pri, 2@24 V ac, 1 A sec (salvaged from old rotator control box).
TB1, TB2—Terminal block, 8 screws, RS 274-670.
U1—Voltage regulator, 7805, RS 276-1770.
U2—Voltage regulator, LM317T, RS 276-1778.

Misc

Gears—2 @ 60 teeth, 48 pitch Delrin gears, Small Parts, Inc GD-4860 (see Note 1).
Enclosure—Project box, RS 270-274.

characters are 0.84 inch high, resulting in an easy to read display. The meter sells for about $14. The scale of the panel meter is set at 199.9 mV at the factory. The control box requires a meter with a full scale of 19.99 V. Shorting pairs of pads on the circuit board sets the scale. Carefully remove the solder on the 200 mV pads, and put a small blob of solder on the 20 V pads. Connect the +5 V power supply to the V+ pad. Connect the GND terminal to ground, and the input voltage to be measured to the V_{IN} terminal. The V– terminal and decimal point selection pads are not used.

The schematic of the control box is shown in Figure 4.

One winding of the transformer supplies the power for the motor, and the other winding is used to build a small dc power supply to excite the potentiometer circuit and supply 5 V to the meter.

The control box is built into a Radio Shack enclosure measuring approximately 3 inches tall × 8 inches wide × 6 inches deep. The enclosure is thin gauge steel, difficult to work, but producing a good-looking result. The front panel was doubled with a 1/16 inch thick piece of aluminum to make the box sturdier when the switches are flipped. The doubler is probably not necessary. It does have the advantage of providing a pattern for the front panel, which could save a misdrill in the expensive box.

The transformer, line cord and motor capacitor were all taken from the original control box. The switches and miscellaneous parts were all purchased new. The switches have large paddles and are easy to flip. However, the paddles are plastic and may not last as long as regular metal bat-handled toggle switches.

The first thing to do before any construction is started is to put two layers of masking tape on the front and rear panels to preserve the paint. Then, drill the panels as shown in Figure 5. Deburr the holes carefully. The cutout for the meter was done with a jigsaw, then dressed with a file. Modify the cutout to fit your meter.

When all the holes are cut and deburred, apply the symbolization label shown in Figure 6. The label is made from clear sheet Mylar label material obtained at an office supply store. Print the label sheet on an ink jet or laser printer. Be careful to get the label positioned perfectly with respect to the holes and free of wrinkles or bubbles. Use a hobby knife to cut away the Mylar in the holes. Mount the front panel and rear panel components first, because they will not fit into place after the transformer is mounted. Terminal strips mounted on the transformer hold most of the miscellaneous electrical parts. See Figure 7. The voltage regulators are mounted on a small angle bracket, also on the transformer.

When everything is mounted, wire the box with #22 AWG or larger stranded wire. Train the wires into cable runs to keep the interior as neat as possible. Use cable ties to secure the wiring. The prototype was wired with military surplus high temperature wire in several colors.

Mounting and Wiring the Rotators

The elevation rotator can be mounted directly to the azimuth rotator by using a small adapter plate made of 1/8 inch aluminum as shown in Figure 8. Build one piece, sand to debur it and paint.

Remove the 1/4 inch mast-mounting studs from the elevation rotator body with a pair of vise grips. Bolt the adapter plate to the rotator with 1/4-20 machine bolts

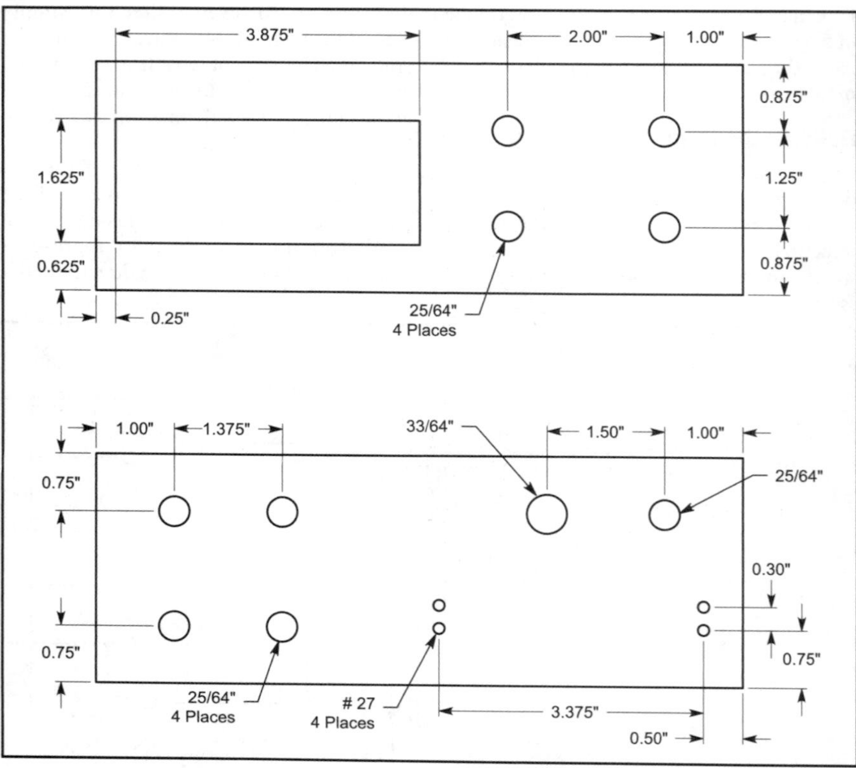

Figure 5—Drilling template for the control box front and back panel.

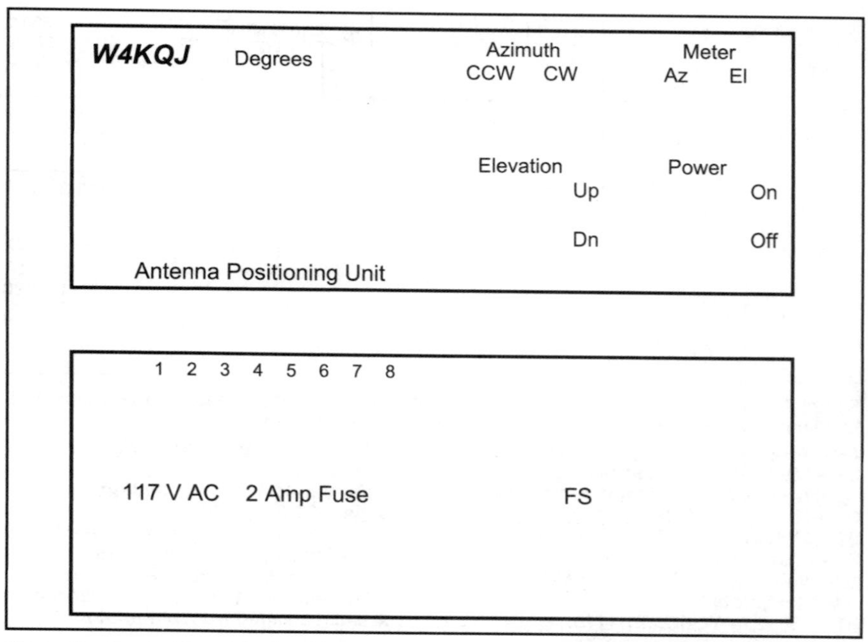

Figure 6—Control box labeling. The label can be printed on a clear sheet and then applied to the box.

½ inch long. Use split-lock washers under the bolt heads and tighten until the washers are compressed.

Cut the U-bolts on the azimuth rotator shorter by ½ inch. Set the mast brackets aside. Mount the elevation rotator on the azimuth rotator shaft by the shortened U-bolts. Start all four nuts, and then tighten the bottom nuts until the split lock washers are compressed fully. Then tighten the top nuts. There will be a slight gap from the plate to the rotator shaft because the shaft is tapered slightly. Do not tighten the nuts so tightly that the plate is deformed.

A 30 inch piece of flat, 5-wire cable was used from the elevation rotator down to the azimuth rotator. Strip ¼ inch of insulation from one end of each of the wires. One wire will be silver colored and the rest will be copper colored. Put a red crimp-on lug on the silver wire and the next two wires. Put a red wire splice on the last two wires. Connect the silver colored wire to terminal 3 in the elevation rotator and the next two wires in order to terminals 2 and 1, respectively. Splice the outside wire farthermost from the silver wire to the wire going to the top of the pot. Splice the remaining wire to the wire going to the wiper. Connect the wire from the bottom terminal of the pot to terminal 3 on the rotator. Bring the 5 wire cable from the elevation rotator through the grommet and then into the azimuth rotator to an 8 screw terminal block mounted to the lid. Put red crimp-on terminals on all five wires. Connect the cable lugs to the correct screws on the terminal block as shown in Figure 9. Connect the wires from the azimuth rotator to the remaining screws. Remember that the elevation rotator will have to rotate through 360°, and leave a slack rotator loop in the cable.

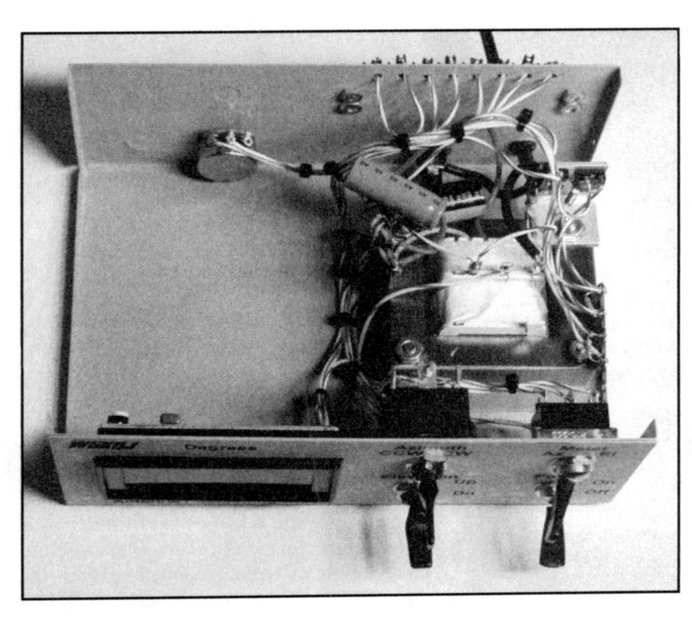

Figure 7—Control box component mounting. The components are mounted on terminal strips secured to the transformer mounting screws. Two screws mount the transformer on diagonal corners; the other corner screws secure the terminal strips.

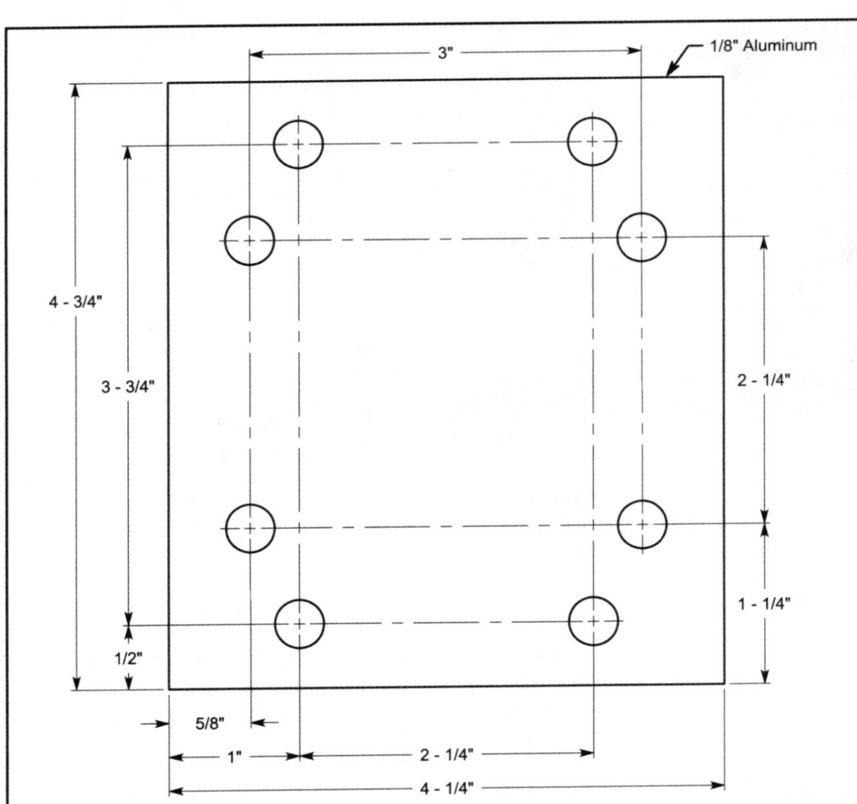

Figure 8— Adapter plate to secure the azimuth rotator to the elevation rotator.

The rotator assembly is connected to the control box by standard 8 wire rotator cable. The rotator cable will have two wires larger than the others. On the CDE rotators these wires were used for the ground and the brake solenoid. On this project, one is used for ground and one for the potentiometer voltage. Connect one of the two large wires, regardless of color, to pin 1 on the block. Connect the other large wire to pin 8. Connect the remainder of the wires in color code order to pins 2 through 7. Run the cable through the slot and replace the lid. The grommet is much too small for both cables, so leave it out and seal the wire entry with a glob of RTV sealant.

Connect the 8 wire cable from the az-el assembly to the control box. Be sure to use the same color code in wiring the box. Test the whole project before mounting the antennas up on the tower or mast.

Rotate the azimuth rotator clockwise to the stop. Do not continue to apply power after the stop is reached. Set the FS potentiometer (R3) for a 360° reading on the display. Rotate the azimuth rotator counterclockwise to the stop. Check the zero reading. If the zero reading is too high, see "That Pesky Zero Error," following. Now rotate the azimuth rotator to 180° and note that the arrows on the rotator housing and rotator shaft line up. These arrows will be used to set the azimuth axis to SOUTH with a compass after the assembly is in place on the tower or the mast. In a similar manner, check the calibration of the elevation axis. There is no independent adjustment for the elevation axis.

Antenna Mounting Options

If the antenna array is small and light, the elevation mast can be mounted directly to the elevation shaft, as shown in Figure 10. The antennas will then be mounted all on one side. If the wind loads reverse-drive the rotator, due to the lack of a brake, the

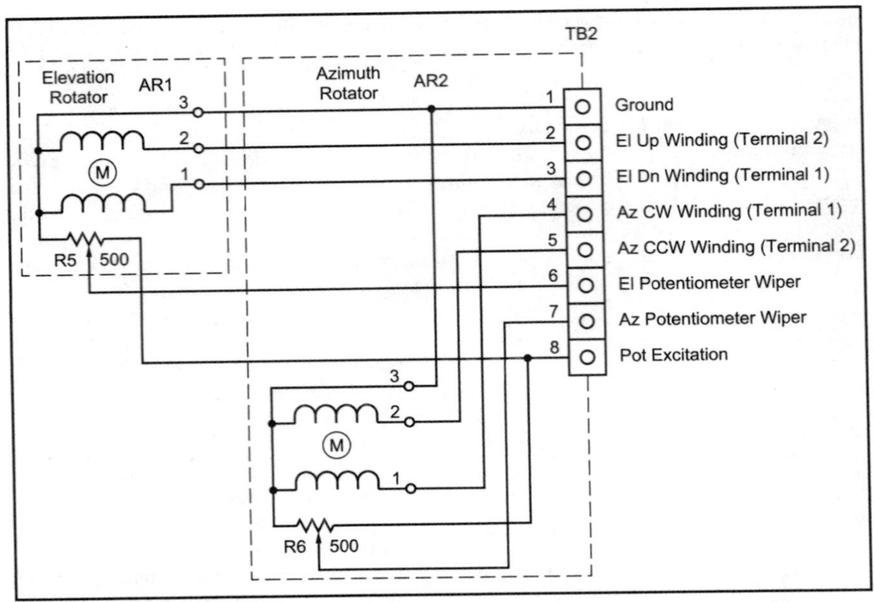

Figure 9—Rotator terminal block wiring. The elevation rotator is wired with 5-wire rotator cable, terminated at a terminal block in the azimuth rotator. The combined unit connects to the control box with 8-wire rotator cable. See Figure 4 parts list.

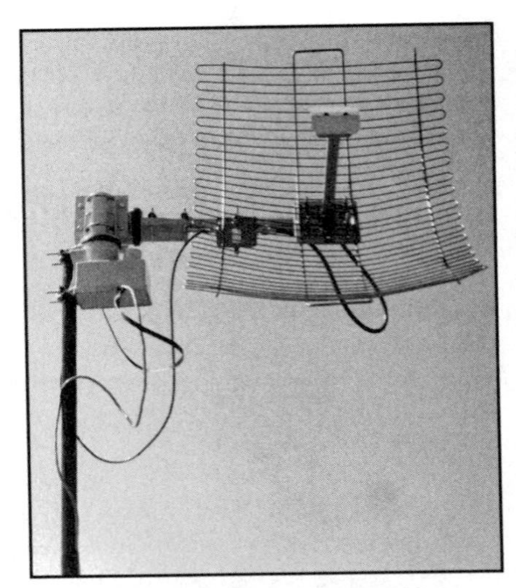

Figure 10—A single antenna mounted to one side. The rotator is ideal for positioning an S band antenna for receiving the AO-40 satellite.

Working Sheet Metal

Aluminum sheet metal is easy to work, but a few tips will aid you in turning out professional looking work. Mark the outline of the work piece with a square and a sharp scribe. If you do not have a metal shear, cut sheet metal with a saw where possible, as hand shears will bend the edges. The best setup is a radial arm saw with a metal cutting blade installed. Lacking that, a hacksaw with a fine-toothed blade can be used.

A couple of pieces of ¼ × 1 inch aluminum stock about a foot long are great tools. Sandwich the sheet metal work piece in a vise between the two tool pieces with the cut line barely exposed. The tools will make the sheet metal rigid and easy to cut. Use a fine file while the sandwich is still clamped to smooth out the saw marks. Sand the edges smooth and straight with a belt sander or a piece of sandpaper laid flat on the workbench. Be very careful to hold the work at 90°. A sloping edge is a dead giveaway of novice work.

When the work piece is sanded to the correct outline, mark the position of the holes with a square and sharp scribe. Lightly center punch the hole locations. Drill a small pilot hole. Use of a small first hole will insure better accuracy. If at all possible, use a small drill press and clamp the work to the table, using a scrap of material between the work piece and the clamp jaw. Clamp brass anytime you work with it. Even a small drill will turn brass sheet into a lethal weapon.

To deburr and finish the work, sand the flat sides of the aluminum sheet with a small power sander. Use ordinary 120 grit wood sandpaper. Finish with 400 grit wet metal sandpaper for a shiny surface. Use a small sanding block if you do not have a power sander. Do not use a larger drill to deburr holes. The marks will show. Sand the flat surface until the burr disappears.

position pot will still continue to indicate correctly and the angle can be reset.

If a larger antenna array is used, the antennas can be mounted on either side of the rotator as shown in Figure 11. Build the mast assembly shown in Figure 12. Use two 1 inch galvanized steel Ts connected with a 3 inch threaded nipple. The mast is then slid through one T and locked in place by the ¼ inch machine bolt tapped into the T. Do not use 1 inch threaded galvanized pipe for the mast. It is too heavy. Obviously, the elevation mast will hit the structure at two rotation angles, but the elevation angle covered is more than adequate. Try to keep the assembly light and as well balanced as possible. Paint the entire assembly. The galvanized pipe will rust where it was threaded.

When the antennas are all mounted, align the azimuth rotator to SOUTH with the indicator reading 180°. Check the setting of the zero and 360° positions and correct as required using the FS potentiometer. Set the elevation angle with a level so that the indicator reads correctly at 0° and at 90°.

That Pesky Zero Error

It was mentioned earlier that a small error at zero exists. This is because the resistance of the ground wire from the rotator to the control box is in series with the end resistance of the pot. The resulting error is only a degree or so, at worse, with a reasonable length of rotator cable and with the pot properly set so that the end resistance is small. However, if the error is deemed unacceptable, two methods of correction are possible. The easiest is to insert a very small resistance in series with the ground of the meter, as shown by the dashed lines in Figure 4. The resistance will be on the order of 0.25 Ω. Parallel wire 1 Ω resistors until the meter reads zero at zero rotation or use a small resistor made from #30 AWG magnet wire wound on a large value ½ W resistor body. The resistor value must be kept very low or the excellent zero stability of the voltmeter will be compromised. The resistor represents positive feedback and if too large in value, the meter will oscillate and never settle down.

A much better but more costly method is to establish a bridge circuit with the position potentiometers on one side and a zero pot on the other. The bridge is balanced at zero rotation and the meter then reads exactly zero. To implement this zero circuit, the plated through hole on pin 30 of the IC must be drilled out so that a wire can be attached to it. The plated through hole is accessible, but extreme care must be used due to the board layout. The schematic of the zero correction circuit is shown in Figure 13. I recommend living with the small error and not implementing either fix.

Growth Potential

These days everyone wants computer-

Figure 11—A balanced antenna mast. Heavier arrays should be balanced on both sides of the rotator to equalize the loading. As the elevation mast can't be run through the rotator, a pair of pipe "T"s are used.

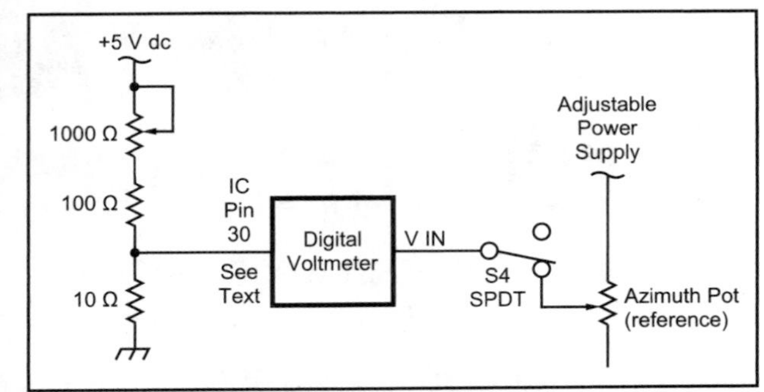

Figure 13—Control box zero error adjustment. If the small zero error is unacceptable a modification is possible. See the text.

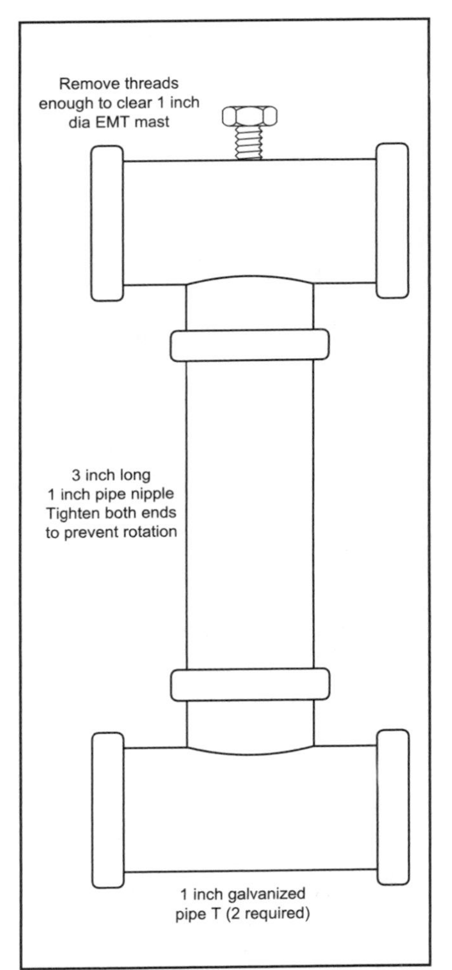

Figure 12—The pipe "T" detail. Two "T"s and a 3 inch nipple allow the mast to be balanced to equalize loading. The top "T" has a machine bolt to lock the mast.

controlled equipment. For AO-40, computer control is not necessary. The satellite apparent motion is so slow that an occasional tweak of the antenna is all that is required. For polar satellites, however, computer control would be a nice feature. To add computer control to this system, one could use any of the popular interfaces already described by others. However, adding even the simplest interface increases the project cost. The simplest, lowest cost interface is the *Fodtrack* designed by Manfred Krohmer, XQ2FOD, and distributed by *AMSAT-CE*. The printed circuit board is available from FAR Circuits.[3]

Several modern tracking programs have a *Fodtrack* driver option, including the software sold by *AMSAT-NA*.[4]

The *Fodtrack* board has four open collector outputs representing each of the directions of rotation of the rotators. Adding four small SPST relays will enable the circuit to drive the rotators. The contacts are simply paralleled across the respective switch contacts. The output of the two position potentiometers will drive the *Fodtrack* board without additional circuitry.

If that interface is added, mount the printed circuit board behind the voltmeter on spacers to the bottom of the box.

The beauty of the interface is that the meter display will continue to function. The angle displayed by the computer can be compared to the angle displayed by the meter, in order to assure that the antenna is really pointed where you think it is.

Notes

[1]The gears used in this project are available from Small Parts Inc, 13980 NW 58th Ct, Miami Lakes, FL 33014-0650; **www.smallparts.com**. Part no. GD-4860. $6.85 each plus shipping and handling.

[2]The C+C voltmeters are available from Circuit Specialists, 220 S Country Club Dr #2, Mesa, AZ 85210; **www.web-tronics.com**. Part no. PM 1029B. $15.95 plus shipping and handling.

[3]The *Fodtrack* printed circuit board is available from FAR Circuits, 18N640 Field Ct, Dundee, IL 60118-9269; **www.farcircuits.net**. Price $4.50 plus shipping and handling.

[4]*AMSAT-NA* and *AMSAT-CE* distribute *Fodtrack* software on the Web as freeware. The distribution package contains instructions for building the interface board. For more information see the extensive discussion of *Fodtrack* interfacing by Jesse Morris, W4MVB, in *The AMSAT Journal*, Mar/Apr 2002.

Lilburn R. Smith, W5KQJ, was first licensed in 1956. He holds an Extra Class license and has a BSEE degree from Texas Tech University. Lilburn has been involved in microwave, VHF and laser design and holds one US patent. He is a past president of the Central States VHF Society and the North Texas Microwave Society, and can be reached at 290 Robinson Rd, Weatherford, TX 76088; or via e-mail at **W5KQJ@arrl.net**.

By Anthony Monteiro, AA2TX

Work OSCAR 40 with Cardboard-Box Antennas!

Are you interested in working the AMSAT-OSCAR 40 satellite but intimidated by the cost and complexity of the special antennas required? Parabolic dishes, axial-mode helices, circularly polarized Yagi arrays...NOT!

Note: Although OSCAR 40 is no longer operational, these antenna designs are still quite useful for other satellites. — *Ed.*

At a recent ARRL convention, several complaints were heard at the AMSAT booth from potential new OSCAR 40 satellite operators. Their concern was that a big, up-front commitment to antennas was required just to try this communications mode. Well, don't despair, potential satellite user...there's help around the corner; if you live near a grocery store, you're in luck! These antennas for working OSCAR 40 are made primarily out of cardboard cartons and aluminum foil and they are designed to be cheap and simple to build. There are no adjustments to make; no test equipment is required and you can leave the soldering iron off. Although they are made out of cardboard, these antennas are no paper tigers when it comes to performance. They perform like the typical antennas used by veteran OSCAR 40 operators.

OSCAR 40 is the satellite formerly known as Phase 3D and it carries lots of transmitters, receivers and transponders in its payload. While not all of these are functional, the mode U/S linear transponder is operating well and it facilitates the most popular operational mode. To access it, you'll need an S band (2.4 GHz) downlink antenna and a UHF (435 MHz) uplink antenna.

Downlink Antenna

OSCAR 40 is in a highly elliptical orbit that, at apogee, has a maximum range beyond 60,000 km. In order to hear the satellite at this distance, a high gain, low-noise antenna is needed. Eighteen to 20 dB of antenna gain is required to get the signal strength high enough to hear over your downconverter's front-end noise. The antenna must also have a clean radiation pattern with minimal sidelobes, coupled with an excellent front-to-back ratio. Otherwise, it will pick up thermal noise from the warm earth as well as interfering signals from cordless telephones and wireless networks that also operate on S band.

Parabolic dish antennas are by far the most popular type used to receive OSCAR 40. They provide the needed gain while maintaining a clean radiation pattern, but are tricky to set up and very sensitive to surface irregularities.

Fortunately, there is a type of antenna that can provide the needed gain and pattern but is much more forgiving of construction tolerances—the *pyramidal horn*, shown in Figure 1. The pyramidal horn looks a lot like a megaphone. If you have ever held a megaphone to your ear, you would have noticed how much louder sounds are through it. That is because the horn shape collects a lot more sound energy than your ear would by itself. The pyramidal horn works in much the same way. It collects lots of microwave signal energy and funnels it to a small antenna, called a probe, at the back of the horn. The probe then feeds the signal to your downconverter.

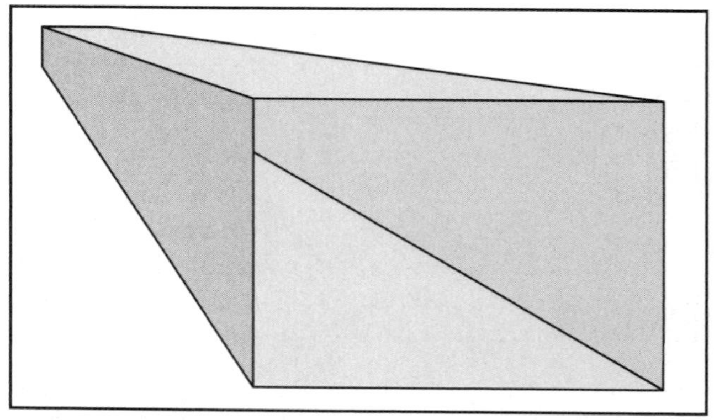

Figure 1—The basic pyramidal horn used for the downlink antenna.

Table 1
Downlink Antenna

Performance Specifications	
Operating frequency	2.4 GHz (S band)
Return loss (SWR < 1.2:1)	> 23 dB @ 2.4 GHz
Gain	20 dBi
Polarization	Vertical
Elevation beamwidth	±9° at –3 dB
Azimuth beamwidth	±8° at –3 dB

List of Materials

(Quantity—2) Shipping carton, 16 × 22 × 22 inches of ⅛ inch cardboard
Roll of 18 inch wide aluminum foil
Roll of 2 inch wide clear packing tape
⅝ × 24 nut
1⅝ inch length solid, bare #14 AWG copper wire

This OSCAR 40 downlink antenna was designed to have a gain of around 20 dBi, which is comparable to the 19-21 dBi gain of popular 2 foot dish antennas. It provides plenty of signal strength, even with the satellite at apogee. The complete technical specifications for the antenna are given in Table 1. Before starting construction on any of these antennas, it's a good idea to familiarize yourself with the tables, specifications and assembly details first—in order to understand the complete procedure.

Constructing the Downlink Antenna

As mentioned previously, the physical structure of the antenna is made of cardboard. A very inexpensive source of cardboard shipping cartons is the corner grocery store; these are generally free for the asking. The large size standard carton is around 16 × 22 × 22 inches and $^1/_8$ inch thick. This is the size you ideally want, although the exact size is not critical. Two cartons are needed. One carton is cut up to make the horn panels and is covered with aluminum foil. The other carton is used to support the horn structure. Packing tape, the type used to wrap packages for mailing, holds everything together.

In order to eliminate the need for any connectors or soldering, the downconverter is directly mounted to the horn using a ⅝ × 24 nut. The nut fits the threads of the female N-connector on the downconverter. This size nut is supplied with single-hole, chassis-mount, coaxial connectors (N or UHF types) and other types of round connectors. If your junk box doesn't have one of these nuts, they are readily available new from coaxial connector suppliers for about 50 cents each.

The only other component required is a straight length of solid, bare, #14 AWG copper wire, used to make the coaxial coupling probe inserted into the center pin of the downconverter's female N-connector. You need to cut this as carefully as possible to 1⅝ inches long. Since the diameter of #14 wire is the same as the center pin of a male N connector, it will fit snugly yet not deform or damage the connector on your downconverter. The complete list of materials is shown in Table 1.

To start construction, carefully unfold one of the cardboard cartons. If you have a carton sized differently than that specified, be sure that the resulting panel pieces have the correct dimensions. You will need all of the sides plus the top or bottom flaps of each side to make the horn panels as shown in Figure 2. Use the 22 inch wide sides to make the top and bottom horn panels and the 16 inch wide sides to make the two side panels. The top and bottom horn panels are the same, except that a ⅝ inch hole in the top panel will be used to mount the downconverter. Draw the horn panels on the cardboard as carefully as possible and then cut them out using either a utility knife or scissors. Be careful to make straight cuts and not bend or crimp the cardboard. After cutting out the horn panels, you may find it helpful to apply some tape along the edges or across the flat surfaces to strengthen them.

Apply the aluminum foil to one side of each horn panel, rolling it over the edges of the panel, and taping it snugly on the back to hold it in place. The foil needs to cover all of the edges of the cardboard panels. Do not tape over the foil on the edges of the panels; the edges of the foil must make electrical contact between panels.

The horn side panels are only 15¼ inches wide, so a single piece of 18 inch wide aluminum foil can be run lengthwise to cover the entire panel. The top and bottom panels are too wide for this, so instead, place one piece of foil across the front 20 inch edge and cover the remainder of the panel with a second piece. Leave an inch or so of overlap between the pieces. You do not need to tape them together. Carefully cut and push back the foil through the ⅝ inch hole in the top panel, again keeping tape off the edges of the hole so that good contact will be made with the connector.

After you have created the four panels, you need to tape them together. The easiest way to do this is to place all four panels on a flat surface, *foil side down*, and align the long edges together. Make sure to place the side panels between the top and bottom panels. This will then look like an unfolded, flat horn. The long edges are all 27 inches long, so all the panel lengths should match. You will be taping the three long sides that are touching. Separate the edges from each other by one cardboard thickness (±⅛ inch) using a scrap piece of cardboard for spacing and then tape the backs, *not the foil side*, of the panels together. This is important, as the long panel edges need to be one cardboard thickness from each other before folding, in order to make good electrical contact when the horn is assembled.

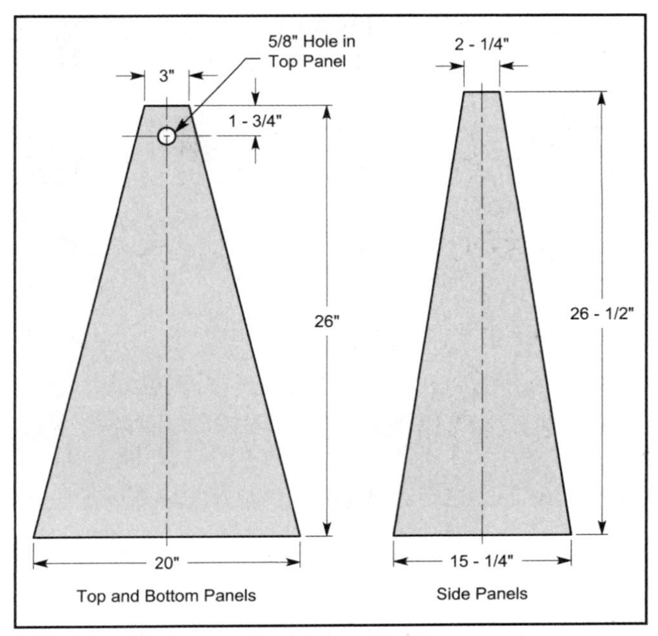

Figure 2—Dimensions for the downlink horn antenna panels.

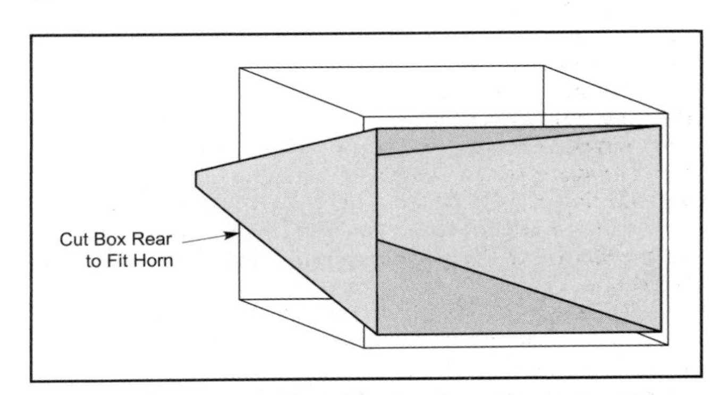

Figure 3—The final downlink antenna assembly showing how the horn fits into the supporting carton.

When the panels are taped together, carefully pick them up and fold them so the aluminum foil is on the inside. Form the pyramid shape, and tape the remaining edge tightly. When you form the pyramid, the cardboard on the edges will crush slightly, holding the aluminum foil covered edges together to make good electrical contact. After taping the fourth edge, hold the horn up to a light to see if there are any gaps between the edges. There will probably be a few. The design can tolerate linear gaps ¼ to ½ inch long without problems. Carefully use tape to pull the larger gaps together. If you have any really large gaps, patch them with foil and tape on the inside.

The remaining cardboard carton will be used to hold the horn, providing both structure and protection, as shown in Figure 3. The mouth of the horn will be centered in the supporting carton. The top and bottom panels of the horn are 15½ inches apart at the mouth and will fit nicely into the 16 inch wide opening on the carton.

The horn is about 25 inches deep from front to back, which is too long to fit inside the carton. Cut a hole in the bottom of the carton to allow the back of the horn to fit through the bottom. The

Figure 4—Front view of the downlink pyramidal horn showing how it is mounted in the support carton. Notice the coax probe at the back of the horn.

Figure 5—Side view of the downlink pyramidal horn showing the elevation control supports and how the downconverter is attached to the horn. The downconverter is pulled forward by the tape to align the probe wire parallel to the rear surface of the horn.

hole should be in the center of the bottom of the carton and will need to be about 5 inches wide by 3¾ inches high. You will need to trim the hole to fit the horn.

Align the front of the horn flush with the open end of the carton and tape the top and bottom edges of the horn to the carton opening. There will be a gap between each horn side and the carton side. Secure the back part of the horn to the hole in the back of the carton with tape.

Now it is time to mount the downconverter. Cut four 2 × 2 inch pieces of cardboard from the left over scrap pieces and cut a ⅝ inch hole in each piece. These are stacked to make a mounting shim for the downconverter. Slip the shim over the N-connector on the downconverter. Next, take the 1⅝ inch piece of #14 AWG wire and gently push it into the center hole of the N connector until it will not go in any further. It should stick out 1¼ inches from the connector, making a probe. Do not bend the wire!

Being careful not to rip the aluminum foil in the hole, push the end of the N connector through the $^5/_8$ inch hole in the top panel at the back end of the horn and secure the downconverter with the ⅝ × 24 nut. You will need to compress the cardboard shim a bit to do this. The top of the N connector should be just about flush with the top of the nut when it is secured. The springiness of the cardboard shim will hold the downconverter in the hole. The connector should not protrude very much into the horn.

The downconverter will still have a fair amount of play in the hole. Hold the downconverter so that the wire probe is parallel with the back plane of the horn and tape the downconverter in place. Finally, use a piece of aluminum foil to cover the back end of the horn and apply tape to hold it taut. Congratulations...your downlink antenna is complete! Figure 4 shows a front view of the completed pyramidal horn and Figure 5 is a side view.

Downlink Operation

The easiest way to test your new downlink antenna is to power-up the downconverter, point the antenna at the satellite and tune around for the beacon. You will need to know the position of AO-40 in order to point the antenna. There are many satellite tracking programs available, including some that are free, on AMSAT's Web site, **www.amsat.org/**; it's a good source for general information on satellites and tracking. Another useful Web site, specifically about AO-40, is **ao40.homestead.com/**.

The antenna has a beamwidth of around 17° in both azimuth and elevation, so pointing is not critical. Use a compass to set the antenna azimuth. You will need to include the magnetic declination in your area, which is the difference between compass north and true north.

For elevation control, it may be helpful to tape some additional pieces of scrap cardboard to tilt up the outer carton as can be seen in the photographs. You can also just prop up the carton with a heavy object. A protractor will be helpful for setting the elevation angle. A certain amount of trial and error will be needed to find the satellite beacon the first time but with a little experience this will be a snap. The satellite beacon is easy to hear and will be a very strong signal, even with the satellite at apogee.

After you find the beacon, you should tune around and listen for CW and SSB signals. You can get a good sense for how to operate on OSCAR 40 by monitoring contacts and nets that use the satellite.

Another activity you can try with your new antenna is decoding the satellite telemetry that is transmitted on the beacon. A convenient way to do this is by using the *ao40rcv* program (available free) at **www.qsl.net/ae4jy/ao40rcv.htm/**. This program uses a PC sound card to demodulate the signal, decode the telemetry and display the results on your PC screen. The program works well with the cardboard-box downlink antenna.

Uplink Antenna

Monitoring the satellite is entertaining, but once you listen a few times you will want to join in the fun and make your own contacts! In order to access the AO-40 satellite on the UHF uplink, an effective radiated power of 100 to 500 W is required. This depends on the mode (CW or SSB) and distance to (or range of) the satellite. Remember that at apogee the range of the satellite can be over 60,000 km. Since the UHF output of a typical transceiver is only 10-50 W, a fair amount of gain is required of the uplink antenna.

Most OSCAR 40 operators use some form of Yagi antenna to get the required gain. A Yagi is, however, hard to construct out of cardboard! Instead, we use a dipole-fed corner reflector. The basic corner reflector antenna is shown in Figure 6.

Corner reflectors are simple to build and are very forgiving of mechanical tolerance errors. The gain of the corner reflector is mostly a function of its size. For more gain, just make it bigger! This means you can get started with a small and really simple corner reflector and make it bigger if you want better performance. With this in mind, two designs are presented here: a simple version and a larger, high-performance version.

The simple version is made from a single cardboard box and has about 9 dBi gain. When fed with 50 W, it will provide a solid SSB uplink signal to OSCAR 40 out to a 30,000-km range and a solid CW signal all the way out to apogee. This is a great way to get started on OSCAR 40, as the antenna is so easy to build.

The high-performance version provides over 14 dBi gain, equivalent to an 8 element optimized Yagi, and will provide solid SSB or CW signals all the way out to apogee. It uses the same construction style, but it is bigger; it takes longer to build, and it requires two to three cardboard boxes. The complete technical specifications for both versions of the uplink antennas are given in Table 2.

Constructing the Uplink Antennas

As with the downlink antenna, you can get all the cardboard you need from grocery store cardboard shipping cartons sized about 16 × 22 × 22 inches and 1/8 inch thick, although exact sizing isn't critical.

The dipole-feed is the same for both versions and is made from ¾ inch aluminum tubing and black PVC insert couplings of the type used in lawn sprinkler systems. The coupling is used as a center insulator for the dipole; two more are used on the ends to hold the dipole in place over the reflector. The dipole presents a 50 Ω impedance when used with the corner reflector. No balun transformer or other matching network is required.

A short length of RG-58 or RG-8X type coaxial cable with a suitable connector is used to connect the dipole to your transmitter. Most UHF transmitters have an N-connector. A pre-made jumper cable can be used…just cut off one end. A minimum of 5 feet of cable is required. The list of materials for the dipole feed can be found in Table 2.

To construct the dipole-feed, cut two pieces of aluminum tubing with a pipe-cutter or a hacksaw, each 5 1/16 inches long. Push the two aluminum tubes onto one of the PVC insert couplings, leaving a ¼ inch gap in the center, as shown in Figure 7. This will require filing down the insert coupling using a grinding tool, a file, or coarse sandpaper. The tubes should fit snugly on the insert coupling. Note that the ¼ inch gap is a critical dimension, so measure this carefully.

With the tubes ¼ inch apart on the insert coupling, drill a 3/16 inch diameter hole through the top of each aluminum tube and into the insert coupling, ⅛ inch from the end of the tube. Insert a #8 sheet-metal screw and thread each hole, backing off two turns. Wrap the center conductor around one screw and the shield around

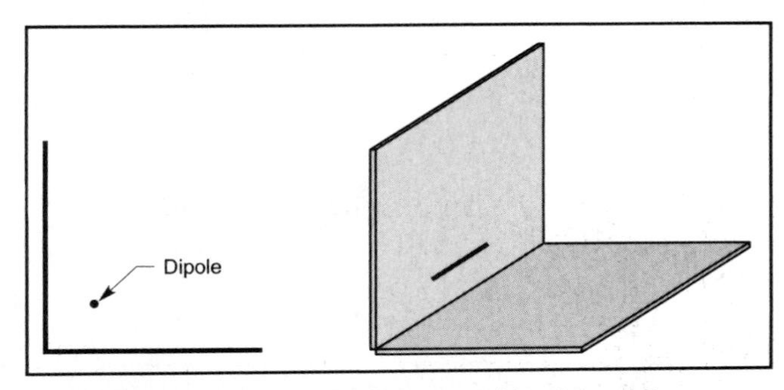

Figure 6—The basic corner reflector used for the uplink antenna. Note the dipole feed.

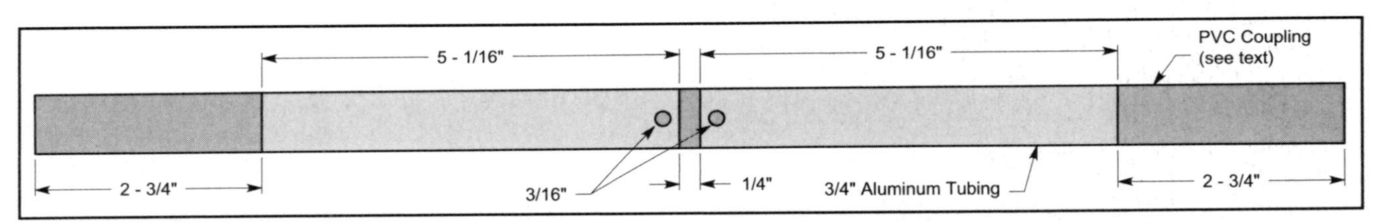

Figure 7—Dimensions for the corner reflector dipole feed.

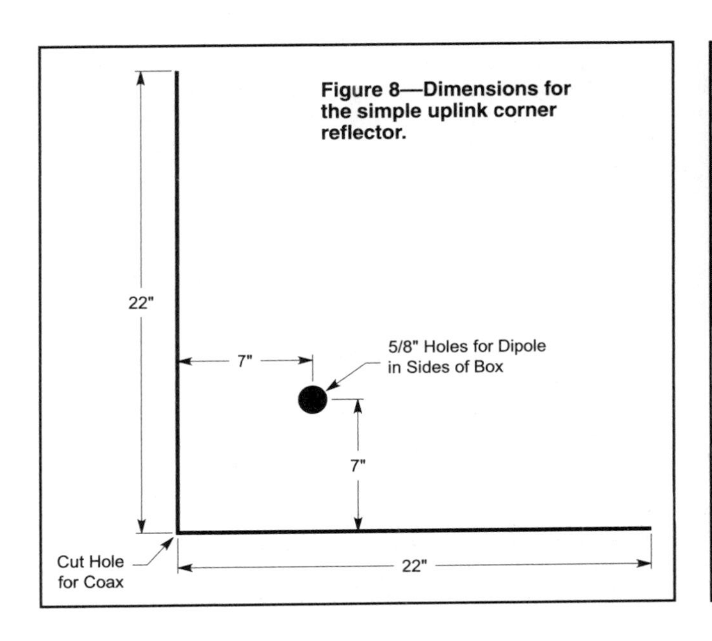

Figure 8—Dimensions for the simple uplink corner reflector.

Figure 9—The completed simple uplink corner reflector.

Table 2
Uplink Antenna Specifications and Materials

Performance Specifications, Simple Version

Operating frequency	425-465 MHz
SWR	< 1.5:1
Reflector size	22 × 22 × 16 inches
Gain	9 dBi
Polarization	Horizontal
Elevation beamwidth	±13° at –1 dB; ±23° at –3 dB
Azimuth beamwidth	±26° at 1 dB; ±43° at 3 dB

Performance Specifications, High-Performance Version

Operating frequency	425-465 MHz
SWR	< 1.5:1
Reflector size	27 × 27 × 38 inches
Gain	14 dBi
Polarization	horizontal
Elevation beamwidth	±8° at –1 dB; ±13° at –3 dB
Azimuth beamwidth	±13° at 1 dB; ±24° at 3 dB

List of Materials

(Quantity 1-3) Cardboard carton, 16 × 22 × 22 inches of ⅛ inch cardboard.
Roll of 18 inch wide aluminum foil.
Roll of 2 inch wide, clear packing tape.
12 inch length of ¾ inch aluminum tubing.
(Quantity 2) #8 × ½ inch sheet-metal screws.
(Quantity 3) ½ inch black PVC insert coupling.
5 ft length of RG-58 or RG-8X with N-connector and ½ inch leads.

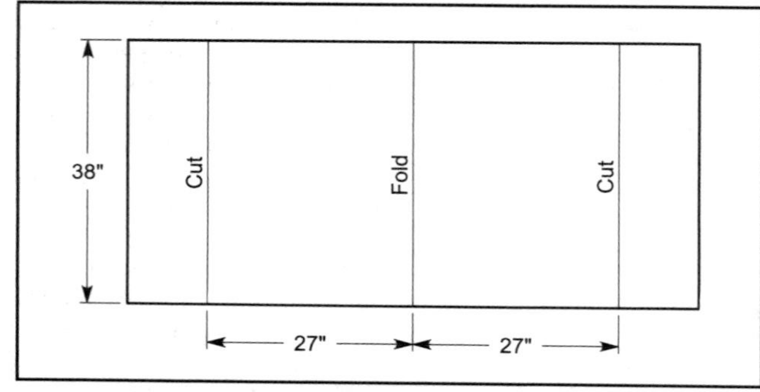

Figure 10—Cutting instructions for the high-performance corner reflector surfaces.

the other and tighten. Make sure there are no short circuits. Push an insert coupling into the other end of each aluminum tube so it sticks out about 2¾ inches and tape it in place. Set this aside while making the reflector.

Simple Corner Reflector

The simple version of the antenna is made from a single cardboard box. The bottom and narrow side of the box should each be about 16 inches wide and 22 inches long. Cover the bottom and one of the narrow sides of the box with a single piece of aluminum foil and tape in place.

To mount the dipole, cut a ⅝ inch hole on each side of the box, 7 inches from the bottom and 7 inches from the foil-covered side, as shown in Figure 8. Cut a hole in the center of the reflector for the coax. Mount the dipole by pushing the insert couplings through the holes in the sides of the box and hold it in place with tape. Push the coax connector and cable through the hole in the center of the reflector and tape the cable to the box to hold it in place. That's it! Your completed antenna should look like the photograph of Figure 9.

High-Performance Corner Reflector

The high-performance version of this antenna is just a bigger version of the simple one. The reflector sides are lengthened, from 22 inches long in the simple version, to 27 inches in this version. The reflector width is expanded from 16 to 38 inches wide. These dimensions are not critical; they were selected based on readily available cardboard box sizes.

To make the high-performance reflector, carefully unfold a 22 × 22 × 16 inch cardboard carton including the top and bottom flaps. This results in a piece of cardboard roughly 38 inches wide by 76 inches long with a fold in the center. Cut 11 inches from each end to make the piece 27 inches on each side of the fold. This will be the reflector surface shown in Figure 10. Cover it with aluminum foil and tape in place.

Flip the reflector over so the aluminum foil side is down. Using the leftover pieces of cardboard and tape, reinforce the reflector so it will hold its shape but make sure the center will still fold. Cut a hole in the center of the reflector surface for the coax connector.

Get another 22 × 22 × 16 inch cardboard box. Remove one of the narrow 22 × 16 inch sides and then cut the box in half down the long dimension, leaving two identical 8 inch wide corners. Bend the reflector surface along the centerfold, so the aluminum foil side is inside and tape it into the two corners to hold the corner reflector shape. You may need to do more reinforcing with cardboard and tape.

Finally, make a holder for the dipole out of cardboard and tape. The holder needs to position the dipole so it is held 7 inches from each of the reflector surfaces just as in the simple version. An easy way to do this is to cut out the corner from yet another box, punching holes in the sides to hold the PVC insulators and, in the bottom, for the coax connector. If you made the simple version first, just peel off the aluminum foil and cut the corner out with the dipole mounting holes already present. Mount the dipole in the holder and fasten with tape, as shown in Figure 11. Push the coax through the hole in the center of the reflector and again fasten with tape. When completed, the antenna should look like the one in Figures 11 and 12.

Uplink Operation

As shown in Table 3, the azimuth beamwidth of the uplink antennas is fairly wide, so the azimuth angle will usually not require frequent adjustment. Elevation positioning is even easier. With one of the reflector sides horizontal and the other one vertical, the antenna radiation pattern points up at about a 45° angle. To lower the radiation angle, raise the back of the antenna. Since the elevation beamwidth is quite high, it may be possible to set the antenna to an angle where no additional elevation positioning is required, depending upon your station location. For example, at my location in Massachusetts, setting the elevation angle to 20° provides good coverage for all AO-40 passes with no other elevation pointing. To set the angle to 20°, raise the back of the antenna until the bottom reflector makes a 25° angle with the ground. This can be done by adding "legs" to the back of the high-performance antenna, as seen in the photograph of Figure 11. Locations that are closer to the equator will require a higher elevation angle.

On the Air

To transmit on the OSCAR 40 uplink you will need to first find your own downlink. This is not as hard as it sounds, after you have done it a couple of times. You need to find the beacon and then tune up or down 100 kHz or so to get away from the popular frequencies and not cause interference. Use your tracking program to tell you the downlink Doppler shift. The uplink shift is roughly one-sixth that of the downlink. Set your VFO to the uplink frequency and send a couple of Morse characters. Tune around your expected downlink frequency until you find the signal. You will need to adjust your transmit power until your signal is 10 dB weaker than the beacon. If

Figure 11—The completed high-performance corner reflector. Note how the box corners hold the reflectors and dipole feed in place. The rear legs set the antenna elevation to 20°—this gives good coverage at the author's latitude.

Figure 13—The complete satellite station with the tracking laptop, FT-847 transceiver, and both downlink and uplink antennas. [Recognize that key? It's a J-38 from 1944, bridging a gap of almost 60 years!–*Ed.*]

Figure 12—A close-up of the high-performance corner reflector dipole feed.

the satellite is not all the way out at apogee, you will probably need to cut back the power quite a bit to prevent triggering LEILA (an uplink power limiting program that ensures that you are not putting an excessive signal into the satellite and thus reducing power available for other contacts). This is true for both versions of the antenna, as even the simple antenna has ample gain when the satellite range is less than 30,000 km.

For more information about working OSCAR 40, check AMSAT's Web site, **www.amsat.org/**. Steve Ford, WB8IMY, has written an excellent article on getting started with OSCAR 40.[1] It includes a list of satellite resources.

The author has used these antennas on several occasions on OSCAR 40. The complete station uses an AIDC-3731AA downconverter and a Yaesu FT-847 transceiver. A laptop computer running *Instant*Track and *Instant*Tune software is used to locate the satellite and auto-tune the FT-847.[2]

As you can see in Figure 13, the complete satellite station easily fits on a picnic table. The laptop computer and the FT-847 are in front of the downlink antenna and the simple version of the uplink antenna. The downconverter can be seen sticking up out of the back end of the downlink antenna and is connected to the FT-847 with a short piece of RG-6 coaxial cable. The uplink antenna is connected directly to the FT-847 UHF output. When the high-performance uplink antenna is used, it is set up next to the picnic table, on the ground.

Do these antennas really work? The answer is a resounding yes! With this simple arrangement, I made several dozen SSB and CW contacts through OSCAR 40. These included DX contacts when the satellite was over the Atlantic. Solid signals can be heard on the downlink antenna and good signal reports were received with both versions of the uplink antenna. Many ragchew sessions ensued when the author mentioned that his antennas were made of cardboard boxes covered with aluminum foil. I'll see you on the bird!

Notes

[1] S. Ford, WB8IMY, "OSCAR 40 on Mode U/S—No Excuses!," *QST*, Sep 2001, pp 38-41.

[2] *Instant*Track and *Instant*Tune are available from AMSAT at **www.amsat.org**.

A version of this article originally appeared in the *Proceedings of the AMSAT-NA 20th Space Symposium, 2002*.

Tony Monteiro, AA2TX, has been a ham since 1973 and is a member of AMSAT, TAPR, ARRL and QCWA. Interested primarily in the technical aspects of Amateur Radio, he can often be found on the satellites. Tony was a member of the technical staff at Bell Laboratories and has held senior management positions at Cisco Systems and several high-tech start-ups. He can be reached at 25 Carriage Chase, North Andover, MA 01845; **aa2tx@amsat.org**.

By L. B. Cebik, W4RNL (SK)

A Simple Fixed Antenna for VHF/UHF Satellite Work

Explore the low-Earth orbiting amateur satellites with this effective antenna system.

When we are just getting interested in amateur satellite operation, the thought of investing in a complex azimuth-elevation rotator system to track satellites across the sky can stop us in our tracks. For starters, we need a simple, reliable, fixed antenna—or set of antennas—to see if we really want to pursue this aspect of Amateur Radio to its limit. We'll look at the basics of fixed antenna satellite work and develop a simple antenna system suited for the home workshop. There will be versions for both 145 and 435 MHz.

Turnstiles and Satellites

For more than decades, many fixed-position satellite antennas for VHF and UHF have used a version of the turnstile. The word "turnstile" actually refers to two different ideas. One is a particular antenna: two crossed dipoles fed 90° out of phase. The other is the principle of obtaining omnidirectional patterns by phasing almost any crossed antennas 90° out of phase. The first idea limits us to a single antenna. The second idea opens the door to adapting many possible antennas to omnidirectional work.

Figure 1 shows one general method of obtaining the 90° phase shift that we need for omnidirectional patterns. Note that the coax center conductor connects to only one of the two crossed elements. A ¼-λ section of transmission line that has the same characteristic impedance as the natural feed point impedance of the first antenna element alone connects one element to the next. The opposing ends of the two elements go to the braid at each end of the transmission line. If the elements happen to be dipoles, then a 70 to 75-Ω transmission line is ideal for the phasing line. However, the resulting impedance at the overall antenna feed point will be exactly half the impedance of one element alone. So we will obtain an impedance of about 35 Ω. For the dipole-based turnstile antenna, we'll either have to accept an SWR of about 1.4:1 or we'll have to use a matching section to bring the antenna to 50 Ω. A parallel set of RG-63 ¼-λ lines will yield about 43 Ω impedance, about right to bring the 35-Ω antenna impedance to 50 Ω for the main coax feed line. For all such systems, we must remember to account for the velocity factor of the transmission line, which will yield a line length that is shorter than a true quarter wavelength.

The dipole-based turnstile is popular for fixed-position satellite work. Figure 2 shows—on the left—one recommended system that has been in *The ARRL Antenna Book* since the 1970s. For 2 meters, a standard dipole-turnstile sits over a large screen that simulates ground. Spacing the elements from the screen by between ¼ and ⅜ of a wavelength is recommended for the best pattern. For satellite operation, the object is to obtain as close to a dome-like pattern overhead as possible. The most desirable condition is to have the dome extend as far down toward the horizon as possible to let us communicate with satellites as long as possible during a pass.

The turnstile-and-screen system, while simple, is fairly bulky and prone to wind damage. However, the turnstile loses performance if we omit the screen. One way to reduce the bulk of our antenna is to find an antenna with its own reflector. However, it must have a good pattern for the desired goal of a transmitting and receiving dome in the sky. The dual Moxon rectangle array, shown in outline form on the right of Figure 2, offers some advantages over the traditional turnstile. First, it yields a somewhat better dome-like pattern. Second, it is relatively easy to build and compact to install.

Almost every fixed satellite antenna shows deep nulls at lower angles, and the number of nulls increases as we raise the antenna too high, thus defeating the desire for communications when satellites are at low angles. Figure 3 shows the elevation patterns of a turnstile-and-screen and of a

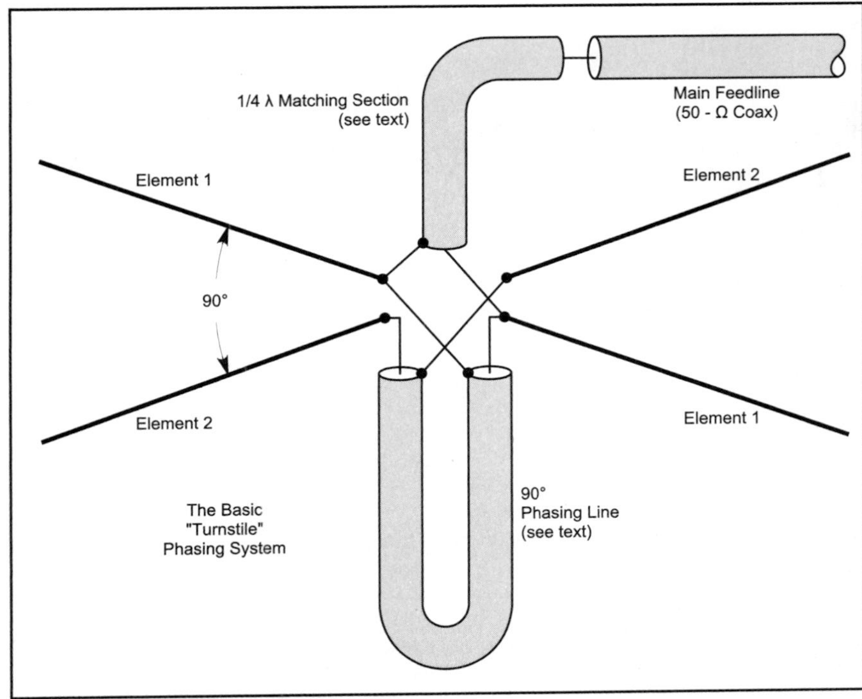

Figure 1—The basic turnstile phasing (and matching) system for any antenna set requiring a 90° phase shift between driven elements in proximity.

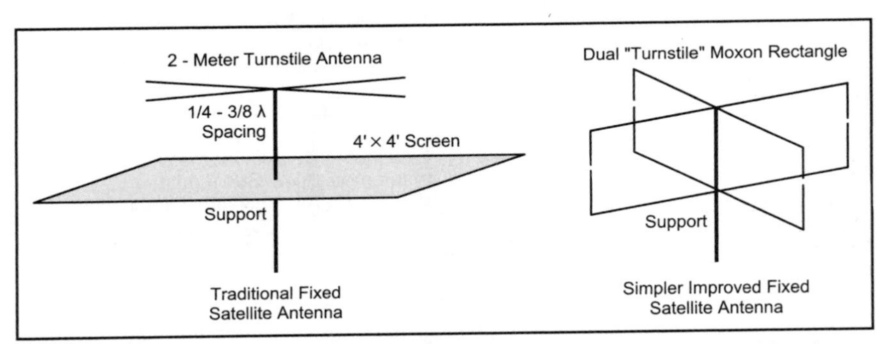

Figure 2—Alternative schemes for fixed-position satellite antennas: the traditional turnstile-and-screen and a pair of "turnstiled" Moxon rectangles.

pair of Moxon rectangles when both are 2λ above the ground. A 1λ height will reduce the low angle ripples even more, if that height is feasible. However, the builder always has to balance the effects of height on the pattern against the effects of ground clutter that may block the horizon.

The elevation patterns show the considerably smoother pattern dome of the Moxon pair over the traditional turnstile. The middle of the turnstile dome has nearly 2 dB less gain than its peaks, while the top valleys are nearly 3 dB lower than the peaks. The peaks and valleys can make the difference between successful communications and broken-up transmissions. So, for the purpose of obtaining a good dome, the Moxon pair may be superior.

A reasonable suggestion offered to me was simply to add reflectors to a standard dipole turnstile and possibly obtain the same freedom from a grid or screen structure. Figure 4 shows the limitation of that solution. The result of placing reflectors behind the dipole turnstile is a pair of crossed 2-element Yagi beams fed 90° out of phase. The pattern is indeed circular and stronger than that of the Moxon pair. However, the beamwidth is reduced to only 56° at the half-power points. The antenna would make an excellent starter for a tracking AZ-EL rotator system, but it does not have the beamwidth for good fixed-position service.

The Moxon pair, with lower but smoother gain across the sky dome, offers the fixed-antenna user the chance to build a successful beginning satellite antenna. The pattern will be circular within under a 0.2-dB difference for 145.5 to 146.5 MHz, and within 0.5 dB for the entire 2-meter band. Since satellite work is concentrated in the 145.8 to 146.0 MHz region, the broadbanded antenna will prove fairly easy to build with success. A 435.6 MHz version, designed to cover the 435 to 436.2 MHz region of satellite activity will have an even larger bandwidth.

Like the dipole-based turnstile, the Moxons will be fed 90° out of phase with a ¼-λ phasing line of 50-Ω coaxial cable. The drivers will be connected just as shown in Figure 1. Since the natural feed-point impedance of a single Moxon rectangle of the design used here is 50 Ω, the pair will show a 25-Ω feed-point impedance. Paralleled ¼-λ sections of 70- to 75-Ω coaxial cable will transform the low impedance to a good match for the main 50-Ω coaxial line to the rig. In short, we have "turnstiled" the Moxon rectangles into a reasonable fixed-position satellite antenna.

Building the Moxon Pairs

The Moxon rectangle is a modification of the reflector-driver Yagi parasitic beam. However, instead of using linear elements, the driver and reflector are bent back toward each other. The coupling between the ends of the elements combined with the coupling between parallel sections of the elements combine to produce a pattern with a broad beamwidth. By carefully selecting the dimensions, we can obtain both good performance (meaning adequate gain and an excellent front-to-back ratio) and a 50-Ω feed point impedance.[1]

In fact, a single Moxon rectangle might be used on each band for reasonably adequate satellite service. When pointed straight up, the Moxon rectangle pattern is a very broad oval, although not a circle. The oval pattern also gives the Moxon another advantage over dipoles in a turnstile configuration. If the phasing-line between dipoles is not accurately cut, the normal turnstile near-circle pattern degrades into an oval fairly quickly because the initial single dipole pattern is a figure **8**. The single Moxon oval pattern allows both dimensional inaccuracies and phasing-line inaccuracies of considerable amounts before degrading from a nearly perfect circle.

Figure 5 shows the critical dimensions for a Moxon rectangle. The lettered references are keys to the dimensions in Table 1. The design frequencies for the two satellite antenna pairs are 145.9 MHz

[1]See "Having a Field Day with the Moxon Rectangle," *QST*, June, 2000, pp 38-42, for further details on the operation of the Moxon rectangle, along with the references in the notes to that article. Also included in the notes is the source for a program to calculate the dimensions for a 50-Ω Moxon rectangle for any HF or VHF frequency using only the design frequency and the element diameter as inputs.

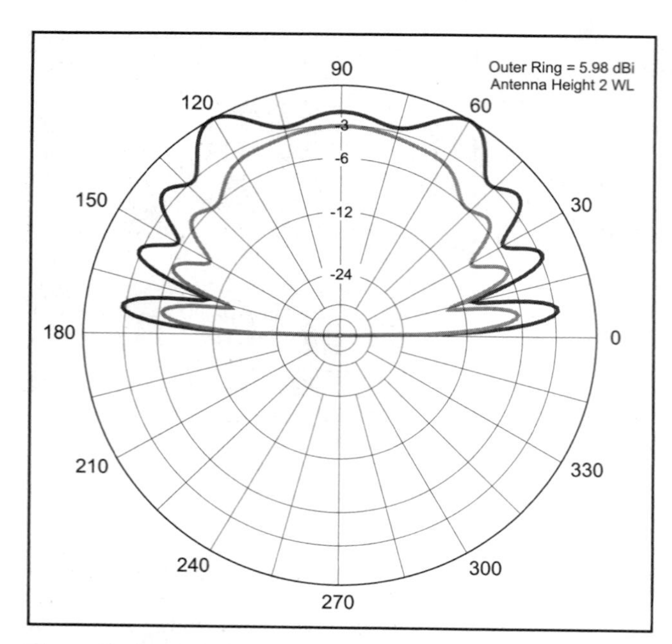

Figure 3—A comparison of elevation patterns for the turnstile-and-screen system (with 3/8λ wavelength spacing, shown in blue) and a Moxon pair (shown in red), both at 2λ height.

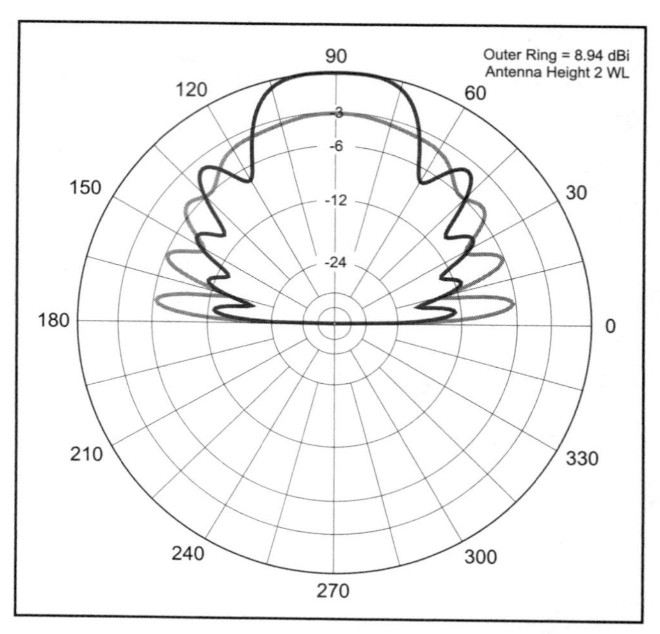

Figure 4—A comparison of elevation patterns for 2-element turnstiles (crossed 2-element Yagis, shown in blue) and a Moxon pair (shown in red), both at 2λ height.

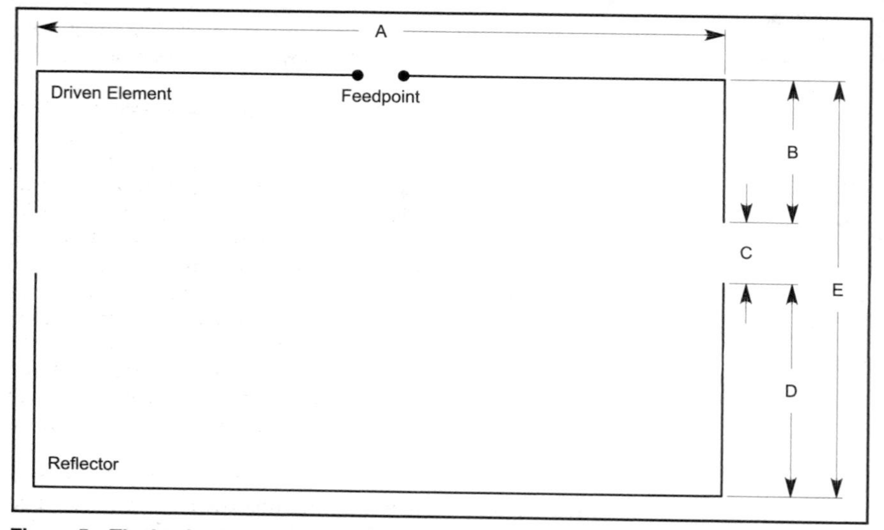

Figure 5—The basic dimensions of a Moxon rectangle. Two identical rectangles are required for each "turnstiled" pair.

Table 1

Dimensions for Moxon Rectangles for Satellite Use

Two are required for each antenna. The phase-line is 50-Ω coaxial cable and the matching line is parallel sections of 75-Ω coaxial cable. Low power cables less than 0.15 inches in outer diameter were used in the prototypes. See Figure 5 for letter references. All dimensions are in inches.

Dimension	*145.9 MHz*	*435.6 MHz*
A	29.05	9.72
B	3.81	1.25
C	1.40	0.49
D	5.59	1.88
E (B + C + D)	10.80	3.62
¼ wavelength	20.22	6.77
0.66 velocity factor phasing and matching lines	13.35	4.47

and 435.5 MHz, the centers of the satellite activity on these two bands. The 2-meter Moxon prototype use 3/16-inch diameter rod, while the 435 MHz version uses #12 AWG wire with a nominal 0.0808-inch diameter. (Single Moxons built to these dimensions would cover all of 2-meters and about 12 MHz of the 432 MHz band.) Going one small step up or down in element diameter will still produce a usable antenna, but major diameter changes will require that the dimensions be recalculated.

The reflectors are constructed from a single piece of wire or rod. I use a small tubing bender to create the corners. The rounding of the corners creates a slight excess of wire for the overall dimensions in the table. I normally arrange the curve so that the excess is split between the side-to-side dimension (A) and the reflector tail (D). Practicing on some scrap house wire may make the task go well the first time with the actual aluminum rod. The total reflector length should be A + (2 × D).

The driver consists of two pieces, since we'll split the element at its center for the feeding and phasing system. I usually make the pieces a bit longer before bending and trim them to size afterwards. The total length of the driver, including the open area for connections, should be A + (2 × B).

Perhaps the most critical dimension is the gap, C. I have found nylon tubing, available at hardware depots, to be very good to keep the rod ends aligned and correctly spaced. When everything has been tested and found correct, a little super-glue on the tubing ends and aluminum stands up to a lot of wind. I usually nick the aluminum just a little to let the glue settle in and lock the junction. For the UHF version, a short length of heat-shrink tubing provides a lock for the size of the gap and the alignment of the element tails.

A close-up view of the 145.9 MHz rectangle pair.

It is one thing to make a single Moxon and another to make a working crossed pair. Figure 6 shows the general scheme that I used for the prototypes, using CPVC. (Standard schedule 40 or thinner PVC or fiberglass tubing can also be used.) The support stock is $^3/_4$ inch nominal. The reflectors go into slots at the bottom of the tube and are locked in two ways. Whether or not the two reflectors make contact at their center points makes no difference to performance, so I ran a very small sheet screw through both 2-meter reflectors to keep their relative positions firm. I soldered the centers of the 435-MHz reflectors. Then I added a coupling to the bottom of the CPVC to support the double reflector assembly and to connect the boom to a support mast. Cementing or pressure fitting the cap is a user option.

The feed point assemblies are attached to solder lugs. The phasing line is routed down one side of the support, while the matching section line is run down the other. Electrical tape holds them in place. For worse weather, the tape may be over-sealed with butylate or other coatings. Likewise, the exposed ends of the coax sections and the contacts themselves should be sealed from the weather. The details can be seen—as built for the experimental prototypes in one of the photos—before sealing, since lumps of butylate or other coatings tend to obscure interesting details.

The overall assembly of the two antennas appears in the second photograph. The PVC from the support **T**s can go to a center Tee that also holds the main support for the two antennas. A series of adapters, made from miscellaneous PVC parts to fit over a standard length of TV mast. Alternatively, the antennas can be separately mounted about 10 feet apart. The 10-foot height of the assembly has proven adequate for general satellite reception, although I live almost at the peak of a hill.

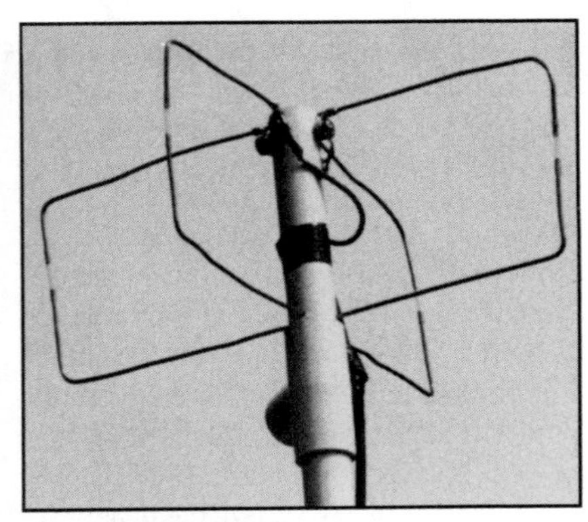

The 435-MHz Moxons.

The antennas can be mounted on the same mast. However, for similar sky-dome patterns, they should each be the same number of wavelengths above ground. For example, if the 2-meter antenna is about two wavelengths up at about 14 feet or so, then the bottom of the 435-MHz antenna should be only about 4.5 feet above the ground. Placing the higher-frequency antenna below the 2-meter assembly will create some small irregularities in the desired dome pattern, but not serious enough to affect general operation.

There is no useful adjustment to these antennas except for making the gap between the drivers and reflectors as accurate as possible. Turnstile antennas show a very broad SWR curve. Across 2 meters, for example, the highest SWR is under 1.1:1. However, serious errors in the phasing line length can result in distortions to the desired circular pattern. There is no substitute for checking the lengths of the phasing line and the matching section several times before cutting. The correct length is from one junction to the next, including the portions of exposed cable interior.

These two little antennas will not compete with tracking AZ-EL rotating systems for horizon-to-horizon satellite activity. For satellite work, however, power is not always the problem (except for using too much) and modern receiver front-ends have enough sensitivity to make communication easy. So when the satellite reaches an angle of about 30° above the horizon, these antennas will give a very reasonable account of themselves. When you become so addicted to satellite communication that you invest in the complete tracking system, these antennas can be used as back-ups while parts of the complex system are down for maintenance!

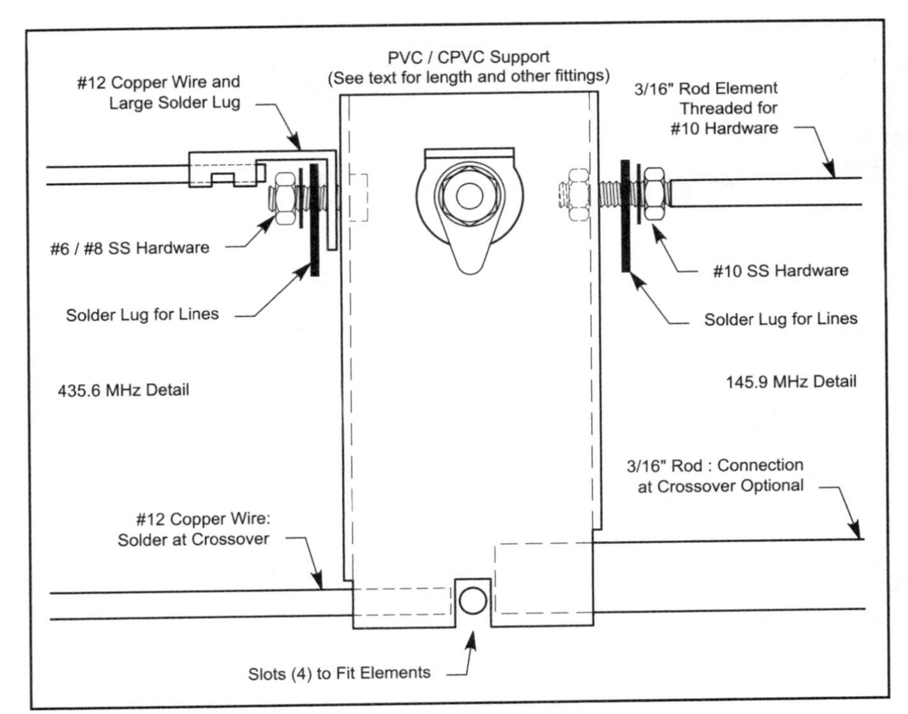

Figure 6—Some construction details for the Moxon pairs constructed as prototypes.

From the *ARRL Antenna Book*

Portable Helix for 435 MHz

Helical antennas for 435 MHz are excellent uplinks for U-band satellite communications. The true circular polarization afforded by the helix minimizes signal *spin fading* that is so predominant in these applications. The antenna shown in **Fig 56** fills the need for an effective portable uplink antenna for OSCAR operation. Speedy assembly and disassembly and light weight are among the benefits of this array. This antenna was designed by Jim McKim, WØCY.

As mentioned previously, the helix is about the most tolerant of any antenna in terms of dimensions. The dimensions given here should be followed as closely as possible, however. Most of the materials specified are available in any well supplied do-it-yourself hardware or building supply store. The materials required to construct the portable helix are listed in **Table 1**.

The portable helix consists of eight turns of ¼-inch soft-copper tubing spaced around a 1-inch fiberglass tube or maple dowel rod 4 feet, 7 inches long. Surplus aluminum jacket Hardline can be used instead of the copper tubing if necessary. The turns of the helix are supported by 5-inch lengths of ¼-inch maple dowel mounted through the 1-inch rod in the center of the antenna. **Fig 57A** shows the overall dimensions of the antenna. Each

Table 1
Parts List for the Portable 435-MHz Helix

Qty	*Item*
1	Type N female chassis mount connector
18 feet	¼-in. soft copper tubing
4 feet	1-inch ID galvanized steel pipe
1	5 feet × 1-inch fiberglass tube or maple dowel
14	5-inch pieces of ¼-inch maple dowel (6 feet total)
1	⅛-inch aluminum plate, 10 inches diameter
3	2 × ¾-inch steel angle brackets
1	30 × 30-inch (round or square) aluminum screen or hardware cloth
8 feet	½ × ½ × ½-inch aluminum channel stock or old TV antenna element stock
3	Small scraps of Teflon or polystyrene rod (spacers for first half turn of helix)
1	⅛ × 5 × 5-inch aluminum plate (boom-to-mast plate)
4	1½-inch U bolts (boom-to-mast mounting)
3 feet	#22 bare copper wire (helix turns to maple spacers)

Assorted hardware for mounting connector, aluminum plate and screen, etc.

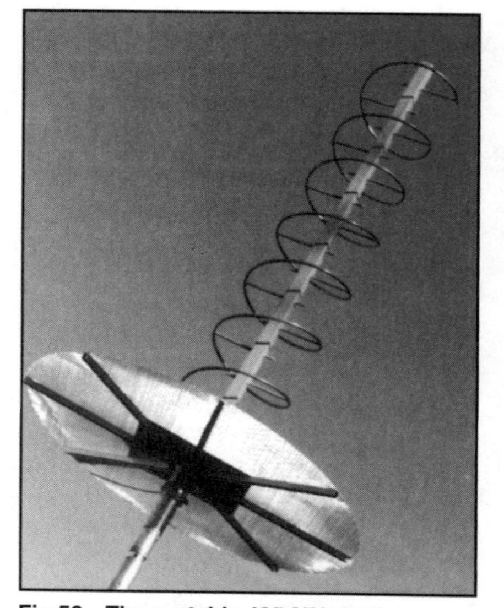

Fig 56—The portable 435-MHz helix assembled and ready for operation. (*WØCY photo.*)

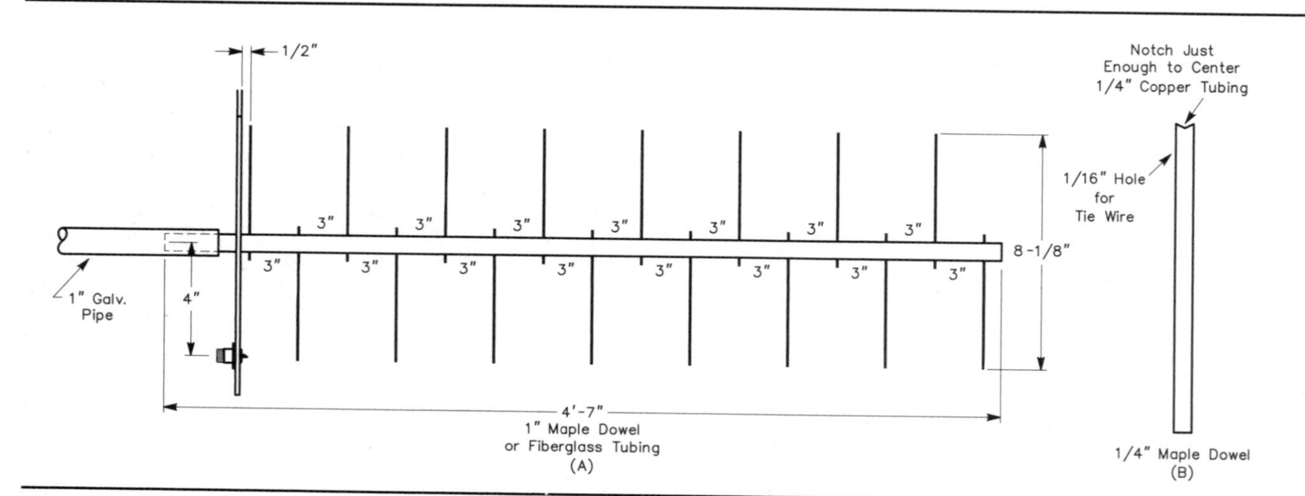

Fig 57—At A, the layout of the portable 435-MHz helix is shown. Spacing between the first 5-inch winding-support dowel and the ground plane is ½ inch; all other dowels are spaced 3 inches apart. At B, the detail of notching the winding-support dowels to accept the tubing is shown. As indicated, drill a ⅟₁₆-inch hole below the notch for a piece of small wire to hold the tubing in place.

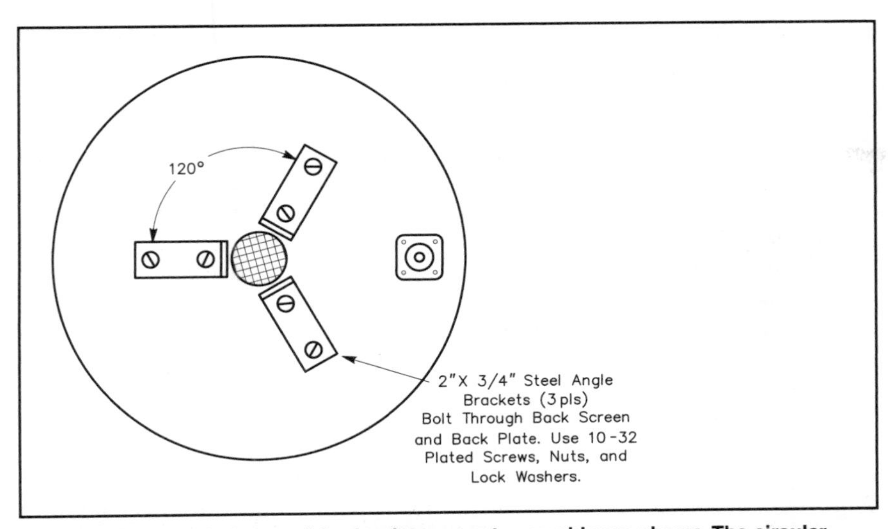

Fig 58—The ground plane and feed-point support assembly are shown. The circular piece is a 10-inch diameter, ⅛-inch thick piece of aluminum sheet. (A square plate may be used instead.) Three 2 × ¾-inch angle brackets are bolted through this plate to the backside of the reflector screen to support the screen on the pipe. The type-N female chassis connector is mounted in the plate 4 inches from the 1-inch diameter center hole.

of these support dowels has a V-shaped notch in the end to locate the tubing, as shown in Fig 57B.

The rod in the center of the antenna terminates at the feed-point end in a 4-foot piece of 1-inch ID galvanized steel pipe. The pipe serves as a counterweight for the heavier end of the antenna. The 1-inch rod material inside the helix must be nonconductive. Near the point where the nonconductive rod and the steel pipe are joined, a piece of aluminum screen or hardware cloth is used as a reflector screen.

If you have trouble locating the ¼-inch soft copper tubing, try a refrigeration supply house. The perforated aluminum screening can be cut easily with tin snips. This material is usually supplied in 30 × 30-inch sheets, making this size convenient for a reflector screen. Galvanized ¼-inch hardware cloth or copper screen could also be used for the screen, but aluminum is easier to work with and is lighter.

A ⅛-inch-thick aluminum sheet is used as the support plate for the helix and the reflector screen. Surplus rack panels provide a good source of this material. **Fig 58** shows the layout of this plate.

Fig 59 shows how aluminum channel stock is used to support the reflector screen. (Aluminum tubing also works well for this. Discarded TV antennas provide plenty of this material if the channel stock is not available.) The screen is mounted on the bottom of the 10-inch aluminum center plate. The center plate, reflector screen and channel stock are connected together with plated hardware or pop rivets. This support structure is very sturdy. Fiberglass tubing is the best choice for the center rod material although maple dowel can be used.

Mount the type-N connector on the bottom of the center plate with appropriate hardware. The center pin should be exposed enough to allow a flattened end of the copper tubing to be soldered to it. Tin the end of the tubing after it is flattened so that no moisture can enter it. If the helix is to be removable from the ground-plane screen, do not solder the copper tubing to the connector. Instead, prepare a small block of brass, drilled and tapped at one side for a 6-32 screw. Drill another hole in the brass block to accept the center pin of the type-N connector, and solder this connection. Now the connection to the copper tubing helix can be made in the field with a 6-32 screw instead of with a soldering iron.

Refer to Fig 57A. Drill the fiberglass or maple rod at the positions indicated to accept the 5-inch lengths of ½-inch dowel. (If maple doweling is used, the wood must be weatherproofed as described below before drilling.) Drill a $^{1}/_{16}$-inch hole near the notch of each 5-inch dowel to accept a piece of #22 bare copper wire. (The wire is used to keep the copper tubing in place in the notch.) Sand the ends of the 5-inch dowels so the glue will adhere properly, and epoxy them into the main support rod.

Begin winding the tubing in a clockwise direction from the reflector screen

Fig 59—At left, the method of reinforcing the reflector screen with aluminum channel stock is shown. In this version of the antenna, the three angle brackets of Fig 58 have been replaced with a surplus aluminum flange assembly. (*WØCY photo.*) At right, this helix view shows the details of a ¼-turn matching transformer, as discussed in the text. (*K9EK photo.*)

end. First drill a hole in the flattened end of the tubing to fit over the center pin of the type-N connector. Solder it to the connector, or put the screw into the brass block described earlier. Carefully proceed to bend the tubing in a circular winding from one support to the next.

See the earlier section entitled "50-Ω Helix Feed" and Figs 19 and 20 to see how the first half-turn of the helix tubing must be positioned close above the reflector assembly. **Fig 59B** shows also an excellent example by K9EK on matching his U-band helical antenna to a 52-Ω feed line. It is important to maintain this spacing, since extra capacitance between the tubing and ground is required for impedance-matching purposes.

Insert a piece of #22 copper wire in the hole in each support as you go. Twist the wire around the tubing and the support dowel. Solder the wire to the tubing and to itself to keep the tubing in the notches. Continue in this way until all eight turns have been wound. After winding the helix, pinch the far end of the tubing together and solder it closed.

Weatherproofing the Wood

A word about preparing the maple doweling is in order. Wood parts must be protected from the weather to ensure long service life. A good way to protect wood is to boil it in paraffin for about half an hour. Any holes to be drilled in the wooden parts should be drilled after the paraffin is applied, since epoxy does not adhere well to wood after it has been coated with paraffin. The small dowels can be boiled in a saucepan. Caution must be exercised here—the wood can be scorched if the paraffin is too hot. Paraffin is sold for canning purposes at most grocery stores. Wood parts can also be protected with three or four coats of spar varnish. Each coat must be allowed to dry fully before another coat is applied.

The fiberglass tube or wood dowel must fit snugly with the steel pipe. The dowel can be sanded or turned down to the appropriate diameter on a lathe. If fiberglass is used, it can be coupled to the pipe with a piece of wood dowel that fits snugly inside the pipe and the tubing. Epoxy the dowel splice into the pipe for a permanent connection.

Drill two holes through the pipe and dowel and bolt them together. The pipe provides a solid mount to the boom of the rotator, as well as most of the weight needed to counterbalance the antenna. More weight can be added to the pipe if the assembly is "front-heavy." (Cut off some of the pipe if the balance is off in the other direction.)

The helix has a nominal impedance of about 105 Ω in this configuration. By varying the spacing of the first half turn of tubing, a good match to 52-Ω coax should be obtainable. When the spacing has been established for the first half turn to provide a good match, add pieces of polystyrene or Teflon rod stock between the tubing and the reflector assembly to maintain the spacing. These can be held in place on the reflector assembly with silicone sealant. Be sure to seal the type-N connector with the same material.

Appendix A
Satellite Orbits

By Dr Martin Davidoff, K2UBC

The satellite-orbit problem (determining the position of a satellite as a function of time and finding its path in space) is essentially the same whether we are studying the motion of the planets around the Sun, the Moon around the Earth, or artificial satellites revolving around either. The similarity arises from the nature of the forces affecting an orbiting body that doesn't have a propulsion system. In the early 17th century Kepler discovered some remarkable properties of planetary motion; they have come to be called *Kepler's Laws*.

I) Each planet moves around the Sun in an ellipse, with the Sun at one focus (motion lies in a plane);

II) The line from the Sun to planet (radius vector, r) sweeps across equal areas in equal intervals of time:

III) The ratio of the square of the period (T) to the cube of the semi-major axis (a) is the same for all planets in our solar system. (T^2/a^3) is constant.

These three properties summarize observations; they say nothing about the forces governing planetary motion. It remained for Newton to deduce the characteristics of the force that would yield Kepler's Laws. The force is the same one that keeps us glued to the surface of the Earth — good old gravity.

Newton showed that Kepler's Second Law would result if the planets were being acted on by an attractive force always directed at a fixed central point: the Sun (central force). To satisfy the First Law, this force would have to vary as the inverse square of the distance between planet and Sun ($1/r^2$). Finally, if Kepler's Third Law were to hold, the force would have to be proportional to the mass of the planet. Actually, Newton went a lot further. He assumed that not only does the Sun attract the planets in this manner, but that every mass (m1) attracts every other mass (m2) with a force directed along the line joining the two masses and having a magnitude (F) given by

$$\mathbf{F} = \frac{\mathbf{G m_1 m_2}}{\mathbf{r^2}} \qquad \text{[Eq 1]}$$

where G is the Universal Gravitational Constant.

The motion of an object results from the forces acting on it. To determine the path of a satellite in space, we will (1) make a number of simplifying assumptions about the forces on the satellite and other aspects of the problem, taking care to keep the most important determinants of the motion intact; (2) solve the simplified model; and then (3) add corrections to our solution, accounting for the initial simplifications.

The Geometry Of The Ellipse

As Kepler noted in his First Law, ellipses take center stage in satellite motion. A brief mathematical look at the geometry of the ellipse is therefore in order. They're related by the formula

$$c^2 = a^2 - b^2 \text{ or } c = \sqrt{(a^2 - b^2)} \qquad \text{[Eq 2]}$$

Using Eq 2, any one of the parameters a, b or c can be computed if the other two are known. In essence, it takes two parameters to completely describe the shape of an ellipse. One could, for example, give the semi-major and semi-minor axes (a and b), the semi-major axis and the distance from the origin to one focus (a and c), or the semi-minor axis and the distance from the origin to one focus (b and c).

There's another convenient parameter, called eccentricity (e), for describing an ellipse. Eccentricity may be thought of as a number describing how closely an ellipse resembles a circle. When the eccentricity is 0, we've got a circle. The larger the eccentricity, the more elongated the ellipse becomes. To be more precise, eccentricity is given by

$$e^2 = 1 - \left(\frac{b}{a}\right)^2 \text{ or } e = \sqrt{\left(1 - \frac{b}{a}\right)^2} \qquad \text{[Eq 3]}$$

Because of its mathematical definition, e must be a dimensionless number between 0 and +1. Using Eqs 2 and 3, we can derive another useful relationship:

$$c = ae \qquad \text{[Eq 4]}$$

It always takes two parameters to describe the shape of an ellipse. Any two of the four parameters, a, b, c or e, will suffice.

Since the Earth is located at a focal point of the ellipse (Kepler's Law 1), it is convenient to introduce two additional parameters that relate to our Earth-bound vantage point: the distances between the center of the Earth and the "high" and "low" points on the orbit.

apogee distance: $r_a = a(l + e)$ [Eq 5a]

perigee distance: $r_p = a(1 - e)$ [Eq 5b]

We now have six parameters, a, b, c, e, r_a and r_p, any two of which can be used to describe an ellipse. With the information we've learned so far, many practical satellite problems can be solved.

When the major and minor axes of an ellipse are equal, the ellipse becomes a circle. From Eq 2 we see that setting a = b gives c = 0. This means that in a circle, both focal points coalesce at the center. Setting a = b in Eq 3 yields e = 0, as we stated earlier.

Since the circular orbit is just a special case of the elliptical orbit, the most general approach to the satellite orbit problem would be to begin by studying elliptical orbits. Circular orbits, however, are often simpler to work with, so we'll look at them separately whenever it makes our work easier.

Our approach to the satellite-orbit problem involves first determining the path of the satellite in space and then looking at the path the sub-satellite point traces on the surface of the Earth. Each of these steps is, in turn, broken down into several smaller steps.

Simplifying Assumptions

We begin by listing the assumptions usually employed to simplify the problem of determining satellite motion in the orbital plane.

1) The Earth is considered stationary and a coordinate system is chosen with its origin at the Earth's center of mass (geocenter).

2) The Earth and satellite are assumed to be spherically symmetric. This enables us to represent each one by a point mass concentrated at its center (M for the Earth, m for the satellite).

3) The satellite is subject to only one force, an attractive one directed at the geocenter; the magnitude of the force varies as the inverse of the square of the distance separating satellite and geocenter ($1/r^2$).

The model just outlined is known as the two-body problem, a detailed solution for which is given in most introductory physics texts. Some of the important results follow.

Solution to the Two-Body Problem

Initial Conditions. Certain initial conditions (the velocity and position of the satellite at the instant the propulsion system is turned off) produce elliptical orbits ($0 \le e < 1$). Other initial conditions produce hyperbolic ($e > 1$) or parabolic ($e = 1$) orbits, which we will not discuss.

The Circle. For a certain subset of the set of initial conditions resulting in elliptical orbits, the ellipse degenerates (simplifies) into a circle ($e = 0$).

Satellite Plane. The orbit of a satellite lies in a plane that always contains the geocenter. The orientation of this plane remains fixed in space (with respect to the fixed stars) after being determined by the initial conditions.

Period and Semi-major Axis. The period (T) of a satellite and the semi-major axis (a) of its orbit are related by the equation

$$T^2 = \frac{4\pi^2}{GM} a^3 \qquad \text{[Eq 6a]}$$

where M is the mass of the Earth and G is the Universal Gravitational Constant. For computations involving a satellite in Earth orbit, the following equations may be used (T in minutes, a in kilometers).

$$T = 165.87 \times 10^{-6} \times a^{\frac{3}{2}} \qquad \text{[Eq 6b]}$$

$$a = 331.25 \times T^{2/3} \qquad \text{[Eq 6c]}$$

Note that the period of an artificial satellite that is orbiting the Earth depends only on the semi-major axis of its orbit. For a circular orbit, a is equal to r, the constant satellite geocenter distance.

The *mean motion*, MM, is defined as the number of revolutions (perigee to perigee) completed by a satellite in a solar day (1440 minutes). A satellite's mean motion is related to its period by Eq 7. (MM in revolutions per solar day, T in minutes.)

$$\mathbf{MM = 1440/T} \qquad \text{[Eq 7]}$$

Since many sources of orbital elements provide the mean motion, it is often necessary to compute period and semi-major axis from it.

Velocity. The magnitude of a satellite's total velocity (v) generally varies along the orbit. It's given by

$$v^2 = GM\left(\frac{2}{r} - \frac{1}{a}\right) = 3.986 \times 10^{14}\left(\frac{2}{r} - \frac{1}{a}\right) \qquad \text{[Eq 8]}$$

where r is the satellite-geocenter distance, r and a are in meters, and v is in meters/sec. Note that for a given orbit, G, M and a are constants, so that v depends only on r. Eq 8 can therefore be used to compute the velocity at any point along the orbit if r is known. The range of velocities is bounded: The maximum velocity occurs at perigee and the minimum velocity occurs at apogee. The direction of motion is always tangent to the orbital ellipse. For a circular orbit r = a and Eq 8 simplifies to (r in meters, v in m/s)

$$v^2 = \frac{GM}{r} = \left(3.986 \times 10^{14}\right)\left(\frac{1}{r}\right)$$

Note that for circular orbits, v is constant.

Position. The satellite position is specified by the polar coordinates r and θ. Note that θ is measured counterclockwise from perigee. Often, it's necessary to know r and θ as a function of the elapsed time, t, since the satellite passed perigee (or some other reference point when a circle is being considered).

For a satellite in a circular orbit moving at constant speed:

$$\theta[\text{in degrees}] = \frac{t}{T}(360^\circ) \text{ or } \theta[\text{in radians}] = 2\pi\frac{t}{T} \qquad \text{[Eq 9]}$$

and the radius is fixed.

The elliptical-orbit problem is considerably more involved. We know (Eq 8) that the satellite moves much more rapidly near perigee. The relation between t and T can be derived from Kepler's Law II.

In an elliptical orbit, time from perigee, t, is given by

$$t = \frac{T}{2\pi\left[E - e \sin E\right]} \qquad \text{[Eq 10]}$$

where the angle E, known as the eccentric anomaly, is defined by the associated equation

$$\mathbf{E = 2\ arctan\left[\left(\frac{1-e}{1+e}\right)^{0.5} tan\frac{\theta}{2}\right] \quad +360^\circ n}$$ Eq 11]

$$\mathbf{n = \begin{cases} 0 \text{ when } -180^\circ \le \theta \le 180^\circ \\ 1 \text{ when } 180^\circ < \theta \le 540^\circ \end{cases}}$$

Eq 11 may also appear in several alternate forms:

$$\mathbf{E - arcsin\left[\frac{\left(1-e^2\right)^{0.5} \sin\theta}{1+e\cos\theta}\right]}$$

or

$$\mathbf{E = arc\cos\left[\frac{e + \cos\theta}{1 + e\cos\theta}\right]}$$

Note that here, “anomaly” just means angle. Eqs 10 and 11, taken together, are commonly referred to as Kepler’s Equation.

There are two common mistakes that people frequently make the first time they try to solve Eqs 10 and 11. Eq 10 contains the first pitfall. Since the expression e sin E is a unitless number, the E term standing by itself inside the brackets *must* be given in *radians*. The second pitfall is encountered when working with the various forms of Eq 11. Although all inverse trigonometric functions are multi-valued, computers and hand calculators are programmed to give only principal values. For example, if sin θ = 0.99, then θ may equal 82° or 98° (or either of these two values ± any integer multiple of 360°), but a calculator only lights up 82°. If the physical situation requires a value outside the principal range, appropriate adjustments must be made. Eq 11 already includes the adjustments needed so it can be used for values of θ in the range –180° to +540°. If the alternate forms of Eq 11 are used, it’s up to you to select the appropriate range. A few hints may help: (1) E/2 and θ/2 must always be in the same quadrant; (2) as θ increases, E must increase; (3) adjustments to the alternate forms of Eq 11 occur when the term in brackets passes through ±1.

We now have a procedure for finding t when θ is known: Plug θ into Eq 11 to compute E, then plug E into Eq 10 to obtain t. The reverse procedure, finding θ when t is known, is more complex. The key step is solving Eq 10 for E when t is known. Unfortunately, there isn’t any way to neatly express E in terms of t. We can, however, find the value of E corresponding to any value of t by drawing a graph of t vs θ, then reading it “backwards,” or by using an iterative approach. An iterative technique is just a systematic way of guessing an answer for E, computing the resulting t to determine how close it is to the desired value, then using the information to make a better guess for E. Although this procedure may sound involved, it’s actually simple. The iterative technique usually employed to solve Kepler’s Equation is known as the Newton-Raphson method.

We now turn to r, the satellite-to-geocenter distance. Rather than attempt to express r as a function of t, it’s simpler and often more useful just to note the relation between r and θ.

$$\mathbf{r = \frac{a(1 - e^2)}{1 + e\cos\theta}}$$ [Eq 12]

Corrections to the Simplified Model

Now that we've looked at the solutions to the two-body problem (the simplified satellite-orbit model), let's examine how a more detailed analysis would modify our results.

1) In the two-body problem, the stationary point is the center of mass of the system, not the geocenter. The mass of the Earth is so much greater than the mass of an artificial satellite that this correction is negligible.

2) Treating the Earth as a point mass implicitly assumes that the shape and the distribution of mass in the Earth are spherically symmetrical. Taking into account the actual asymmetry of the Earth (most notably the bulge at the equator) produces additional central force terms acting on the satellite. These forces vary as higher orders of 1/r (for example, $1/r^3$, $1/r^4$, and so on). They cause (i) the major axis of the orbital ellipse to rotate slowly in the plane of the satellite and (ii) the plane of the satellite to rotate about the Earth's N-S axis. Both of these effects are observed readily, and we'll return to them shortly.

3) The satellite is affected by a number of other forces in addition to gravitational attraction by the Earth. For example, such forces as gravitational attraction by the Sun, Moon and other planets; friction from the atmosphere (atmospheric drag); radiation pressure from the Sun, and so on enter into the system. We turn now to the effects of some of these forces.

Atmospheric Drag. At low altitudes the most prominent perturbing force acting on a satellite is drag caused by collisions with atoms and ions in the Earth's atmosphere. Let us consider the effect of drag in two cases: (i) elliptical orbits with high apogee and low perigee and (ii) low-altitude circular orbits. In the elliptical-orbit case, drag acts mainly near perigee, reducing the satellite velocity and causing the altitude at the following apogee to be lowered (perigee altitude initially tends to remain constant). Atmospheric drag therefore tends to reduce the eccentricity of elliptical orbits having a low perigee (makes them more circular) by lowering the apogee.

In the low-altitude circular orbit case, drag is of consequence during the entire orbit. It causes the satellite to spiral in toward the Earth at an ever-*increasing* velocity. This is not a misprint. Contrary to intuition, drag causes the velocity of a satellite to *increase*. As the satellite loses energy through collisions it falls to a lower orbit; Eq 8 shows that velocity increases as the height decreases.

A satellite's lifetime in space (before burning up on reentry) depends on the initial orbit, the geometry and mass of the spacecraft, and the composition of the Earth's ionosphere (which varies a great deal from day to day and year to year).

Solar activity has a very big effect on the composition of the Earth's atmosphere at altitudes between 300 and 600 km. High solar activity results in increased atmospheric density and greater drag on spacecraft. Early predictions, for a satellite lifetime of three to five years, had to be revised because of the very low level of solar activity recorded during the 1984-87 time period.

Gravitational Effects. Gravitational attraction by the Sun and Moon can affect the orbit of Earth satellites that have a large (roughly greater than one Earth radii) apogee distance such as those in the Phase 3 series. In many cases these small perturbations (forces) average out to zero over long time periods so their impact is minor. However, in some instances these forces tend to exert their effects during the same part of the orbit for months or even years. In such cases the effects do not average out — they are cumulative. The result, called a resonant perturbation, can produce major changes in an orbit. An effect of this type caused OSCAR 13 to reenter after about 8.5 years in orbit.

In the mid 1980s, when possible orbits for OSCAR 13 were being evaluated, the existing small computers that AMSAT engineers had access to didn't have the power to fully investigate the long term effects that slightly different launch times or initial orbits would have on the spacecraft. Today we have access to far more powerful computers and, thanks to the development work of James Miller, G3RUH, and others, the analysis software needed to prevent another OSCAR 13 type scenario.

Now that the motion of the satellite in space has been described, we turn to the problem of relating this motion to an observer on the surface of the Earth.

Satellite Motion Viewed From Earth

Terrestrial Reference Frame

To describe a satellite's movement as seen by an observer on the Earth, we have to establish a terrestrial reference frame. Once again we simplify the situation by treating the Earth as a sphere. The rotational axis of the Earth (N-S axis) provides a unique line through the geocenter that intersects the surface of the Earth at two points that are designated the *north* (N) and *south* (S) geographic *poles*. The intersection of the surface of the Earth and any plane containing the geocenter is called a *great circle*. The great circle formed from the *equatorial plane*, that plane containing the geocenter that also is perpendicular to the N-S axis, is called the equator. The set of great circles formed by planes containing the N-S axis are also of special interest. Each is divided into two *meridians* (half circles), connecting north and south poles.

Points on the surface of the Earth are specified by two angular coordinates, *latitude* and *longitude*.

Latitude. Given any point on the surface of the Earth, the latitude is determined by (i) drawing a line from the given point to the geocenter, (ii) dropping a perpendicular from the given point to the N-S axis and (iii) measuring the included angle. A more colloquial, but equivalent, definition for latitude is the angle between the line drawn from the given point to the geocenter and the equatorial plane. To prevent ambiguity, an N or S is appended to the latitude to indicate whether the given point lies in the Northern or Southern Hemisphere. The set of all points having a given latitude lies on a plane perpendicular to the N-S axis. Although these latitude curves form circles on the surface of the Earth, most are *not* great circles. The equator (latitude = 0°) is the only one to qualify as a great circle, since the equatorial plane contains the geocenter. The significance of great circles will become apparent later in this discussion when we look at spherical trigonometry. Better models of the Earth take the equatorial bulge and other asymmetries into account when latitude is defined. This leads to a distinction between geodetic, geocentric and astronomical latitude. We won't bother with such refinements.

Longitude. All points on a given meridian are assigned the same longitude. To specify longitude one chooses a reference or "prime" meridian (the original site of the Royal Greenwich Observatory in England is used). The longitude of a given point is then obtained by measuring the angle between the lines joining the geocenter to (i) the point where the equator and prime meridian intersect and (ii) the point where the equator and the meridian containing the given point intersect. For convenience, longitude is given a suffix, E or W, to designate whether one is measuring the angle east or west of the prime meridian.

Inclination

As the Earth rotates about its N-S axis and moves around the Sun, the orientation of both the plane containing the equator (*equatorial plane*) and, to a first approximation, the plane containing the satellite (*orbital plane*) remain fixed in space relative to the fixed stars. The line of intersection of the two planes is called the *line of nodes*, since it joins the ascending and descending nodes. The relative orientation of these two planes is very important to satellite users. It is partially specified by giving the inclination. The *inclination*, i, is the angle between the line joining the geocenter and north pole and the line through the geocenter perpendicular to the orbital plane (to avoid ambiguity, the half-line in the direction of advance of a right-hand screw following satellite motion is used).

The inclination can vary from 0° to 180°. To first order, none of the perturbations to the simplified model we discussed earlier cause the inclination to change, but higher-order effects result in small oscillations about a mean value. A quick analysis of these three cases yields the following information. When the inclination is 0°, the satellite will always be directly above the equator. When the inclination is non-zero the satellite passes over the equator twice each orbit, once heading north and once heading south. When the inclination is 90°, the satellite passes over the north pole and over the south pole during each orbit.

Orbits are sometimes classified as being polar (near polar) when their inclination is 90° (near 90°), or equatorial (near equatorial) when their inclination is 0° (near 0° or 180°). The maximum latitude (ϕ_{max})' north or south, that the subsatellite point will reach equals (i) the inclination when the inclination is between 0° and 90° or (ii) 180° less the inclination when the inclination is between 90° and 180°.

Argument of Perigee

The angle between the line of nodes (the segment joining the geocenter to the ascending node) and the major axis of the ellipse (the segment joining the geocenter and perigee) is known as the argument of perigee. In the simplified two-body model of satellite motion, the argument of perigee is constant. In reality however, it does vary with time, mainly as a result of the Earth's equatorial bulge. The rate of precession (variation) is given by

$$\mathbf{w} = 4.97\left(\frac{R_{eq}}{a}\right)^{3.5}\frac{\left(5\cos^2 i - 1\right)}{\left(1 - e^2\right)^2} \qquad \text{[Eq 13a]}$$

where
$\dot{w}$ = rate of change of argument of perigee in degrees per day
R_{eq} = mean equatorial radius of Earth in same units as a
a = semi-major axis
i = inclination
e = eccentricity

Focusing on the (5 $\cos^2 i$ – 1) term, we see that no matter what the values of a and e, when i = 63.4° the argument of perigee will be constant. The position of the perigee rotates in the same direction as the satellite when i < 63.4° or i > 116.6°, and in the opposite direction when 63.4° < i < 116.6°.

Let w_o represent the value of w at a specific time. Future values of w can be obtained from

$$\mathbf{w(t) = w^{o} + \dot{w}t} \qquad \text{[Eq 13b]}$$

where t is the elapsed time in days.

Nodal vs Anomalistic Period

Once we've seen how the Earth's equatorial bulge affects the argument of perigee, we have to introduce a new term, *nodal period*, to refer to the elapsed time as a satellite travels from one ascending node to the next. The period that we've been referring to up to this point in this discussion is the *anomalistic period* (elapsed time from perigee to perigee). The adjectives "nodal" and "anomalistic" are often omitted in technical literature and conversation when the meaning is clear from the context. In the equations in this appendix, we'll be explicit when we refer to nodal period. The term period by itself will refer to anomalistic period.

The numerical differences between anomalistic period and nodal period are generally quite small. However, if one is making long-term predictions using the wrong period, the error is cumulative. After a few weeks, the predictions will be useless. As a result, it's sometimes necessary to calculate nodal period from the information distributed with classical orbital elements which refers to anomalistic period.

Solar and Sidereal Time

Living on Earth we quite naturally keep time by the sun. So when we say the Earth undergoes one complete rotation about its N-S axis each day, we're actually referring to a mean *solar day*, which is arbitrarily divided into exactly 24 hours (1440 minutes).

The time for the Earth to rotate exactly 360° is known as the *sidereal day*. When the word "day" is used by itself, solar day is meant. For example, orbital elements distributed by NASA give the units for mean motion as revolutions/day. The day referred to is the solar day of 1440 minutes.

Precession: Circular Orbits

Although the two-body model predicts that the orbital plane will remain stationary, we've already noted that when the Earth's equatorial bulge is taken into account, the plane precesses about the Earth's N-S axis.

$$\Omega = -\,\mathbf{9.95}\left(\frac{\mathbf{R_{eq}}}{\mathbf{r}}\right)^{\mathbf{3.5}} \mathbf{\cos i} \qquad \text{[Eq 14]}$$

(circular Earth orbits only)

where
$\dot{\Omega}$ = orbital plane precession rate in °/day.
R_{eq} = mean equatorial radius of Earth = 6378 km
R = satellite-geocenter distance in same units as R_{eq}
i = orbital inclination

Sun Synchronous Orbits

By choosing the altitude and inclination of a satellite, we can vary $\dot{\Omega}$ over a considerable range of values. An orbit that precesses very nearly 360° per year is called

Sun-synchronous. Such orbits pass over the same part of the Earth at roughly the same time each day, making communication and various forms of data collection convenient. They can also provide nearly continuous sunlight for solar cells or good Sun angles for satellite photos. Because of all these factors, orbits are often carefully designed to be Sun-synchronous.

To obtain an orbital precession of 360° per year, we need a precession rate of 0.986°/day (360°/365.25 days). Substituting this value in Eq 14 and solving for i we obtain

$$i* = \arccos\left[-(0.09910)\left(\frac{r}{6378}\right)^{3.5}\right] \quad \text{[Eq 15]}$$

where i* is the inclination needed to produce a Sun-synchronous circular orbit. In this form, we can plug in values of r and calculate the inclination which will produce a Sun-synchronous orbit. Graphing Eq 15, we see that for low-altitude satellites Sun-synchronous orbits will be near polar. You may have noted that the 0.986°/day precession rate needed to produce a Sun synchronous orbit exactly corresponds to the amount in excess of 360° that the Earth rotates each solar day. This is no accident; the precession rate was chosen precisely for this purpose.

Precession: Elliptical Orbits

The precession of the orbital plane about the Earth's N-S axis for elliptical orbits is given by

$$\dot{\Omega} = -9.95\left(\frac{R_{eq}}{a}\right)^{3.5} \frac{\cos(i)}{(1-e^2)^2} \quad \text{[Eq 16]}$$

If a = r and e = 0 (that is, when the ellipse becomes a circle), Eq 16 simplifies to Eq 14.

Longitude Increment

We now know how the satellite moves in the orbital plane and how the orientation of the orbital plane changes with time. Our next objective is to relate this information to the longitude increment. The *longitude increment* (I), or simply increment, is defined as the change in longitude between two successive ascending nodes. In mathematical terms

$$I = \lambda_{n+1} - \lambda_n \quad \text{[Eq 17]}$$

where λ_{n+1} is the longitude at any ascending node in degrees east of Greenwich [°E], λ_n is the longitude at the preceding ascending node in °E, and I is in degrees east per revolution [°E/rev].

There are two ways to obtain the increment: experimentally by averaging observations over a long period of time, or theoretically by calculating it from a model. Though the best numbers are obtained experimentally, the calculation approach is needed; we, after all, want a value for I before a spacecraft is launched, and in the early weeks or months of its stay in orbit when observations haven't accumulated over a long time period.

The computations that follow will use the sign convention specified with Eq 16. However, when convenient, the final result may be expressed in degrees west by simply changing the sign.

If we neglect precession of the orbital plane, the increment can be estimated by computing how much the Earth rotates during the time it takes for the satellite to complete one revolution from ascending node to ascending node. (In this section, period refers to nodal period.)

$$\bar{I} = \left(\frac{T}{1440}\right)(-360.98563°E) \qquad \text{[Eq 18a]}$$

$$\bar{I} = -(0.250684°E)T \qquad \text{[Eq 18b]}$$

The period, T, must be in minutes; the negative sign means that each succeeding node is further west; and 360.98563° is the angular rotation of the Earth about its axis during a solar day (1440 minutes) From Eq 18b we see that it's easy to get a quick estimate of I by computing T/4 and expressing the result in degrees west per revolution.

The value for the orbital increment provided by Eq 18 can be improved by taking into account the fact that the precession of the orbital plane (Eq 14 or Eq 16) will affect the apparent rotation of the Earth during one solar day. The result is given in Eq 19.

$$\bar{I} = (T/1440)\left(-360.98563 + \dot{\Omega}\right) \qquad \text{[Eq 19]}$$

Once the increment is known we can compute the longitude of any ascending node, λ_m, given the longitude of any other ascending node, λ_n. The orbit reference integers, m and n, may either be the standard ones beginning with the first orbit after launch, or any other convenient serial set.

$$\lambda_m = \lambda_n + (m - n)I \qquad \text{[Eq 20]}$$

This formula works either forward or backward in time. When future orbits are being predicted, m> n. The right side of Eq 20 must be brought into the range of 0-360° by successive subtractions or additions of 360° if necessary.

Ground Track

To study the ground track, we have to (1) look at the geometry involved when the orbital plane intersects the surface of the Earth, (2) consider the motion of the satellite about its orbit and (3) take into account the rotation of the Earth. The best way to handle complex problems of this type is to use mathematical objects known as vectors. All advanced texts in orbital mechanics proceed in this manner. However, many simple problems can be treated using spherical trigonometry. If you have a reasonable background in plane trigonometry, a brief introduction to spherical trigonometry will provide you with all the information needed to understand how the ground-track equations are derived. Since many more readers have experience with trigonometry than with vectors, we'll use the spherical trigonometry approach.

A barebones introduction to spherical trigonometry follows. The results for circular orbits are then generalized and summarized. (Readers who just need access to the ground-track equations for programming a computer can skip the spherical trigonometry and derivation sections and jump right to the summary.) We then go on to derive and summarize the ground-track equations for elliptical orbits.

Spherical Trigonometry Basics

A triangle drawn on the surface of a sphere is called a spherical triangle *only* if all three sides are arcs of great circles. A great circle is formed *only* when a plane containing the center of a sphere intersects the surface. The Earth's equator is a great circle; other latitude lines are not. The intersection of a satellite's orbital plane and the surface of a static (nonrotating) Earth is a great circle. Range circles drawn around a ground station are not great circles.

Spherical trigonometry is the study of the relations between sides and angles in spherical triangles. The notation of spherical trigonometry closely follows that of plane trigonometry. Surface angles and vertices in a triangle are labeled with capital letters A, B and C, and the side opposite each angle is labeled with the corresponding lowercase letter. Note that the arc length of each side is proportional to the central angle formed by joining its end points to the center of the sphere. For example, side b is proportional to angle AOC. The proportionality constant is the radius of the sphere, but because it cancels out in the computations we'll be interested in, the length of a side will often be referred to by its angular measure.

The rules governing the relationships between sides and angles in spherical triangles differ from those in plane triangles. In spherical trigonometry, the internal angles in a triangle do not usually add up to 180° and the square of the hypotenuse does *not* generally equal the sum of the squares of the other two sides in a right triangle.

A spherical triangle that has at least one 90° angle is called a right spherical triangle.

Recall how in plane trigonometry the rules for right triangles were simpler than those for oblique triangles. In spherical trigonometry the situation is similar: The rules for right spherical triangles are simpler than those for general spherical triangles. Fortunately, since the spherical triangles we'll be working with have at least one right angle, we need only consider the laws for right spherical triangles.

Two major pitfalls await newcomers attempting to apply spherical trigonometry for the first time. The first pitfall, the degree-radian trap, comes from overlooking the fact that angles must be expressed in units appropriate to a given equation *and* a given computing machine. For example, focus on the angle $\theta = 30° = \pi/6$ radians. Consider the machine-dependent aspect first. To evaluate $\sin(\theta)$ on most simple scientific calculators, you must input "30" since the calculator expects θ to be in degrees unless you've been instructed otherwise. To evaluate $\sin(\theta)$ in BASIC on a computer, you must input $\pi/6$ (or 0.52360), because the BASIC language expects θ to be in radians. In some situations, especially in cases where θ is not the argument of a trigonometric function, the form of the equation determines whether θ must be in degrees or radians. Consider a radio station at 30° N latitude trying to use the equation $S = R\theta$ to find the surface distance (S) along a meridian (Earth radius = R) to the equator. The equation only holds for θ in radians, so the input must be $\pi/6$.

The second trap awaiting spherical trigonometry novices is using a latitude line as one side of a spherical triangle. The only latitude line that will serve in this manner is the equator. All other latitude lines do not work, since they are not arcs of great circles.

Circular Orbits: Derivation

The most important step in deriving the ground-track equations for circular orbits is drawing a clear picture. Our object is to compute the latitude and longitude of the subsatellite point (SSP) — Φ_s and λ_s —when the spacecraft reaches S, t minutes after the most recent ascending node. We assume that the period T, orbit inclination i, and the longitude of the ascending node, λ_o, are known.

Since arc AS, along the actual ground track, is not a section of a great circle, we first consider the situation for a static Earth (one not rotating about its N-S axis). On such an Earth, the SSP would be at point B at t minutes after the ascending node. Triangle ABC is a right spherical triangle. Angle A is given by 180° – i. Arc AB (side c of the spherical triangle) is a section of the circular orbit with

$$\mathbf{c = 2\pi \frac{t}{T}} \qquad \text{[see Eq 9]}$$

By definition, the latitude of point B is equal to a.

The problem of finding the latitude of point B, ϕ_B in terms of i, t and T is identical to the problem of finding a in terms of A and c.

$$\mathbf{a = \arcsin [\sin (c) \sin (A)]}$$

Substituting the variables ϕ_B, T, i and t we obtain

$$\boldsymbol{\phi_B = \arcsin\left[\sin\left(2\pi\frac{t}{T}\right)\sin\left(180^\circ - i\right)\right]}$$

Using the symmetry of the sine function, this simplifies to

$$\phi_B = \arcsin\left[\sin\left(2\pi\frac{t}{T}\right)\sin(i)\right] \quad \text{(nonrotating Earth)}$$

If for computations we wish to specify c in degrees, we would replace

$$2\pi\frac{t}{T} \text{ by } 360^\circ\frac{t}{T}$$

To solve for the longitude at B, λ_B, we note that $b = \lambda_o - \lambda_B$. So, our problem of solving for λ_B in terms of ϕ_B, t and T is equivalent to solving for b in terms of a and c.

$$\mathbf{b = \arccos\left[\frac{\cos (c)}{\cos(a)}\right]}$$

Making the appropriate substitutions, this yields

$$\boldsymbol{\lambda_0 - \lambda_B = \text{ar}\cos\left[\frac{\cos\left(\frac{2\pi t}{t}\right)}{\cos\left(\phi_B\right)}\right]} \quad \textbf{(nonrotating Earth)}$$

The effect of the Earth's rotation is to move the SSP from B to S. The latitude remains constant ($\phi_S = \phi_B$); only the longitude changes. The longitude change is simply the angular rotation of the Earth during the time t. To a first approximation, the rotation rate of the Earth is 0.25°/minute, so if we measure t in minutes, $\lambda_s = \lambda_B - t/4$. For long-term predictions a more accurate figure for the Earth's rotation should be used (as we saw when the longitude increment was discussed). This value is 0.25068°/minute. (See Eq 18b.)

A more complete derivation would consider several additional cases: satellites in the southern hemisphere, inclination between 0° and 90°, spacecraft headed south, and so on. As the approach is similar, we'll just summarize the results in the next section.

Circular Orbits: Summary

Latitude of SSP:

$$\phi(t) = \arcsin\left[\sin(i)\left(360^0 \frac{t}{T}\right)\right] \qquad \text{[Eq 21]}$$

Note: "$\phi(t)$" should be read "latitude as a function of time"; it does *not* mean ϕ times t.

Longitude of SSP:

$$\lambda(t) = \lambda o - (0.250684)T + (s1)(s2)\arccos\left[\left(\frac{\cos\left(\frac{360°t}{T}\right)}{\cos(\phi(t))}\right)\right] \qquad \text{[Eq 22]}$$

$$S1 = \begin{Bmatrix} +1 \text{ when } 0° \le i \le 90° \\ -1 \text{ when } 90° \le i \le 180° \end{Bmatrix}$$

$$S2 = \begin{Bmatrix} +1 \text{ when } \phi(t) \ge 0°(\text{Northern Hemisphere}) \\ -1 \text{ when } \phi(t) > 0°(\textit{Southern Hemisphere}) \end{Bmatrix}$$

SIGN CONVENTIONS

Latitude
North: positive
South: negative

Longitude
East: positive
West: negative

All angles are in degrees and time is in minutes
i = inclination of orbit
T = period
t = elapsed time since most recent ascending node
λ_o = longitude of SSP at most recent ascending node

Please note the sign conventions for east and west longitudes. Most maps used by radio amateurs in the US are labeled in degrees west of Greenwich. This is equivalent to calling west longitudes positive. Because there are important computational advantages to using a right-hand coordinate system, however, almost all physics and mathematics books refer to east as positive, a custom that we follow for computations. When calculations are completed it's a simple matter to re-label longitudes in degrees west. This has been done for all user-oriented data in this book.

Eq 22 should only be applied to a single orbit. At the end of each orbit, the best available longitude increment should be used to compute a new longitude of ascending node. Eq 22 can then be reapplied.

Eq 21 and Eq 22 can be solved at any time, t, if i, λ_0,~ and T are known. In other words, it takes four parameters to specify the location of the SSP for a circular orbit. The four we've used are known as the "classical orbital elements." They were chosen because each has a clear physical meaning. There are several other sets of orbital elements that may also be employed.

Elliptical Orbits: Derivation

Now that we've seen how the ground-track equations for a circular orbit are derived, we go on to look at the additional parameters and steps required for elliptical orbits.

We assume that the following parameters are known: T (period in minutes), i (inclination in degrees), λ_p (SSP longitude at perigee), w (argument of perigee) and e (eccentricity). Our object is to solve for the latitude and longitude of the SSP — $\phi(t)$ and $\lambda(t)$ — at any time t. We will measure t from perigee.

The actual ground track is not a great circle, so our strategy will again be to focus first on a static Earth model where the principles of spherical trigonometry can be applied. The results will then be adjusted to take into account the rotation of the Earth.

Step 1. Our object here is to relate our perigee-based parameters to the ascending node. More specifically, we wish to calculate (a) elapsed time as the satellite moves from D to P, (b) the latitude at perigee and (c) the longitude at the ascending node.

la) Consider the static-Earth model and focus on spherical triangle CPD. The arc PD is, by definition, equal to the argument of perigee, w. Using Kepler's Equation (Eqs 10 and 11), we can plug the value of w in for θ and calculate the elapsed time between perigee and the ascending node, which we call t_p.

lb)The latitude at perigee is, by definition, the length of arc PC. Angle PDC is equal to 180° – i. Knowing angle PDC and arc PD, we use Napier Rule II to solve for arc PC.

$$\phi_p = \arcsin[\sin(i)\sin(w)]$$

1c) To obtain the longitude at point D, we again apply Napier Rule II.

$$\dot{\lambda}_o = \lambda_p + \arccos\left[\frac{\cos(w)}{\cos(\phi_p)}\right]$$

1d) The actual longitude at the ascending node is found by computing how far the Earth rotated as the satellite traveled from the ascending node to perigee and adding this to the preceding static-Earth result. To simplify the following equations we approximate the rotation of the Earth by 0.25°/ mm.

$$\lambda_o = \lambda_p + \arccos\left[\frac{\cos(w)}{\cos(\phi_p)}\right] + \left|\frac{t_p}{4}\right|$$

Step 2. We now turn to the problem of locating the SSP at S, any time, before or after perigee. We again begin by focusing on the static-Earth mode to find the latitude and longitude of point B. To do this we use spherical triangle BDE.

2a) To emphasize that 0 changes with time, we write this term as (θ(t) + w). Using Napier Rule II we obtain the latitude of point B, which is also the actual latitude of SSP at S.

$$\phi(t) = \arcsin[\sin(i)\sin(\phi(t) + w)]$$ [Eq 23]

2b) Applying Napier Rule II once again we obtain the longitude of point B.

$$\dot{\lambda}\ (t) = \dot{\lambda}_o - \arccos\left[\frac{\cos(\theta(t) + w)}{\cos(\theta(t))}\right]$$

2c) Finally, correcting for the rotation of the Earth we obtain the actual longitude of the SSP at S.

$$\lambda(t) = \lambda_o - \arccos\left[\frac{\cos(\theta(t) + w)}{\cos(\theta(t))}\right] - \frac{t}{4} - \frac{t_p}{4}$$ [Eq 24]

Eq 24 only gives the correct results when the spacecraft is in the northern hemisphere, $\theta > w$, and both are less than 90°.

Right Ascension Declination Coordinate System

We've now discussed all classical orbital elements except for the term *right ascension*. Calculations by astronomers and those working with satellites are best carried out in an inertial coordinate system (one that has fixed directions with respect to the distant stars). The right ascension-declination coordinate system is often used for this purpose. The position of the center of the system isn't important. For convenience we take it to be the geocenter. We now imagine a sphere of infinite radius, called the *celestial sphere*, surrounding the Earth. When the Earth's equatorial plane is extended in all directions, it becomes the celestial equator. When the Earth's North-South axis is extended, it becomes the celestial polar axis. To locate points in the right ascension-declination coordinate system, we need a set of three perpendicular axes. The extended north polar axis is one. The other two lie in the equatorial plane. We take one of these to be parallel to the directed line from the center of the Sun to the geocenter on the first day of spring. This is frequently referred to as the direction of the vernal equinox or first point of Aries. The position of an object in space is described by the two angles, called right ascension and declination and, if necessary, its distance from the Earth.

This coordinate system is particularly convenient for keeping track of the orientation of a satellite's orbital plane as, the plane rotates slowly about the celestial spheres polar axis. Let Ω_o be the angle (at a particular time) in the celestial equator between (1) the line from the geocenter to the vernal equinox and (2) the line from the geocenter to the ascending node of a satellite orbital plane. This angle is referred to as the right ascension of ascending node (RAAN) at the epoch time. When Ω_o is known, we can predict the RAAN at any future time with the equation

$$\Omega(t) = \dot{\Omega}_o + \Omega t$$ [Eq 25]

By adopting the right ascension-declination coordinate system, we can separate the problems of keeping track of (1) the direction of the line of nodes and (2) the position of a terrestrial observer, as the Earth rotates about its polar axis and revolves around the Sun. At any time these two pieces of information can be combined with the location of the spacecraft in the orbital plane to give the longitude of the SSP.

Bahn Latitude And Longitude

The orientation of a satellite in its orbital plane is specified by two angles, *Bahn latitude* and *Bahn longitude*. Bahn latitude is the angle between the spacecraft + z axis and the orbital plane. Suppose we rotate the spacecraft through this angle so that the Z axis is now in the orbital plane and then rotate the spacecraft a second time so that its + Z axis is aligned with the directed line running from the geocenter to the spacecraft at apogee. This second rotation angle is the Bahn longitude.

Declination, rotation of a line into a plane, is essentially a latitude. Ascension, rotation in a plane, is essentially a longitude.

Orbital Elements

The parameters used to describe the position and motion of a satellite, rocket, planet or other heavenly body are called *orbital elements*. There's a lot of flexibility in selecting particular parameters to serve as orbital elements. The choice depends on the characteristics of the problem being examined, the information available and the coordinate system being used. If we focus our attention on satellites that do not contain propulsion systems and are not affected by atmospheric drag, we find that it requires a set of six parameters specified at a particular time (called the epoch time) to specify the current spacecraft location and to accurately predict its future positions (the basic satellite tracking problem).

From a computational viewpoint, a desirable set of six elements consists of three position coordinates and three velocity components expressed in an inertial Cartesian coordinate system. One shortcoming of this set of elements is that it doesn't provide any immediate clues as to what the orbit looks like. In contrast, classical orbital elements, which involve parameters like eccentricity, inclination, argument of perigee, mean motion (or semi-major axis), RAAN and time since last perigee are extremely helpful for visualizing an orbit.

The orbital elements used by amateurs in tracking problems are almost always a variation of the classical set. In many instances, more than six parameters are provided by the source (or requested by the program) at a particular epoch time. One reason that many programs require more than six elements is that they do more than solve the basic problem. For example, they may take drag into account or keep track of the spacecraft orientation in the orbital plane so that squint angle can be determined. Some element distributors provide redundant data to allow for the fact that one tracking program may, for example, request mean motion as an input while another might request semi-major axis. In any event, a set of orbital elements must include at last six parameters specified at a particular time.

The Oblate Earth

As a first approximation, it is reasonable to treat the Earth as a sphere having a mean radius of 6371 km. A model of satellite motion based on a spherical Earth predicts that the orientation (right ascension and inclination) of a satellite's orbital plane will remain fixed in space and that the position of a satellite's perigee in the orbital plane (argument of perigee) will not change. Such a model might be acceptable for a single orbit, but it is not adequate for long-term predictions.

In order to obtain equations for important factors such as rate of change of perigee (Eq 13) and precession of the line of nodes (Eq 16), one has to adopt a more complex model. The next step in complexity is to use an ellipsoidal Earth (an ellipse rotated about the polar axis). Higher-order models are possible, but it turns out that an oblate Earth model based on the ellipse gives very good results. The semi-major axis of the ellipse is the equatorial radius of the Earth: 6378 km; the semi-minor axis is the polar radius: 6357 km; and the Earth's eccentricity is 0.08 182.

An oblate Earth model enters our calculations two ways. First, it affects the motion of the satellite in space. We have taken these effects into account by our use of Eqs 13 and 16. Second, it affects calculations that involve the position of a ground station on the surface of the Earth, such as antenna aiming parameters and range. If one compares antenna pointing predictions (azimuth and elevation) using a spherical Earth model to those using an oblate Earth model, the differences turn out to be small fractions of a degree, an amount that will never be visible with any amateur antenna. Using an oblate Earth model for these calculations does lead to complications, such as a need to distinguish between geocentric latitude, geodetic latitude and astronomical latitude.

A reasonable approach to designing tracking programs for amateurs would therefore be to incorporate the oblate Earth by using Eqs 13 and 16, but to treat the Earth as a sphere for calculations involving range and antenna aiming. This would suffice for 99.9% of amateur needs. However, for the critical orbit determination process leading to the rocket burns used to boost Phase 3 satellites from a transfer orbit to an operating orbit, it's desirable to use the most accurate model possible. Since key early tracking programs were specifically developed for the orbit transfer process, they incorporated an oblate Earth model at all steps. This refinement is included in most programs being distributed today.

Special Orbits

With at least four parameters to vary—eccentricity, inclination, semi-major axis and argument of perigee—there are many possible ways of classifying orbits. We've already paid considerable attention to low Earth circular orbits with inclinations near 90°. We've also discussed the special characteristics of Sun-synchronous orbits. In this section, we briefly look at two additional types of orbits of special interest to radio amateurs.

The Geostationary Orbit

A satellite launched into an orbit with an inclination of zero degrees will always remain directly above the equator. If such a satellite is in a circular orbit (constant velocity), traveling west to east, at a carefully selected height (35,800 km), its angular velocity will equal that of the Earth about its axis (period = 24 hours). As a result, to an observer on the surface of the Earth the spacecraft will appear to be hanging motionless in the sky. Satellites in such orbits are called *geostationary* (or stationary for short).

The geostationary orbit has a number of features that make it nearly ideal for a communications satellite. Of prime importance, Doppler shift on the radio links is nonexistent, and ground stations can forget about orbit calendars and tracking. These features have not gone unnoticed — so many commercial spacecraft are spaced along the geostationary arc above the equator that a severe "parking" problem exists. From an Amateur Radio point of view, a geostationary satellite is not without problems. The biggest shortcoming is that a single spacecraft can only serve slightly less than half the Earth. It's sometimes stated that a geostationary satellite provides poor east-west communications coverage to radio amateurs at medium to high latitudes. This may be true when Molniya-type orbits (see next section) are the standard of comparison.

If the orbital inclination of a satellite is not zero, the spacecraft cannot appear stationary; stationary satellites can only be located above the equator. A 24 hour-period circular orbit of non-zero inclination will have a ground track like a symmetrical figure eight. Note that the ascending and descending nodes of such an orbit coincide and the longitude of ascending node is constant (the increment is nearly zero). The 24-hour circular path is known as a *synchronous orbit*. The geostationary orbit is a special type of synchronous orbit, one with a zero-degree inclination.

Note that some authors apply the term synchronous (or geosynchronous) to other types of orbits, ones that are circular or elliptical with periods that are an exact divisor of 24 hours, such as 8 hours or 12 hours. Because this might lead to unnecessary confusion, we'll avoid this use of the term synchronous.

Molniya Type Orbit

Looking at elliptical orbits earlier, we noted that the position of the perigee in the orbital plane (the argument of perigee) changes from day to day at a rate given by Eq 13. An interesting feature of this equation is that when i = 63.4°, the argument of perigee remains constant regardless of the values of the period and eccentricity. As a consequence, the argument of perigee, period and eccentricity can be chosen independently to satisfy other mission requirements.

Orbits with i = 63.4°, eccentricities in the 0.6 to 0.7 range and periods of 8 to 12 hours have a number of features that make them attractive communications satellites. Spacecraft in the Russian Molniya series were designed to take advantage of this type of orbit.

Let's take a brief look at a Molniya II series communications satellite of the type once used for the Moscow-Washington Hotline. (The Hotline used redundant Molniya and Intelsat links.) The spacecraft is maintained in an orbit with an inclination of 63.4°, and argument of perigee constant at 270° and a period of 12 hours. Because of the 12-hour period, the ground track tends to repeat on a daily basis (there is a slow drift in the longitude of ascending node). Apogee, where the satellite moves slowly, always occurs over 63.4° N latitude. At apogee nearly half the Earth, most in the northern hemisphere, is in view. The primary Washington-Moscow mutual visibility window lasts 8 to 9 hours. A single spacecraft is accessible to the Washington station about 16 hours per day, and simultaneously to both stations about 12 hours each day. Thus, a three-satellite Molniya system can provide a reliable Washington-Moscow link 24 hours a day.

From an Amateur Radio point of view, the Molniya-type orbit has a number of attractive features. The orbits selected for Phase III spacecraft have been variations on the Molniya theme.

There has been a great deal of discussion as to the specific orbit desired. Some amateurs preferred an orbit with i = 63.4°. With this inclination, the apogee can be set to favor the northern hemisphere continuously. Other amateurs preferred a different value of inclination, since a changing argument of perigee eventually gives one access to a considerably larger portion of the world. Thinking in terms of a long-term Phase III system, such a satellite would begin to favor the southern hemisphere six or seven years after launch as the apogee drifted south of the equator. A new spacecraft would then be launched to take over in the northern hemisphere.

Meanwhile, spacecraft design engineers must also investigate the trade-offs involving the thrust required to reach a high inclination orbit and the Sun angles the spacecraft would encounter during the year. The Sun-angle numbers have a big impact on the transmitter power available on the spacecraft.

When compared to a geostationary orbit, the Molniya-type orbit has several advantages, which we've been focusing on, and several shortcomings.

Most of the shortcomings are minor. Greater attention must be given to antenna aiming and Doppler shifts, but to a lesser degree than with low-altitude spacecraft.

The major problem is one for the AMSAT spacecraft engineers. A satellite in a Moloiya orbit traverses the Van Allen radiation belts twice each orbit, subjecting many of the onboard electronic subsystems, especially those associated with the central computer, to damage from the high-energy particles that may be encountered. Extensive shielding of the computer chip is necessary, but this shielding increases the weight, restricting access to desirable orbits.

The trade-off involved here is so important that AMSAT undertook a special research program to look into the effects of radiation on the RCA CMOS integrated circuits used on OSCAR 10. Chips of the type which were to be flown were exposed to radiation under conditions that simulated the anticipated space environment. These failure-rate studies, performed at Argonne and Brookhaven National Laboratories, used various amounts and types of shielding. The results provided the data used to design OSCAR 10. Using every option possible, the engineers were able to project a three- to five-year lifespan for the spacecraft computer. The projections were remarkably accurate. When OSCAR 13 was built, integrated circuits having a much higher resistance to radiation damage were available. With respect to radiation damage, OSCAR 13 had a projected lifetime of nearly a century.

Appendix B
Other Satellite Subsystems: Structural, Environmental, Power Energy, Attitude, Propulsion and Control

By Dr Martin Davidoff, K2UBC

Structural

The spacecraft structural subsystem, the frame that holds it all together, serves a number of functions including physical support of antennas, solar cells and internal electronic modules; protection of onboard subsystems from the environment during launch and while in space; conduction of heat into and out of the satellite interior; mating to the launch vehicle and so on. Structural design (size, shape and materials) is influenced by launch vehicle constraints, by the spacecraft's mission, and by the orbit and attitude stabilization system employed. OSCARs 6, 7 and 8 had masses in the 18 to 30 kg range. The larger early Phase III spacecraft containing either a kick motor or liquid fuel engine were in the 90 to 125 kg range at launch. Spacecraft in the MicroSAT series have a mass of approximately 10kg contained in a cube roughly 23 cm on edge.

The prominent features one observes when looking at a satellite are the attach fitting used to mate the satellite to the launch vehicle, antennas for the various radio links, solar cells, the heat-radiative coating designed to achieve the desired spacecraft thermal equilibrium and, for Phase III, the nozzle of the apogee kick motor. Insofar as possible, AMSAT satellite structures are fabricated from sheet aluminum to minimize the complexity of machining operations.

The MicroSAT structure is unique. It consists of five modules (trays) formed into a composite stack held together with stainless steel tie bolts. See **Figure B1**. Each frame is approximately 210×210×40 mm. Aluminum side panels, about 5 mm thick, holding the solar panels cover the sides. The entire structure is a rectangular solid 230×230×213 mm. The trays form an extremely sturdy structure; no additional spacecraft frame is needed.

Each tray contains an electronic subassembly. The basic modules, needed by every MicroSAT, are the (1) battery/BCR unit, (2) receiver unit, (3) CPU/RAM and (4) transmitter unit. The fifth module is the TSFR tray. TSFR stands for "this space for rent." The TSFR tray is essentially space set aside for experimental mission-specific subsystems.

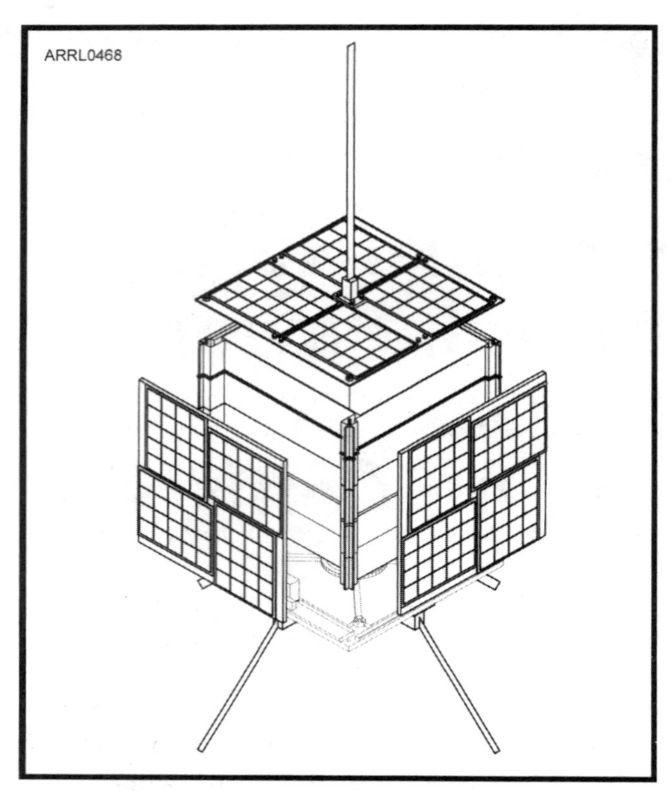

Figure B1—MicroSat structure.

One significant aspect of the MicroSAT structure is that, by its very nature, it invites small amateur groups around the world to build a TSFR module for inclusion on future spacecraft. It also establishes a standard structure that greatly simplifies the work of groups who wish to build their own MicroSAT and reduces the cost of future MicroSATs.

The OSCAR 40 structure was designed to accommodate two ESA requirements. When ESA launches two large primary satellites they are placed one above the other and separated by a conical adapter. The bottom of the lower satellite is attached to the rocket. The bottom of the upper satellite is attached to the adapter ring that is in turn attached to the top of the lower satellite. The adapter ring must therefore be capable of supporting the upper satellite during the entire launch phase. And, the upper spacecraft can weigh up to 10,350 pounds.

AMSAT studied the standard adapter ring to see if it could be incorporated into a P3 satellite spaceframe. However, this approach did not lead to promising results. In discussions with ESA, AMSAT suggested an alternative approach that would use a cylindrical Specific Bearing Structure (SBS) as a spacer between the two primary spacecraft with the OSCAR 40 spaceframe contained inside the SBS. After extensive consultations, ESA agreed to the approach. The design, construction and testing of the SBS was a major technical accomplishment for AMSAT.

The spaceframe for OSCAR 40 had a completely different shape from that of previous Phase III spacecraft. It was the first AMSAT satellite to employ solar panels that extend out after launch. The central structure was hexagonal and weighs about 60kg, nearly 10 times the weight of OSCARs 10 and 13. The entire spacecraft including the spaceframe, everything mounted to it, fuel and the SBS weighed about 400 kg.

Environmental Control

The function of a spacecraft environmental control subsystem is to regulate temperatures at various points, shield against high-energy particles and protect the onboard electronics from RF interference. We'll focus on thermal control.

The temperature of a satellite is determined by the inflow and outflow of energy. More specifically, the satellite temperature will adjust itself so heat inflow equals heat outflow. Although we talk about the "temperature" of the satellite, different parts of a spacecraft are at different temperatures and these temperatures are constantly varying.

The goal of the spacecraft designer is to create a model of the spacecraft and its environment that will accurately predict the average and extreme temperatures that each unit of the spacecraft will exhibit during all phases of satellite operation. This includes prelaunch, where the satellite may sit atop a rocket more than a week baking under a hot tropical Sun; the launch and orbit-insertion sequence of events; and the final orbit where, during certain seasons of the year, the spacecraft may go for months without any eclipse time. (An eclipse occurs when the Earth passes between the spacecraft and the Sun. During other seasons it may be eclipsed for several hours each day.) Since excessive temperature extremes, either too hot or too cold, may damage the battery or electronic or mechanical subsystems permanently, the thermal design must keep the temperatures of susceptible components within bounds at all times.

Once the satellite is in the vacuum of space, heat is transferred only by radiation and conduction; convection need not be considered. The complete energy balance model of Phase IIIA depicted the satellite as being composed of 121 subunits, each connected via conduction and radiation links to several other subunits. To solve the resulting energy balance equations mathematically, the designers had to manipulate 121 nonlinear coupled equations, each consisting of about three or four terms. This was not a job for pad and pencil. A fairly large computer was needed. Even with sophisticated computer models, achieving a precision of ±10 K is difficult; commercial satellite builders usually resort to testing the thermal balance of full scale models in space simulators. The Phase IIIA thermal design problem was handled by Dick Jansson, KD1K, using computer time donated by the Martin Marietta Corp.

Dick also handled the OSCAR 40 design. This time it was done on a powerful personal computer using a model involving 450 nodes (temperature points) and 3750 heat conduction links.

Earlier OSCARs used a far simpler and less accurate approach to thermal design that nevertheless provided reasonable results. Since the details of the simple approach provide a good introduction to the science of thermal design, we'll go through an example.

Thermal Design: A Simple Example

The Sun is the sole source of energy input to the satellite. Quantitatively we can write:

$$P_{in} = P_o \langle A \rangle \alpha\beta\sim \qquad \text{(Eq 1)}$$

where

P_{in} = energy input to the satellite

P_o = solar constant = incident energy per unit time on a surface of unit area (perpendicular to direction of radiation) at 1.49×10^{11} m (earth-Sun distance) from the Sun.

P_o= 1380 W/m^2

$\langle A \rangle$ = effective capture area of the satellite for solar radiation

α = absorptivity (fraction of time satellite is exposed to the Sun during each complete orbit)

β = eclipse factor (fraction of time satellite is exposed to the Sun during each complete orbit)

Power output from the satellite consists of blackbody radiation at temperature T, and the radio emissions. Since blackbody radiation is very much greater than the radio emissions, we can ignore the latter.

$$P_{out} = A\sigma eT^4 \qquad \text{(Eq 2)}$$

where

P_{out} = energy radiated by satellite

A = surface area of satellite

σ = Stefan-Boltzmann constant = $\frac{5.67 \times 10^{-8} \text{joules}}{K^4 m^2 s}$

e = average emissivity factor for satellite surface

T = temperature (K)

For equilibrium, incoming and outgoing radiation must balance,

$P_{in} = P_{out}$ or

$P_o \langle A \rangle \alpha\beta = A\sigma eT^4$ (energy balance equation) (Eq 3)

Solving for temperature, we obtain

$$T = \left(\frac{P_o ab < A >}{\sigma eA}\right)^{1/4}$$

Reasonable average values for the various parameters are $\alpha = 0.8$, $\beta = 0.8$ and e = 0.5. For AMSAT-OSCAR 7, A = 7770 cm^2 and <A> = 1870 cm^2. Inserting these values we obtain T = 294 K. This is equivalent to 21°C, which is close to the observed equilibrium temperature of OSCAR 7. Over the course of a year, as β varied from 0.8 to 1.0, the temperature of OSCAR 7 varied between 275 K and 290 K. The transponder final amplifier, of course, ran considerably hotter.

Passive methods used to achieve a desired spacecraft operating temperature include adjusting surface absorptivity (α) and emissivity (e) by roughening or painting and taking thermal conductivity of structural components into account when the spacecraft is designed. Active techniques for temperature control include fitting the spacecraft with shutters, or louvers, that are controlled by bimetallic strips or conducting pipes that can, by ground command, be filled with Helium gas or evacuated. To a certain extent active temperature control has been employed on OSCAR spacecraft; in several instances specific subsystems have been activated primarily because of their effect on spacecraft temperature.

Heat pipe technology was an important component in the design of OSCAR 40. A heat pipe consists of a long, thin evacuated tube partially filled with a liquid and a material that serves as a wick (supports capillary action). The fluid is vaporized at the hot end. The vapors flow through the hollow core of the heat pipe and condense at the cold end. The fluid then circulates back to the hot end via capillary action through the wicking material. A well designed heat pipe can, kg for kg, transport many times more heat than the best conducting materials. The process requires no power and works in a zero-gravity environment.

OSCAR 40 contained four heat pipes using anhydrous ammonia as the transport liquid. Each forms a hexagonal ring just inside the spacecraft structure.

Energy-Supply Subsystems

Communications satellites can be classified as active or passive. An example of a passive satellite is a big balloon (Echo I, launched August 12, 1960, was 30 meters in diameter when fully inflated) coated with conductive material that reflects radio signals. Used as a passive reflector, such a satellite does not need any electronic components or power source. While such a satellite is appealingly simple, the radio power it reflects back to Earth is less, by a factor of 10 million (70 dB), than the signal transmitted by a transponder aboard an active satellite (assuming equal uplink signal strength and a comparison based on equal satellite masses in the 50-kg range).

An active satellite (one with a transponder) needs power. The energy source supplying the power should be reliable, efficient, low-cost and long-lived. By efficient we mean that the ratio of available electrical power to weight and the ratio of available electrical power to waste heat should be large. We examine three energy sources that have been studied extensively: chemical, nuclear and solar.

Chemical Power Sources

Chemical power sources include primary cells, secondary cells and fuel cells. Early satellites such as Sputnik I, Explorer I and the first few OSCARs were flown with primary cells. When the batteries ran down, the satellite "died." Spacecraft of this type usually had lifetimes of a few weeks, although Explorer I with low-power transmitters (about 70 mW total), ran almost four months on mercury (Hg) batteries. These early experimental spacecraft demonstrated the feasibility of using satellites for communications and scientific exploration and thereby provided the impetus for the development of longer-lived power systems. Today, batteries (secondary cells in this case) are used mainly to store energy aboard satellites to cushion against peak loads and to power the spacecraft during eclipse periods; they are no longer used as a primary power source.

Batteries for early OSCAR missions were donated by companies or provided by government agencies from spares remaining when programs ended. In the early 1980s AMSAT realized that a more stable source was necessary to support the activities of the growing amateur program. AMSAT therefore initiated an ongoing battery qualification program to provide cells for future missions. Larry Kayser (VE3PAZ) looked into the procedures used to "space qualify" batteries. He concluded that carefully screened commercial grade NiCad's were likely to perform as reliably. Kayser and a group of amateurs in Ottawa purchased a large supply of 6 Ah GE aviation NiCds and put them through a qualification procedure that involved X-raying to look for internal flaws and extensive computer controlled charge-discharge cycling with extremely detailed computer monitoring. All anomalous cells, whether better or worse than the others, were eliminated. The remaining cells were matched, potted, and stored in a freezer to prevent deterioration.

OSCAR 11 used cells from this batch and it has performed flawlessly since March 1984. The MicroSATs and UoSATs launched in late 1990 were also powered by cells from this supply. AMSAT continues to monitor emerging technologies such as sealed nickel-hydrogen batteries that have energy densities about five times the value of NiCd cells. However, cost and reliability remain important selection criteria so there's good reason to stick to proven technology that's served us well in the past.

Another chemical power system, the fuel cell has been used as a source of energy on manned space missions where large amounts of power are required over relatively short time spans. Fuel cells do not appear to be appropriate for current OSCAR missions.

Nuclear Power Sources

One nuclear power source to be flight-tested is the radio-isotopic-thermoelectric power plant. In devices of this type, heat from decaying radioisotopes is converted directly to electricity by thermoelectric couples. Some early US Transit navigation satellites, the SNAP 3B and SNAP 9A (25 W), have flown generators of this type. The US is not currently using nuclear power in Earth orbit but development work on reactors for "Star Wars" related projects continues. The USSR currently flies nuclear reactors on Radar Ocean Reconnaissance Satellites (RORSATs). In 1989 some 34 deactivated but still radioactive reactors were orbiting the Earth.

Nuclear power sources have a high available-power-to-weight ratio, a very long operational lifetime and the ability to function in a high radiation environment. But they generate large amounts of waste heat and have a high hardware cost per watt because of the fuel. (They also have an extremely high cost due to related safety concerns and insurance). Nuclear power is most useful on missions where one of the following conditions holds: solar intensity is greatly reduced (deep-space), the radiation environment would quickly destroy solar cells (orbits inside the Van Allen Belts) or very large amounts of power are required.

There are strong pressures in the US and USSR to put an end to the use of nuclear power in Earth orbit. The primary reason is the danger of nuclear contamination in case of launch failure or satellite reentry. Satellites with nuclear power are often placed in near Earth orbit with the intention of boosting them to a higher orbit when they reach the end of their useful life. However, many believe the likelihood of a spacecraft failure before the satellite is moved to a higher orbit is unacceptably high. This has been clearly demonstrated since three nuclear powered spacecraft have reentered and spewed radioactive material in either the atmosphere or on the ground (SNAP-9A [1964], COSMOS 954 [1978], COSMOS 1402 [1983]). The growing problem of space debris also poses a risk to these spacecraft. A second reason for outlawing nuclear powered spacecraft is that they cause severe gamma ray pollution and are having a serious impact on gamma-ray astronomy.

Nuclear power is clearly out of the question for OSCAR satellites. If the previous facts don't convince you then practical concerns like the cost of insurance and the facilities, security precautions, and paperwork associated with handling nuclear material certainly will. This conclusion does not, of course, imply that nuclear power is not suitable for scientific deep-space missions. The Voyager mission to the outer planets would have been impossible without the 400 W nuclear generator aboard.

Solar Power Sources

The third power source we consider is solar. The first solar cells were built in 1954 using silicon. Solar cells quickly became the dominant supplier of power to spacecraft. However, they are far from ideal for this application. They compete for mounting space on the outer surface of the satellite with antennas and heat-radiating coatings. Their efficiency decreases with time, especially when the satellite orbit passes through the Van Allen radiation belts (roughly at altitudes between 1600 and 8000 km). They work most efficiently below 0°C (most electronics systems perform best at about 10° C). They call for a spacecraft orientation that may conflict with mission objectives. Finally, they produce no output when eclipsed from the Sun.

Despite these shortcomings, power sources that use solar cells to produce electrical energy and secondary cells to store energy are by far the simplest for long lifetime spacecraft. They're affordable; they generate little waste heat and they provide acceptable ratios of available-electric power to weight.

In Earth orbit a 1-meter-square solar panel, oriented perpendicular to incoming solar radiation intercepts about 1380 W. The amount of this power that can be used on a spacecraft depends on solar cell efficiency. The efficiency of cells used on OSCAR satellites has nearly doubled in the past 15 years. The silicon solar cells used on the initial group of MicroSATs employed back-surface reflector technology to produce an efficiency of more than 15%. In the near future we'll probably use GaAs solar cells having even higher efficiencies. In 1988 researchers reported the production of two layer cells using Si and GaAs that had an efficiency of 31%. While these are one-of-a-kind laboratory devices they do demonstrate solar cell technology is continuing to move forward. A few years ago a satellite the size of a MicroSAT would not have been capable of producing enough power to supply a useful mission.

Once a satellite is launched, its solar cell efficiency begins to decrease due to radiation damage. To minimize the rate of decrease the solar cells are usually covered by glass cover slides. These cover slides tend to reduce initial efficiency and increase spacecraft weight. One of the mission payloads on UoSAT E was the Solar-Cell Experiment designed to evaluate the long term performance of various new solar cell technologies in the space environment. Cells tested included: gallium arsenide, indium phosphide and new silicon designs. The study focused on issues related to radiation degradation over time using several cover slide geometries.

Practical Energy Subsystem

The typical AMSAT satellite energy system consists of a source, a storage device and conditioning equipment (shown in **Figure B2**). The source consists of silicon solar cells (future missions may use GaAs or lithium cells). A storage unit is needed because of eclipses (satellite in Earth's shadow) and the varying load.

Power conditioning equipment typically flown on AMSAT spacecraft includes a battery charge regulator (BCR) and at least one instrument switching regulator (ISR) to provide dc-to-dc conversion with changes of voltage, regulation and protection. Because failures in the energy subsystem could totally disable the spacecraft, special attention is paid to ensuring continuity of operation. BCRs and ISRs usually are built as redundant twin units with switch-over between units controlled automatically, in case of internal failure, or by ground command.

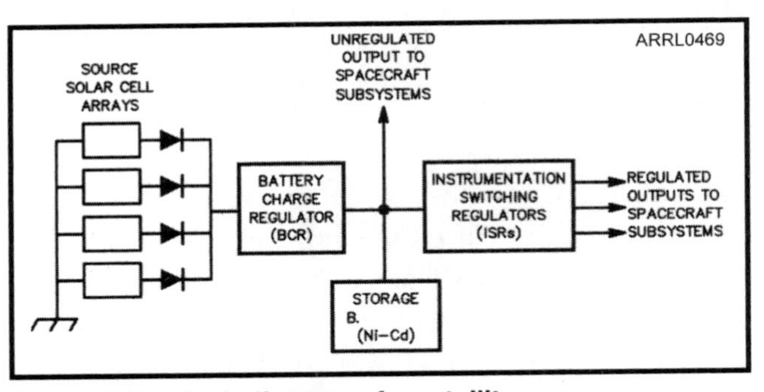

Figure B2—Block diagram of a satellite energy subsystem.

Phase III spacecraft carry two separate batteries: a lower capacity backup capable of providing full operation through all but the longest eclipses, is kept in cold storage. Solar cell strings are isolated by diodes, so a failure in one string will lower total output capacity but will not otherwise affect spacecraft operation. These diodes also prevent the battery from discharging through the cells when the satellite is in darkness. The MicroSAT approach to reliability does not include the use of redundant power systems.

The MicroSATs use an interesting method for battery charge regulation. Spacecraft transmitters are designed to operate efficiently over a wide range of power levels running from a fraction of a watt up to about 4 W. The onboard computer selects a power level that places minimum strain on the battery system. Software is the key element in a feedback loop that operates in an overdamped condition. The software is periodically refined to maximize spacecraft longevity.

When the energy supply subsystem provides sufficient energy to operate a satellite's major systems on a continuous basis we say the spacecraft has a *positive power budget*. If some subsystems must be turned off periodically for the storage batteries to be recharged, we say the spacecraft has a *negative power budget*. An illustration of how spacecraft geometry can be taken into account when estimating the average power output of a solar cell array covering a spacecraft can be found in the references.

Attitude-Control Subsystem

The orientation of a satellite (its attitude) with respect to the Earth and Sun greatly affects the effective antenna gain, solar cell power production, thermal equilibrium and scientific instrument operation. Attitude-control subsystems vary widely in complexity. A simple system might consist of a frame-mounted bar magnet that tends to align itself parallel to the Earth's magnetic field; a complex system might use cold gas jets, solid rockets and inertia wheels, all operating under computer control in conjunction with a sophisticated system of sensors. Attitude-control systems can be used to provide three-axis stabilization, or to point a selected satellite axis in a particular direction—toward the Earth, in a fixed direction in inertial space (with respect to the fixed stars), or parallel to the Earth's local magnetic field. Fixing a spacecraft's orientation in inertial space is generally accomplished by spinning the spacecraft about its major axis (spin-stabilized).

Attitude-control systems are classified as *active* or *passive*. Passive systems do not require power or sensor signals for their operation. Consequently, they are simpler and more reliable, but also less flexible. Some of the attitude control systems in general use are described in the following paragraphs.

Mass Expulsors

Devices of this type are based on the rocket principle and are classified as active and relatively complex. Examples are cold gas jets, solid-propellant rockets and ion-thrust engines. Mass expulsors are often used to spin a satellite around its principal axis. The resulting angular momentum of the satellite is then parallel to the spin axis. As a result of conservation of angular momentum the spin axis will tend to maintain a fixed direction in inertial space.

Angular Momentum Reservoirs

This category includes devices based on the inertia (fly) wheel principle. Assume that a spacecraft contains a flywheel as part of a dc motor that can be powered up on ground command. If the angular momentum of the flywheel is changed, then the angular momentum of the rest of the satellite must change in an equal and opposite direction (conservation of angular momentum). These systems are classified as active.

Environmental-Force Coupler

Every satellite is coupled to (affected by) its environment in a number of ways. In the two-body central force model the satellite and Earth were first treated as point masses at their respective centers of mass. Further analysis showed that the departure of the Earth from spherical symmetry causes readily observable perturbations of the satellite's path. The departure of the satellite's mass distribution from spherical symmetry likewise causes readily observable effects. An analysis of the mass distribution in the spacecraft defines a specific axis that tends to line up pointing toward the geocenter as a result of the Earth's gravity gradient. Gravity-gradient devices exploit this tendency. Anyone who's been on a sailboat, however, knows that gravity can produce two stable states. The gravity gradient effect is greatly accentuated when the spacecraft is in a very low orbit and if one of the satellite dimensions is much longer than the others. Attaching a long boom, with a weight at the far end, to the spacecraft is one way of achieving this configuration.

Another environmental factor that can be tapped for attitude control is the Earth's magnetic field. A strong bar magnet carried by the satellite will tend to align itself parallel to the local direction of this field (passive attitude control).

At any point in space the Earth's magnetic field can be characterized by its magnitude and direction. A simple model for the Earth's magnetic field employs a dipole offset somewhat from the Earth's rotational axis. The magnitude of a dipole field decreases as $1/r^3$, where r is the distance from the center of the Earth, so attitude control systems that depend on this field are most efficient at low altitudes. The direction of the magnetic field is often specified in terms of bearing and inclination (dip) angle. To describe bearing and dip we imagine a sphere concentric with the Earth drawn through the point of interest. Bearing and dip play the same role on this sphere that bearing and elevation play when describing a direction on the surface of the Earth.

The Earth's magnetic field can also be exploited for attitude control via an active system based on electromagnets consisting of coils of wire. By passing current through these coils, one forms a temporary magnet. With proper timing, the coils can produce torques in any desired direction. Devices of this type are often called torquing coils.

Note that even if a satellite designer does not exploit magnetic or gravity-based environmental couplers for attitude control, these forces are always present and their effect on the satellite must be taken into account.

Energy Absorbers

Energy absorbers or dampers are used to convert undesired motional energy into heat. They are needed in conjunction with many of the previously mentioned attitude control schemes. For example, if dissipative forces did not exist, gravity gradient forces would cause the satellite's principal axis to swing pendulum-like about the local vertical instead of pointing toward the geocenter. Similarly, a bar magnet carried on a satellite would oscillate about the local magnetic field direction instead of lining up parallel to it. Dampers may consist of passive devices such as springs, viscous fluids or hysteresis rods (eddy-current brakes). At times, torquing coils are used to obtain similar results.

Practical Attitude Control

Passive magnetic stabilization was first tried on OSCAR 5 and it has since been used on numerous other Phase II spacecraft. When passive magnetic stabilization is used, Permalloy hysteresis damping rods are generally employed to reduce rotation about the spacecraft axis aligned parallel to the Earth's local magnetic field and to damp out small oscillations. Note that the principal axis of a spacecraft in a near polar orbit using this type of stabilization will rotate 720° in inertial space during each revolution of the Earth.

Because of temperature regulation concerns we don't want to allow one side of a spacecraft to face the Sun for too long a time. Part of this spin was introduced purposely for temperature regulation. The technique was a novel one wherein the elements of the canted turnstile antenna were painted with reflective paint on one side and absorbent paint on the other. Solar radiation pressure then produced a radiometer-like rotation dubbed by users at the time as the "barbecue rotisserie" technique.

Because the camera on UoSAT must be pointed directly at the Earth, a passive magnetic stabilization system wasn't sufficient. To accomplish their mission objectives, UoSAT engineers chose a complex gravity-gradient stabilization system used in conjunction with torquing coils. Equipment modifications over the course of the UoSAT series, coupled with several generations of software optimization, have resulted in major performance improvements of this active control system. The evolution of the system hardware and control software represents an important contribution to spacecraft stabilization technology.

OSCARs 10 and 13 used spin stabilization (approximately 20 to 30 RPM). The spin axis is ideally aimed at the geocenter when the spacecraft is at apogee. However, for adequate spacecraft illumination the orientation must often depart from this ideal state. When the orbital inclination is near 57° these departures will be relatively small (on the order of 20°). The need for off-aiming occurs periodically. On average, a spacecraft like OSCAR 13 will need to have its attitude adjusted about every three months. Attitude information is obtained by Sun and Earth sensors under the control of the spacecraft computer.

To produce attitude changes the torquing coils must be pulsed at precisely the correct time. Software loaded by a ground station directs the satellite computer to monitor Sun and Earth sensors and pulse the torquing coils when the proper conditions are met.

OSCARs 10 and 13 employed viscous fluid dampers to discourage nutation (small oscillations in the direction of the spin axis). These dampers consist of a mixture of glycerine and water (about 50/50) contained in thin tubes (about 0.2 cm in diameter and 40 cm long) that run along the far edge of each arm of the spacecraft.

The attitude control system on OSCAR 40 involved a number of major changes that allowed this satellite to point the antennas directly at the center of the Earth throughout the entire orbit. It consisted of a set of three magnetically suspended, orthogonally mounted reaction wheels, two torquing coils, and six nutation dampers. When outside forces were absent the angular momentum of the spacecraft remained constant, but one could change the orientation of the body of the spacecraft by transferring momentum from one reaction wheel to another. The reaction wheels were magnetically suspended. Therefore there is no friction and no lubricants were needed. As a result, the estimated lifetime of this system was estimated to be greater than the 10 to 15 year lifetime expected for the spacecraft. The torquing coils will be used to "dump" accumulated momentum so that the velocity of the reaction wheels can be kept in a desired range. The nutation dampers are similar to those flown on OSCAR 10 and 13. Their main function would have been during orbit transfer when OSCAR 40 was spin stabilized

Propulsion Subsystem

The simplest type of space propulsion system consists of a small solid-propellant rocket which, once ignited, burns until the fuel is exhausted. Rockets of this type are often used to boost a satellite from a near-Earth orbit into an elliptical orbit with an apogee close to geostationary altitude (35,800 km) or to shift a satellite from this type of elliptical orbit into a circular orbit near geostationary altitude. Such rockets are known as "apogee kick motors" or simply "kick motors."

Liquid propellant motors are more complex to construct and they require on-site fueling involving hazardous materials. However, they make multiple burns possible. This greatly increases mission planning flexibility.

Both kick motors and liquid propellant motors are extremely dangerous. Their use, handling, shipping and storage must conform to rigid safety procedures.

Phase 3A: Solid Fuel Rocket

The only AMSAT satellite to use a kick motor was Phase IIIA. The kick motor was intended to shift the spacecraft from the planned transfer orbit (roughly 300×35,800 km, 10° inclination) to the target operating orbit (1500×35,800 km, 57° inclination).

The kick motor was a solid-propellant Thiokol TEM 345-12 containing approximately 35 kg of a mixture of powdered aluminum and organic chemicals in a spherical shell (17 cm radius) with a single exit nozzle. These units were originally designed as retro rockets for the Gemini spacecraft. The TEM 345 was capable of producing a velocity change of 1600 m/s during its single 20-second burn. Because of the launch failure this kick motor was never fired.

OSCARs 10, 13, 40: Liquid Fuel Rocket

OSCARs 10 and 13 used liquid-fuel rockets donated by the German aerospace firm Messerschmitt-BoelkowBlohm. These units were significantly more powerful (400 N thrust) than the solid propellant motor used on P3A. The added thrust enabled AMSAT to fly a heavier spacecraft (thicker shielding, more electronics modules, and so on) and to compensate for the lower inclination and perigee of the transfer orbit being provided. The fuels used were Unsymmetrical DiMethyl Hydrazine (UDMH) and Nitrogen Tetroxide (N_20_4). Both of these chemicals are extremely toxic. Those involved in loading fuel had to wear protective suits and breathe filtered air.

While the rocket motor and several of the associated valves were a donation much of the "plumbing" needed for the fuel system had to be devised and constructed by AMSAT personnel. There were a number of components required: filling valves, mixing valves, pressure regulators, check valves to prevent backflow, explosive (pyro) valves to prevent the system from accidentally firing before it was in orbit. Construction of the two chamber fuel tank and the high pressure helium bottle presented significant challenges.

The orbit transfer strategy planned for OSCAR 10 included two burns. There were important advantages to a two burn transfer over a single burn maneuver. With a single burn transfer the spacecraft velocity passes through a danger zone where premature termination will cause the spacecraft to reenter at the next perigee. Another advantage of the two burn approach is that it gives AMSAT an opportunity to compensate for unexpected performance characteristics of the rocket unit; the initial burn serves in part as a calibration run. The first burn went relatively well — there was a small deviation from the expected burn duration that placed the spacecraft perigee somewhat higher than planned. However, the engine could not be reignited for a second burn due to the fact that a slow leak in the high pressure helium system during the week the spacecraft was being reoriented prevented the opening of valves feeding fuel to the thrust assembly.

OSCAR 13 used a liquid fuel kick motor similar to the one employed on OSCAR 10. MBB again agreed to donate a rocket but this time very little support hardware was available. For this mission AMSAT designed a new propellant flow/storage assembly that used available hardware and made the system more robust. A two burn strategy was again planned. This time the motor firings went perfectly. Each burn was accomplished exactly as planned. OSCAR 40 carried an updated liquid fuel motor, but it developed problems that resulted in a damaging explosion.

Other Propulsion Systems

Water fuel rockets. AMSAT has evaluated the potential utility of a propulsion system using water as a fuel. The inherent safety of these systems is a key factor if AMSAT hopes to attain a useful orbit from a shuttle launch where it's unlikely that we could obtain permission to fly a solid fuel or chemical fuel kick motor. The problem with "water rockets" is that they take a considerable amount of time to affect orbit transfer. For example, it's been estimated that it would require approximately one year to reach a Phase III type orbit from a shuttle launch. The economic consequences of the time involved in orbit change make water rockets impractical for commercial satellites so this method has never been developed. However, now that AMSAT is constructing spacecraft with long expected lifetimes, this approach may be well matched to our needs. There are two types of water rockets. Hughes Aircraft Company developed a working model of a rocket based on the electrolysis of water to produce hydrogen and oxygen. A second method involves the production of steam, which is released through a specially designed thruster. The steam method produces less impulse per unit of water than the electrolysis method but system complexity is extremely low and less external energy is required for operation. The reliability of such a system should therefore be extremely high.

Arc-jet thruster. In addition to the main propulsion system, OSCAR 40 carried an arc-jet thruster (aka an ion-jet engine) to provide for station-keeping and minor adjustments once the spacecraft is in its near final orbit. Using ammonia as a fuel, the arc-jet can provide 100 milli-Newtons of thrust, far less than the 400 Newtons of the primary system. Operating the arc-jet requires about 1 kW of power to heat the ammonia. The system was designed at the University of Stuttgart.

Computer Guidance And Control Subsystems

On OSCARs 5-8, hard-wired logic was used to interface the various spacecraft modules to both the telemetry system and the command system. As overall spacecraft complexity grows, at some point it becomes simpler and more reliable to use a central computing facility in place of hardwired logic. Once the decision to incorporate a computer is made, the design of the spacecraft must be reevaluated totally to take advantage of the incredible flexibility provided by this approach. Ground stations need no longer be located in position to send immediate commands; they uplink pretested computer programs. After correct reception is confirmed, these programs take control of the spacecraft and the uploaded directions are executed at designated times or when needed. Phase III spacecraft feed data from Sun and Earth sensors to the computer. Using a simple model of the Earth's magnetic field the computer pulses the torquing coils at the appropriate times to maintain the correct spacecraft attitude. Firing the apogee kick motor on a Phase III spacecraft is also handled by the computer.

Computer programs control telemetry content and format. If we want to change the scale used to monitor a particular telemetry channel or to sample it more frequently, we simply add a couple of bytes to the computer program and it's done. Want to send out a daily Codestore message at 0000 GMT? No problem; uplink the message and control program whenever it's convenient and the message will be broadcast on schedule.

The Phase III Series IHU

Each Phase III satellite contains a module composed of a central processing unit (CPU) board, a random access memory (RAM) board, and a multiplexer (MUX) and command detector (CMD) board. The entire module is known as the Integrated Housekeeping Unit (IHU). The IHU combines a traditional multi-tasking computer (CPU and RAM) with a telemetry encoder and command decoder so that a single unit handles all guidance, control and telemetry functions.

The CPU. The 8-bit RCA COSMAC CDP18O2 microprocessor was selected for the Phase IIIA CPU back in the mid l970s for a simple reason. It was the only suitable device available when the spacecraft was being designed. The choice proved to be a good one since this processor has proved powerful and flexible enough to meet the more complex demands of later missions. A radiation hardened version (one more resistant to radiation damage) later became available.

A novel feature of the spacecraft 1802 CPU design is that it does not use any read-only-memory. This is considered an important attribute since radiation damage to ROM is considered a serious threat to spacecraft longevity. When the spacecraft IHU recognizes a particular sequence of bits on the command link a reset is sent to the computer. The next 128 bytes uplinked are fed into sequential locations in low memory. When the last of the 128 bytes are received the processor is automatically toggled into the run mode. A bootstrap loader contained in the 128 bytes then controls the loading of the rest of the operating system.

CPU Language. The CPU runs a high-level language called IPS (Interpreter for Process Structures), a threaded code language similar to Forth. IPS was developed by Dr. K. Meinzer, DJ4ZC, for multi-tasking industrial control type operations. IPS is fast, powerful, flexible, and extremely efficient in terms of memory usage. Some say it's also nearly incomprehensible for anyone not brought up using an HP Reverse Polish Notation hand calculator. In any event, users are not required to know IPS to recover data from the downhink telemetry beacons.

RAM. Initial plans (1975) were to fly Phase IIIA with 2 kbytes of NMOS RAM. By the time Phase IIIA was flown it was possible to include 16 k. The unit flown on OSCAR 13 contained 32k of radiation hardened CMOS RAM. This may not seem like much when compared to today's personal computers that typically contain 640 k or more of RAM. However, the Voyager mission to the outer planets ran fine using 32k of memory and memory size didn't limit OSCAR 13 in any important way. Each byte of 8-bit memory is backed up by 4 additional bits in an error detection and correction (EDAC) arrangement. The CPU, as a background task, constantly cycles through RAM checking each memory cell. This is called a memory wash. If an error (radiation induced bit-flip) is detected it's corrected. The memory is thus protected against soft errors (radiation-induced bit-flips), provided that no more than one occurs in a byte in the time it takes for the CPU to check the entire memory, typically less than one minute. The EDAC circuitry is based on a Hamming code.

MUXJCMD Board. The CMD unit just sits monitoring the uphink. When it identifies a unique bit sequence it passes data to the CPU. The MUX is an electronic 64 pole switch. When used in conjunction with the analog-to-digital converter on the CPU board it forms a 64 channel scanning voltmeter with a 0 to 2 V range. Each parameter to be measured must provide voltage in this range. Temperatures are measured using thermistors. An ingenious technique is used to measure currents without incurring the losses and reliability problems that series resistors would introduce. The system works as follows: A symmetrical low level ac signal is impressed on a coil wound on a small toroid. A dc measurement across the coil normally reads zero. If a wire carrying a dc current passes through the center of the toroid, however, a small dc offset voltage will be superimposed on the ac signal. The magnitude of the offset signal is proportional to the current flowing in the wire passing through the center. Small toroids are placed about each wire carrying a current to be measured.

Radiation Shielding. The integrated circuits comprising the electronics modules aboard the spacecraft are susceptible to radiation damage from high energy particles.

This problem is especially acute with Phase III spacecraft because these craft make two passes through the Van Allen radiation belts during each orbit and the high energy particle density in these belts is a severe threat. Soft errors, ones that simply cause a bit-flip in a memory cell, are not a major problem. The EDAC circuitry will take care of these. The problem is hard errors — permanent destruction of a memory cell. When this occurs in RAM, sections can be placed off limits. However, if too many memory cells are eliminated, or if certain key cells are destroyed, the computer cannot be rescued. There are lots of strategies for minimizing the susceptibility of the spacecraft to such damage and AMSAT used them all: choosing chips with the best possible radiation properties, placing individual shields on chips, placing shields over groups of chips, and so on.

With OSCAR 10 susceptibility to radiation damage was the acknowledged Achilles heel. After slightly more than three years in orbit the IHU did succumb. When OSCAR 13 was being readied for launch, chips with a much higher resistance to radiation became available and these were used on the new spacecraft. Radiation damage was not a limiting factor on OSCAR 13. OSCAR 40 carried several computers including the old faithful 1802 and a complex local area network.

UoSAT Series IHU

Early UoSAT spacecraft were controlled by CDP1 802 microprocessors running the IPS language. OSCAR 9 and OSCAR 11 supported the 1802 with 48 k of dynamic RAM and carried secondary computers. On OSCAR 9 the backup computer was a Ferranti 16-bit F100L supported by 32 k of static CMOS RAM. On OSCAR 10 was an NSC-800 with 128 k of CMOS RAM used in the digital communications experiment. UoSAT D carries three computers (1802, 80C3 1, 80C 186) and more than 4 Mbytes of RAM. More recent UoSAT spacecraft have used computers having architectures similar to the Intel 186 series for overall spacecraft and communications system control and transputer parallel processing microcomputers for image processing.

One of the primary objectives of the UoSAT series has been to test and evaluate new hardware and systems approaches. Since Phase IIIA never attained orbit, OSCAR 9 gave amateurs their first opportunity to control a computer in space. Many of the techniques used for packet radio satellites were first tested on UoSAT spacecraft.

MicroSAT Series IHU

It has been said, somewhat seriously, that a MicroSAT is a compact, low power IBM computer clone masquerading as a spacecraft. The MicroSAT CPU uses an NEC CMOS V40 (similar to the 80C 188) and 2 k of ROM for a bootstrap loader. EDAC is used for 256 k of memory that holds the operating system software. An additional 8 Megabytes of static RAM is used to hold messages. MicroSAT software is written in assembler and Microsoft *C*, linked with Microsoft LINK.

The spacecraft computer control system on the MicroSATs represented a significant change in direction over the 8-bit 1802 architecture running IPS used on early Phase III and UoSAT spacecraft. The reason for the change is straightforward. The development of the primary MicroSAT mission subsystem (the mailbox) required a microprocessor more powerful than the 1802. It was decided to use one from the Intel series so that development work could be done on IBM clones that a great many amateurs have access to. The V-40 had the desired characteristics. It would have been possible to use an 1802 for overall spacecraft control and use the V-40 to manage the mailbox but this would require that extra circuitry be placed on the satellite and that the spacecraft command and development teams work in IPS in addition to the more familiar languages used for Intel microprocessors. It was simpler to place the V-40 in overall control and treat spacecraft management as one of many tasks the V-40 was responsible for.

A major innovation of the MicroSAT series is the introduction of a standard spacecraft bus (interconnection scheme) for linking the onboard computer and the various electronic modules. In the past, each satellite required a unique and extremely elaborate wiring harness. These harnesses had to be designed and constructed to provide all the links needed between the various modules. As spacecraft became more complex the number of interconnections that the harness had to handle grew rapidly and construction difficulty increased at an even faster rate. This has a negative effect on spacecraft reliability. In the MicroSATs the function of the wiring harness is mainly handled by a ribbon cable that plugs into each module. The setup is similar to a simple local area network with the on board computer (OBC) acting as the master and the modules as slaves. Each module contains an AART (Addressable Asynchronous Receiver/Transmitter) chip and associated components for communications with the OBC and telemetry sensor measurements.

The bus-communications orientation provides several additional advantages. With small engineering groups spread around the world, each working on a different spacecraft module, communication has always been a problem. A detailed bus definition greatly reduces this problem and facilitates distributed engineering. The bus orientation also makes it easier for new groups to become involved in spacecraft construction. Finally, it provides a design and control approach that can be efficiently applied to spacecraft of widely varying complexity and with all types of mission objectives.

Sensors

A satellite guidance and control subsystem includes components and software involved in the measurement of spacecraft position and orientation, attitude adjustment and in control of all other onboard systems in response to orders issued by telecommand or the spacecraft computer. On OSCAR spacecraft the command receiver generally uses elements of the transponder (linear or digital) front end. A tap goes to a dedicated IF strip, demodulator and decoder. Data from the decoder is routed to the IHU. An active attitude control system requires sensors. Since most of the elements of the guidance and control subsystem, except for the sensors, have already been discussed we'll focus our attention on sensors. The sensors on OSCAR 13 serve as a good example.

OSCAR 13 contained three sets of sensors designed to provide attitude information: a Sun sensor, an Earth sensor, and a top/bottom sensor. In the following discussion it will help to picture the shape of OSCAR 13 and the fact that it's rotating about its symmetry axis that is aligned in a fixed direction in space. The attitude determination strategy consists of first finding the relative orientation or the spacecraft with respect to two celestial bodies, the Sun and Earth, whose positions are accurately known and then mathematically reducing this data to absolute orientation. When this is done there are frequently two solutions, only one of which is correct. The top/bottom sensor is used to eliminate the ambiguity.

The top/bottom sensor consists of a few solar cells mounted on the top and bottom of the spacecraft. When the spacecraft is in sunlight only one set will be illuminated. This provides a crude estimate of the satellite's orientation, sufficient to choose between the solutions provided by the Sun and Earth sensors. The Sun and Earth sensors are mounted at the end of arm two of the spacecraft.

Because the Sun is extremely bright and virtually a point source, construction of the

Sun sensor is relatively simple. It consists of two slits and two photodiodes. Because of the satellite's spin the Sun sensors will scan a region ±60° from the spacecraft's equator once each revolution (about 20 times per minute). If the Sun is in this region we will get a "pip" from each photodiode. The time between these two pips and knowledge of which one pips first provide important information about satellite orientation. There are some real world complications to this simple model caused by the fact that extraneous pips may be introduced by sunlight reflecting off an antenna.

The second body we choose to focus on is the Earth. The Earth is rather dim and its diameter as seen from OSCAR 13 varies from about 18° at apogee to 90° at perigee. Light enters the Earth sensor through an anti-glare shield and is focused by a lens on a photodiode. The threshold sensitivity of the diode electronics is set so that a step-like change in output will occur when the diode field of view changes from dark space to sunlight reflected off the Earth. Temporarily assume that the Earth is completely bathed in sunlight (of course this is impossible). During most of the satellite's orbit the Earth sensor will not scan through the Earth as the spacecraft rotates. However, there will be two periods, one slightly before perigee and one slightly after perigee, when the sensor will view the Earth. As we enter one of these periods the sensor will scan through the Earth very quickly, but the time during which the sensor is focused on the Earth will increase, reaching a maximum slightly further along the satellite's orbit, and then decrease back down to zero.

The mathematics needed to convert data from this simple model into information on satellite orientation is not too horrendous. However, there are complications. The most serious one occurs because only part of the Earth may be illuminated. The Earth sensors will therefore be reporting illuminated crescent acquisition and loss. Additional uncertainty is introduced by the fact that the transition from light to dark is not very sharp when the Earth is partially illuminated. To help with data interpretation the Earth sensor actually includes two photodiodes each having a beamwidth of about 2°. One points about 4° above the spacecraft's equator and the other points about 4° below.

Using OSCAR 13 telemetry values containing the sensor data, a personal computer can be used to determine the direction of the spacecraft spin axis to about 1°. The absolute spin angle (needed for determining when torquing coils should be pulsed) can be determined to within about 0.1°. This accuracy is completely adequate for orbital transfer maneuvers and for attitude adjustment via the torquing coils.

OSCAR 40 was actually the second OSCAR satellite to carry a Global Positioning Satellite receiver for use in orbit and attitude determination. The first such unit was on PoSAT but amateurs were not responsible for its construction or operation. However, OSCAR 40 was the first satellite in any service to experiment with the use of GPS in highly elliptical orbits. GPS receivers on OSCAR 40 were designed to be used to determine the satellite's orbit (to within 20 meter accuracy) and back up the attitude determination system. OSCAR 40 carried eight GPS antennas. Information from the system will be converted to Keplerian orbital elements on the satellite and made available to users on the downlink.

Notes

Notes

Notes

INDEX

P

U

V

W

Y

About the ARRL

The seed for Amateur Radio was planted in the 1890s, when Guglielmo Marconi began his experiments in wireless telegraphy. Soon he was joined by dozens, then hundreds, of others who were enthusiastic about sending and receiving messages through the air — some with a commercial interest, but others solely out of a love for this new communications medium. The United States government began licensing Amateur Radio operators in 1912.

By 1914, there were thousands of Amateur Radio operators — hams — in the United States. Hiram Percy Maxim, a leading Hartford, Connecticut inventor and industrialist, saw the need for an organization to band together this fledgling group of radio experimenters. In May 1914 he founded the American Radio Relay League (ARRL) to meet that need.

Today ARRL, with more than 150,000 members, is the largest organization of radio amateurs in the United States. The ARRL is a not-for-profit organization that:

- promotes interest in Amateur Radio communications and experimentation
- represents US radio amateurs in legislative matters, and
- maintains fraternalism and a high standard of conduct among Amateur Radio operators.

At ARRL headquarters in the Hartford suburb of Newington, the staff helps serve the needs of members. ARRL is also International Secretariat for the International Amateur Radio Union, which is made up of similar societies in 150 countries around the world.

ARRL publishes the monthly journal *QST*, as well as newsletters and many publications covering all aspects of Amateur Radio. Its headquarters station, W1AW, transmits bulletins of interest to radio amateurs and Morse code practice sessions. The ARRL also coordinates an extensive field organization, which includes volunteers who provide technical information and other support services for radio amateurs as well as communications for public-service activities. In addition, ARRL represents US amateurs with the Federal Communications Commission and other government agencies in the US and abroad.

Membership in ARRL means much more than receiving *QST* each month. In addition to the services already described, ARRL offers membership services on a personal level, such as the ARRL Volunteer Examiner Coordinator Program and a QSL bureau.

Full ARRL membership (available only to licensed radio amateurs) gives you a voice in how the affairs of the organization are governed. ARRL policy is set by a Board of Directors (one from each of 15 Divisions) elected by the membership. The day-to-day operation of ARRL HQ is managed by a Chief Executive Officer.

No matter what aspect of Amateur Radio attracts you, ARRL membership is relevant and important. There would be no Amateur Radio as we know it today were it not for the ARRL. We would be happy to welcome you as a member! (An Amateur Radio license is not required for Associate Membership.) For more information about ARRL and answers to any questions you may have about Amateur Radio, write or call:

ARRL—the national association for Amateur Radio
225 Main Street
Newington CT 06111-1494
Voice: 860-594-0200
Fax: 860-594-0259
E-mail: **hq@arrl.org**
Internet: **www.arrl.org/**

Prospective new amateurs call (toll-free):
800-32-NEW HAM (800-326-3942)
You can also contact us via e-mail at **newham@arrl.org**
or check out ARRLWeb at **http://www.arrl.org/**

FEEDBACK

Please use this form to give us your comments on this book and what you'd like to see in future editions, or e-mail us at **pubsfdbk@arrl.org** (publications feedback). If you use e-mail, please include your name, call, e-mail address and the book title, edition and printing in the body of your message. Also indicate whether or not you are an ARRL member.

Where did you purchase this book?

☐ From ARRL directly ☐ From an ARRL dealer

Is there a dealer who carries ARRL publications within:

☐ 5 miles ☐ 15 miles ☐ 30 miles of your location? ☐ Not sure.

License class:

☐ Novice ☐ Technician ☐ Technician Plus ☐ General ☐ Advanced ☐ Amateur Extra

---

Name ____________________ ARRL member? ☐ Yes ☐ No

Call Sign ____________________

Daytime Phone ( ) ____________________ Age ____________________

Address ____________________

City, State/Province, ZIP/Postal Code ____________________

If licensed, how long? ____________________ E-mail ____________________

Other hobbies ____________________

____

Occupation ____________________

For ARRL use only	SAT HBK
Edition	1 2 3 4 5 6 7 8 9 10 11 12
Printing	1 2 3 4 5 6 7 8 9 10 11 12

From ______________________________

____

____

Please affix postage. Post Office will not deliver without postage.

EDITOR, SATELLITE HANDBOOK
AMERICAN RADIO RELAY LEAGUE
225 MAIN STREET
NEWINGTON CT 06111-1494

- - - - - - - - - - - - - - - - please fold and tape - - - - - - - - - - - - - - - -